JACARANDA GEOGRAPHY ALIVE 10

VICTORIAN CURRICULUM | SECOND EDITION

JILL PRICE

CATHY BEDSON

DENISE MILES

KINGSLEY HEAD

BENJAMIN ROOD

JANE WILSON

CONTRIBUTING AUTHORS

ALEX ROSSIMEL | TRISH DOUGLAS

jacaranda
A Wiley Brand

Second edition published 2020 by
John Wiley & Sons Australia, Ltd
42 McDougall Street, Milton, Qld 4064

First edition published 2017

Typeset in 11/14 pt TimesLTStd

ISBN: 978-0-7303-7472-5

Front cover image: © Franck Metois/Alamy Australia Pty Ltd

Illustrated by various artists, diacriTech and Wiley Composition Services

Typeset in India by diacriTech

Printed in Singapore by
Markono Print Media Pte Ltd

A catalogue record for this book is available from the National Library of Australia

10 9 8 7 6 5 4 3 2

CONTENTS

Fieldwork inquiry: Comparing wellbeing in the local area   online only

HOW TO USE
the *Jacaranda Geography Alive* resource suite

The ever-popular *Jacaranda Geography Alive for the Victorian Curriculum* is available as a standalone Geography series or as part of the *Jacaranda Humanities Alive* series, which incorporates Geography, History, Civics and Citizenship, and Economics and Business in a 4-in-1 title. The series is available across a number of digital formats: learnON, eBookPLUS, eGuidePLUS, PDF and iPad app.

Skills development is integrated throughout, and explicitly targeted through SkillBuilders and a dedicated Geographical skills and concepts topic for each year level.

This suite of resources is designed to allow for differentiation, flexible teaching and multiple entry and exit points so teachers can *teach their class their way*.

Features

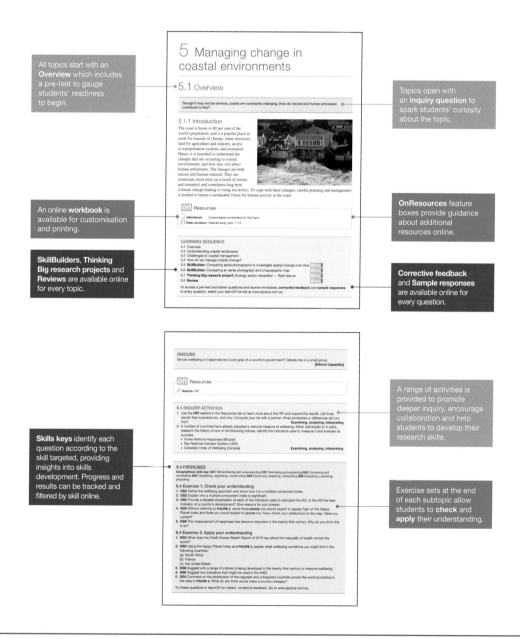

All topics start with an **Overview** which includes a pre-test to gauge students' readiness to begin.

Topics open with an **inquiry question** to spark students' curiosity about the topic.

An online **workbook** is available for customisation and printing.

OnResources feature boxes provide guidance about additional resources online.

SkillBuilders, Thinking Big research projects and **Reviews** are available online for every topic.

Corrective feedback and **Sample responses** are available online for every question.

A range of activities is provided to promote deeper inquiry, encourage collaboration and help students to develop their research skills.

Skills keys identify each question according to the skill targeted, providing insights into skills development. Progress and results can be tracked and filtered by skill online.

Exercise sets at the end of each subtopic allow students to **check** and **apply** their understanding.

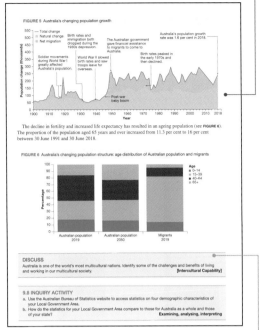

Content is presented using age-appropriate language, and a wide range of engaging sources, diagrams and images support concept learning.

Skillbuilders model and develop key skills in context.

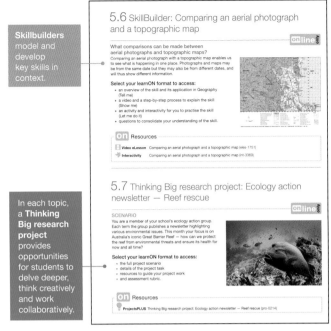

In each topic, a **Thinking Big research project** provides opportunities for students to delve deeper, think creatively and work collaboratively.

Discuss features explicitly address Curriculum Capabilities.

Links to the **myWorld Atlas** are provided throughout.

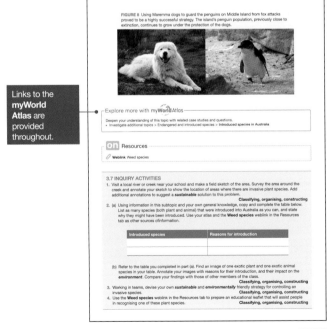

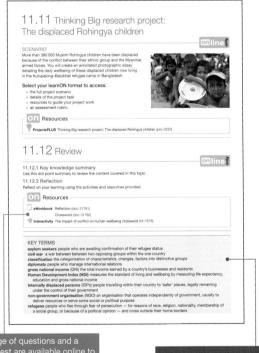

A range of questions and a post-test are available online to test students' understanding of the topic.

Key terms are available in every topic review.

learn**on**

Jacaranda Geography Alive learnON is an immersive digital learning platform that enables student and teacher connections, and tracks, monitors and reports progress for immediate insights into student learning and understanding.

It includes:

- a wide variety of embedded videos and interactivities
- questions that can be answered online, with sample responses and immediate, corrective feedback
- additional resources such as activities, an eWorkbook, worksheets, and more
- Thinking Big research projects
- SkillBuilders
- teachON, providing teachers with practical teaching advice, teacher-led videos and lesson plans.

teach**on**

Conveniently situated within the learnON format, teachON includes practical teaching advice, teacher-led videos and lesson plans, designed to support, save time and provide inspiration for teachers.

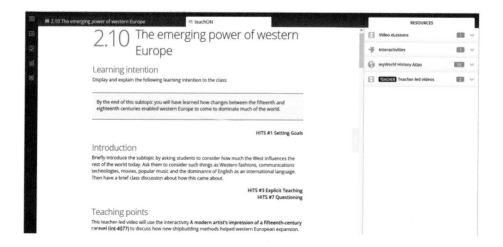

ACKNOWLEDGEMENTS

The authors and publisher would like to thank the following copyright holders, organisations and individuals for their assistance and for permission to reproduce copyright material in this book.

The Victorian Curriculum F–10 content elements are © VCAA, reproduced by permission. The VCAA does not endorse or make any warranties regarding this resource. The Victorian Curriculum F–10 and related content can be accessed directly at the VCAA website. VCAA does not endorse nor verify the accuracy of the information provided and accepts no responsibility for incomplete or inaccurate information. You can find the most up to date version of the Victorian Curriculum at http://victoriancurriculum.vcaa.vic.edu.au.

Images

• AAP Newswire: **167** (left, right)/Gerald Herbert; **195**/Luigi Costantini; **302** (top)/Alexander Kots/Shutterstock; **304** (right)/Thanassis Stavrakis; **306** (c)/EPA/Darko Dozet • Alamy Australia Pty Ltd: **169** (top right); **56** (top)/Pete Titmuss; **58** (top)/Genevieve Vallee; **65** (bottom)/Mike Patterson; **71** (bottom)/Green Eyes; **81** (left)/J C Clamp; **81** (right)/Paul Dymond; **93**, **252** (bottom), **301** (bottom), **302** (bottom), **306** (b)/© EPA European Pressphoto Agency b.v.; **97** (bottom)/SCPhotos; **132**/imageBROKER; **156** (bottom right)/F. Bettex – Mysterra.org; **188** (bottom)/National Geographic Image Collection; **192** (bottom)/Roy Garner; **193** (Wittenoom)/Paul Mayall Australia; **248** (top right)/Bill Bachman; **276**/Dinodia Photos; **278**/Susanna Bennett; **265** (right)/Jan Sochor; **306** (a)/dpa picture alliance • Ashden: **96** • Australian Bureau of Statistics: **277** • Cartoon Movement: **218** (bottom)/Global Food Consumption — Richard & Slavomir Svitalsky/Cartoon Movement • Climate Council of Australia: **34**/Climate Council 2019 • Country Womens Association of Australia: **274**/© Country Womens Association of Australia • Creative Commons: **21** (top, bottom), **33**, **77**, **145** (top), **191** (bottom), **216**, **249** (bottom), **261**; **7**/Department of Environment, Land, Water & Planning; **72** (left)/Allan Fox & DSEWPAC Australia © Commonwealth of Australia 2013; **165**/Reefbase; **174**/© Bureau of Meteorology; **186**, **187**/Department of Environment, Land, Water & Planning; **187**/Department of Environment, Land, Water & Planning; **215**/© AusAID Photolibrary; **218** (top)/FAO, IFAD, UNICEF, WFP and WHO. 2018. The State of Food Security and Nutrition in the World 2018. Building climate resilience for food security and nutrition. Rome, FAO; **225**/Copyright © International Labour Organization 2017; **207**, **254**/© World Bank; **272**/© Australian Institute of Health and Welfare; **248** (bottom), **249** (top), **279** (top)/© Australian Bureau of Statistics; **273** (top), **279** (bottom)/© Australian Institute of Health and Welfare; **287** (top)/International Institute for Strategic Studies; **287** (bottom)/DoD photo by Spc. DeYonte Mosley, U.S. Army/Released • Credit Suisse Wealth Report: **210** • CSIRO Land and Water: **47** (natural forest); **47** (irrigation agriculture)/Willem van Aken; **47** (sheep grazing)/Greg Heath; **47** (vacant land)/John Coppi • Dept of Agriculture and Food: **71** (top)/© Western Australia Agriculture Authority Department of Primary Industries and Regional Development • Economist Intelligence Unit: **301** (top)/EIU Data Liveability Index 2018 • Freedom House: **223** • Getty Images: **160** (top)/Klaus Vedfelt; **258**/Xia Yuan; **47** (algal bloom)/Science Photo Library; **83**/edelmar; **92** (bottom)/Mario Tama; **105**/China Photos; **117**/Vince Streano; **118**/Peter Harrison; **123** (bottom left)/Paul A. Souders/Corbis/VCG; **123** (bottom right)/Rodney Hyett; **129**/Ashley Cooper; **134**/STR/AFP; **137**/The Asahi Shimbun; **139**/David Wall Photo; **188** (top)/Matt Mawson; **189**/Sam Panthaky; **193** (Kowloon)/Jodi Cobb/National Geographic Image Collection; **194** (bottom)/Jonathan Blair; **285**/Abdullah Doma; **297**/Per-Anders Pettersson; **273** (bottom)/Auscape; **290** (bottom)/MOHAMMED ABED/AFP; **294** (top)/Sebastian Meyer Australia; **304** (left)/BULENT KILIC/AFP; **306** (e)/Denis Charlet/AFP • GhostNets Australia: **162**, **163** • Global Footprint Network: **19** (bottom) • Internal Displacement Monitoring Centre IDMC: **299** • International Organizing Committee for the World Mining Congresses: **294** (bottom) • John Ivo Rasic: **49** (middle) • John Wiley & Son Australia: **286** (bottom) • MAPgraphics Pty Ltd Brisbane: **44**, **128** (bottom) • Matter Of Trust: **169** (top left)/Amanda Bacon • Medecins Sans Frontieres Aus: **306** (d), **311** • Murray Darling Wetlands Working Group Ltd.: **113** (a, b, c) • NASA: **11** (left), **108** (left), **144** (left)/NASA Earth Observatory image created by Jesse Allen and Robert Simmon, using Landsat data provided by the United States Geological Survey; **11** (right), **108** (right), **144** (right)/NASA Earth Observatory images by Joshua Stevens, using MODIS data from the Land Atmosphere Near real-time Capability for EOS LANCE and Landsat data from the U.S. Geological Survey. Caption by Adam Voiland, with information from Amin Eimanifar University of Florida and Mohammed Tourian University of Stuttgart; **65** (top)/Image courtesy Jacques Descloitres, MODIS Rapid Response Team at NASA GSFC/NASA Earth Observatory; **86**/NASA Earth Observatory; **232**/NASA/Goddard Space Flight Center Scientific Visualization Studio • Newspix: **35** (top)/Zak Simmonds; **75**/Mike Keating; **140** (top)/Brad Wagner • NSW Department of Primary Industries: **114** • NSW Land and Property: **142** • Ocean Conservancy: **156** (top) • Outback stores: **280** • Oz Aerial Photography: **141**/© 2012 Oz Aerial Photos Australia • Public Domain: **20** (graph), **28**, **177** (bottom), **206**, **213**, **219–21**, **238**, **253**, **260** (right); **10** (left, right)/US Census Bureau; **128** (top)/Mohammed Seeneen/AAP Newswire; **234**/United Nations Inter-agency Group for Child Mortality Estimation UN IGME, 'Levels & Trends in Child Mortality: Report 2017, Estimates Developed by the UN Inter-agency Group for Child Mortality Estimation', United Nations Children's Fund, New York, 2017; **181** (top, bottom)/Demographia World Uban Areas http://www.demographia.com/db-worldua.pdf; **242** (top left, top right), **243** (left, right), **245** (top left, top right)/Based on information from United Nations, Department of Economic and Social Affairs, Population Division. *World Population Prospects: The 2015 Revision.*; **255**/Global Burden of Disease Collaborative Network. *Global Burden of Disease Study 2016* GBD 2016 Results. Seattle, United States: Institute for Health Metrics and Evaluation IHME, 2017 • Peter Coyne: **72** (right) • Pauline English: **43** • Shutterstock: **1**/NicoElNino; **2** (bottom)/goodluz; **2** (middle)/Free Wind 2014; **2** (top)/Goodluz; **12**/Caleb Holder; **15**/Tony Campbell; **16** (bottom)/Vladimir Wrangel; **14** (bottom)/goodluz; **14** (top)/Ioannis Pantzi; **35** (bottom)/Pete Niesen; **37**/idiz; **39**/Dirk Ercken; **46**/Valentin Valkov; **47** (saline scald)/Alberto Loyo; **48** (top)/HHelene; **48** (middle)/Mark Winfrey; **48** (bottom)/Neil Bradfield; **49** (bottom)/Caleb Holder; **56** (bottom)/Ashley Whitworth; **58** (bottom)/doraclub; **63**/SJ Allen; **67**/velvetweb; **68**/YuliiaKas; **70** (bottom (a))/mrfotos; **73** (left)/Grezova Olga; **73** (right)/Susan Flashman; **79**/Sam DCruz; **85**/Mark Schwettmann; **87**/Warren Price Photography; **107**/Sherrianne Talon; **115**/xpixel; **119**/Martin Maun; **127**/Seaphotoart; **131**/Mohamed Shareef; **145** (bottom)/Tanya Puntti, **147**/Mikadun; **152**/Antonio V. Oquias; **161**/Maria Pomelnikova; **175**/Rich Carey; **176**/chungking; **177** (top)/Nataliya Hora; **178** (top)/FCG; **178** (bottom)/Alexandr Vlassyuk; **179**/VIGO-S; **182**/donsimon; **183**/View Factor Images; **192** (top)/Vladimir Korostyshevskiy; **193** (Bodie)/Zack Frank; **193** (Oradour-sur-Glane)/Ivonne Wierink; **193** (Pripyat)/SvedOliver; **193** (top)/Esin Deniz; **198**/Paolo Bona; **200**/Yury Birukov; **202** (bottom)/2xSamara.com; **202** (top)/Denis Cristo; **203**/Anton_Ivanov; **214**/GNEs; **227**/Masum Ali; **229**/© Milles Stud; **231**/Aaron Gekoski; **235**/Andrei Shumskiy; **241**/meunierd; **239** (bottom)/wizdata1; **242** (bottom)/mykeyruna; **244** (bottom)/Matyas Rehak; **248** (top left)/Brisbane; **250**/hvostik; **251**/Amir Ridhwan; **256**/Monkey Business Images; **260** (left)/Lucian Coman; **262** (bottom)/singh_lens; **265** (left)/Publio Furbino; **271**/Nils Versemann; **275**/JHMimaging; **282**/Rawpixel.com; **283** (bottom)/John Wollwerth; **298**/Sadik Gulec; **305** (centre)/Robert Biedermann; **305** (left)/vectorlight; **305** (right)/Q STOCK; **310** (bottom)/john austin; **314**/Sk Hasan Ali • Snowy Monaro Regional Council: **70** (bottom (b)) • Spatial Vision: **8**, **16** (top), **45**, **54**, **91**, **92** (top), **104**, **111**, **130**, **140** (bottom), **149**, **153**, **170**, **191** (top), **197** (bottom), **197** (top), **205**, **236** (bottom), **237**, **252** (top), **270**, **281**, **286** (top), **289**, **293**, **310** (top); **13**/Government of India, Ministry of Home Affairs, Office of Registrar General; **57**/© Commonwealth of Australia Geoscience Australia 2013; **50**/© The State of Victoria, Department of Environment and Primary Industries 2013; **109**/Vector Map Level 0 Digital Chart of the World; **61**/UNEP World Conservation Monitoring Centre; **64**/United States Department of Agriculture, Natural Resources Conservation Service, Soil Survey Division, World Soil Resources; Paul Reich, Geographer. 1998. Global Desertification Vulnerability Map. Washington, D.C.; **69**, **70** (top)/© Commonwealth of Australia Geoscience Australia 2013. © Commonwealth of Australia Department of Sustainability, Environment, Water, Population and Communities 2013; **89**/University of New Hampshire UNH/Global Runoff Data Centre GRDC http://www.grdc.sr.unh.edu; **97** (top)/World Climate – http://www.worldclim.org; **101**/BGR & UNESCO 2008: Groundwater Resources of the World 1 : 25 000 000. Hannover, Paris; **102**/UNEP Global Environmental Alert Service GEAS. Made with Natural Earth. Vector Map Level 0 Digital Chart of the World; **103**/BBC News, http://news.bbc.co.uk/2/hi/8545321.stm; **109** (bottom)/United Nations Environment Programme. Vector Map Level 0 Digital Chart of the World; **112**/© Commonwealth of Australia Geoscience Australia 2013. Murray Darling Basin Commission; **136**/Based on information from the United Nations Environment Programme; **154** (bottom)/Greenpeace International; **155**/© Commonwealth of Australia Geoscience Australia 2013. Ghost Nets Australia, http://www.ghostnets.com.au/index.html; **159**/Data derived from ReuseThisBag.com and G. Cabrera, August 28, 2017 with source data by UNEP, Greenpeace, and national governments; **169** (bottom)/Conservation Biology Institute, Publication Date: 7/5/2010; **172** (bottom)/National Oceanic and Atmospheric Administration, Office of Ocean Exploration and Research, U.S. Department of Commerce. Adapted by Spatial Vision; **180** (top)/Data sourced from: United Nations, Population Division 2017, *World Population Prospects: The 2017*

Revision, Medium Variant; **180** (bottom)/Data sourced from United Nations Department of Economics and Social Affairs; Worldatlas.com; **184**/.idplacemaker © The State of Victoria, Department of Environment and Primary Industries 2013 © Commonwealth of Australia Geoscience Australia 2013; **185**/© Commonwealth of Australia Geoscience Australia 2013. © The State of Victoria, Department of Environment and Primary Industries 2013; **194** (top)/© OpenStreetMap contributors; **217**/FAO 2013; **230**/NASA Socioeconomic Data and Applications Center SEDAC. http://sedac.ciesin.columbia.edu/data/collection/gpw-v3; **236** (top)/World Bank Data; **247**/Australian Bureau of Statistics; **259**/Data Source: World Health Organization Map Production: Health Statistics Information Systems HSI World Health Organization Total population: United Nations Population Division. Regional aggregates calculated by UNFPA based on data from United Nations Population Division; **264**/Instituto Brasileiro de Geografia e Estatística; **266**/UNEP-WCMC 2012. *Data Standards for the World Database on Protected Areas.* UNEP-WCMC: Cambridge, UK; **296** (bottom right)/ttps://www.unhcr.org/en-au/statistics/unhcrstats/ 5b27be547/unhcr-global-trends-2017.html; **300**/Based on data from UNHCR, Government of Turkey; **245** (bottom)/Government of India, Ministry of Home Affairs, Office of Registrar General; **246** (left)/Percentage of Population Below Poverty Line URP Consumption, ChartsBin.com, viewed 27th August, 2013, http://chartsbin.com/view/2797; **246** (right), **262** (top)/Government of India, Ministry of Home Affairs, Office of Registrar General; **290** (top)/From 'The Humanitarian Impact of Restrictions on the access to land near the perimeter fence in the Gaza Strip' August 2018 ©United Nations Office for the Coordination of Humanitarian Affairs in the occupied Palestinian territory. Reprinted with the permission of the United Nations; **296** (top)/UNHCR Statistics; **309** (top)/Institute for Economics & Peace. Global Peace Index 2018: Measuring Peace in a Complex World, Sydney, June 2018. Available from: http://visionofhumanity.org/reports • State of Victoria, Department of Primary Industries: **49** (top)/Victorian Resources Online www.dpi.vic.gov.au/vro, Photo by Clem Sturmfels; **51** (top, bottom), **52**/© State of Victoria Department of Environment and Primary Industries 2013. Reproduced with permission. Photograph by Stuart Boucher • Surfrider Foundation: **158** • Susan Middleton: **156** (bottom left) • UNDP: **209**/© UNDP 2012 • UNHCR: **296** (bottom left) • UNICEF: **267**/WHO/*UNICEF JMP report Progress on household drinking water, sanitation and hygiene 2000–2017: Special focus on inequalities assesses progress made at national, regional and global levels in reducing inequalities in WASH services.* New York: United Nations Children's Fund UNICEF and World Health Organization WHO, 2019; **268**/© UNICEF and World Health Organization; United Nations/Populations Division; **233**/© United Nations Publications • Cynthia Wardle: **82** (Newcastle)/© Cynthia Wardle • Worldmapper: **211**/Happy Planet Index

Text

• Amnesty International: **224** • Creative Commons: **27** (table)/GARNAUTCLIMATECHANGE2008; **27–8** (Figure 2)/© Australian Conservation Foundation; **33**, **277**/© Commonwealth of Australia; **271** (table)/© Australian Institute of Health and Welfare; **274** (table), **275** (table)/© State of Victoria Department of Environment, Land, Water and Planning • Fraser Island Tourist Guide: **20**/Fraser Island Travel Guide • IPCC: **30–1** (table)/IPCC_Mitigation of Climate Change 2007 • NOAA: **171** (table)/© NOAA Fisheries • Ocean Conservancy: **151** (bottom) • Public Domain: **196**, **219–21**; **44** (table)/World Bank Data; **154**(top)/Reprinted with permission from 'Export of Plastic Debris by Rivers into the Sea' by Christian Schmidt, Tobias Krauth, and Stephan Wagner, published in *Environmental Science & Technology* 2017 American Chemical Society; **187** (table)/Government of India Census; **241** (table)/Based on information taken from the Population Reference Bureau • UNHCR: **305** (Figure 3 text) • World Bank: **209** (table)

1 Geographical skills and concepts

1.1 Overview

1.1.1 Introduction

As a student of Geography, you are building knowledge and skills that will be needed by you and your community now and into the future. The concepts and skills that you use in Geography can also be applied to everyday situations, such as finding your way from one place to another. Studying Geography may even help you in a future career here in Australia or somewhere overseas.

Throughout your study of Geography, you will cover topics that will give you a better understanding of the social and physical aspects of the world around you at both the local and global scale. You will investigate issues that need to be addressed now and for the future.

LEARNING SEQUENCE

1.1 Overview
1.2 Work and careers in Geography
1.3 Concepts and skills used in Geography
1.4 **Review**

To access interactivities and resources, select your learnON format at www.jacplus.com.au.

1.2 Work and careers in Geography

1.2.1 Careers that will help the Earth

As the world's population increases and the impacts of environmental changes affect living conditions, people and organisations will need to adapt and develop strategies to manage and sustain fragile environments and resources. Land degradation, marine pollution and feeding the future world population are just three environmental challenges that will be the focus for many occupations in the future. Which careers will be helpful in managing environmental change?

TABLE 1 Careers that will manage environmental change

Conservationists	Conservationists will work to find solutions to land degradation. They will work for governments, in national parks and on policy development, and with local communities on environmental protection projects.

Oceanographers	Work for oceanographers will mainly involve research and monitoring of the marine environment. They may work for governments, providing data and advice on pollution levels, or they may work for private or not-for-profit organisations, helping to suggest and implement plans for cleaning up the oceans.

Agricultural scientists	Agricultural scientists will be employed by the government, and agricultural and horticultural producers. They will work with farmer groups and agribusiness to carry out research, and with mining companies, working on regeneration projects.

1.2.2 Profile of a geographer

Geographers have a love of learning. They are the explorers of the modern world. Geographers are lifelong learners; they expand their knowledge to adapt their skills to the tasks required.

Expansion of knowledge requires a willingness to learn. How many of these skills and attributes have you developed?

- Willingness to learn
- Curiosity and adaptability
- Active listening
- Good communication
- Critical thinking
- Time management
- Problem solving

You can develop your skills and work attributes by undertaking work experience or volunteering activities while you are still at school.

1.2.3 The importance of work experience

The activities you undertake and skills that you develop in Geography will be useful to you in many aspects of your life and your career. In building and managing your career options, it is also helpful to have an understanding of the interconnections between various careers. One way of building your knowledge of these interconnections, particularly in relation to Geography and the career paths that lead from it, is to undertake work experience in the field. Work experience can help you to understand the tasks involved in various roles and the training required to specialise in a particular area. You can gain first-hand experience through observation of and participation in the day-to-day tasks of workplaces.

Volunteering

Volunteering in your community is a great way to find out about different work environments and the things that affect the delivery of the services or programs. Volunteering your time to support local communities and businesses also demonstrates your willingness to learn and support others, and it can provide a great boost to your self-confidence, as well as important skill development. **FIGURE 1** provides some ideas to get you thinking!

FIGURE 1 Some options for volunteering

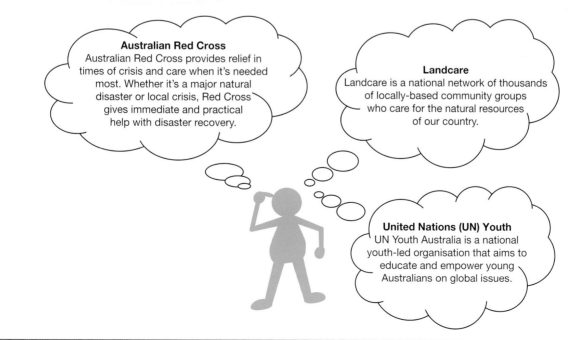

Australian Red Cross
Australian Red Cross provides relief in times of crisis and care when it's needed most. Whether it's a major natural disaster or local crisis, Red Cross gives immediate and practical help with disaster recovery.

Landcare
Landcare is a national network of thousands of locally-based community groups who care for the natural resources of our country.

United Nations (UN) Youth
UN Youth Australia is a national youth-led organisation that aims to educate and empower young Australians on global issues.

Learning directly from industry experts through volunteering can help you to consolidate your interests while also picking up valuable core skills for work (refer to **TABLE 2**). These core skills are considered the most important component of a career portfolio. The study of Geography also assists in the development of these skills.

TABLE 2 The core skills for work

Communication	Ability to use effective listening and speaking skills
Teamwork	Ability to connect and work with others
Learning	Ability to recognise and utilise diverse perspectives
Planning and organisation	Ability to develop plans and see things through to completion
Self-management	Ability to make decisions
Problem solving	Ability to identify and solve problems
Initiative and enterprise	Ability to create and innovate through new ideas
Use of technology	Ability to work in a digital world

How many of the core skills for work have you developed? Use the **FIGURE 2** chart to help you think about your own skills. You may find you have particular strengths and other areas you need to improve upon. If you do this periodically, you can monitor your progress in this area. **FIGURE 3** is an example of a completed chart.

FIGURE 2 Evaluating my core skills for work

Legend
1 = Poor
2 = Fair
3 = Good
4 = Very good
5 = Excellent

FIGURE 3 Core skills analysis — March 2020

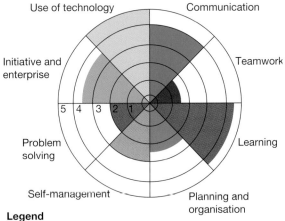

Legend
1 = Poor
2 = Fair
3 = Good
4 = Very good
5 = Excellent

1.2.4 Future careers and Geography

Studies in Geography, along with other Social Science subjects and evidence of your work experience or volunteering, can demonstrate your adaptability, creativity and enterprise skills for future work.

In the future, the type of work that will be available will change in response to the impact of climate change, population growth and decline, and technological innovation. The rapid expansion of world economies will mean that industries will adapt their workforces. Migration and a borderless world will mean that individuals will become global citizens working in large teams around the world. Many of the occupations of this century are yet to be created, while others have been imagined and offer a glimpse into the future.

The hypothetical job advertisement in **FIGURE 4** outlines some of the skills that will be needed to tackle these future roles.

FIGURE 4 Agroecologist — a career of the future

SEEKING AN AGROECOLOGIST...	Job requirements/skills
Agroecologists help restore ecological balance while feeding and fuelling the planet. Agroecologists work with farmers to design and manage agricultural ecosystems whose parts (plants, water, nutrients and insects) work together to create an effective and sustainable means of producing the food and environmentally-friendly biofuel crops of the future.	You will need an undergraduate degree in Agroecology, in which you'll have learned how plants, soil, insects, animals, nutrients, water and weather interact with one another to create the living systems in which crop-based foods are grown. You'll also have learned about the technologies and methods involved in growing food in a sustainable way.
Agroecologists also work with Ecosystem Managers to reintroduce native species and biodiversity to repair the damage done by the ecosystem-disruptive farming techniques of the past.	To be successful in this role, you will need to be responsive to change, demonstrate adaptability by working as part of a global team, and be creative and enterprising in all elements of the business to ensure that business growth is sustainable.

 Resources

 **Weblinks** ACTU Worksite
Careers 2030

1.2 INQUIRY ACTIVITIES

1. Part-time, casual or vacation work are all useful ways to build your core skills for work. Use the **ACTU Worksite** weblink in the Resources tab to locate information on work experience, volunteering and being ready for your first job. **Examining, analysing, interpreting**

2. Over the coming decades, new careers in geography will emerge. Ecosystem auditors, localisers and rewilders will become commonplace in the future. Exploring these careers today can provide an insight into the type of studies and further training you will need to undertake to ensure that you are ready for the workforce of tomorrow.
   (a) Use the **Careers 2030** weblink in the Resources tab to learn about the work of an ecosystem auditor, a localiser or a rewilder.
   (b) Develop a career profile for this emerging career. Include the following details in your profile:
      i. a definition for this occupation
      ii. the core skills needed in this field
      iii. the study or training required to successfully carry out the tasks of the role
      iv. the industries that will employ these occupations. **Examining, analysing, interpreting**

1.3 Concepts and skills used in Geography

1.3.1 Skills used in studying Geography

As you work through each of the topics in this title, you'll complete a range of exercises to check and apply your understanding of concepts covered. In each of these exercises, you'll use a variety of skills, which are identified using the Geographical skills (GS) key provided at the start of each exercise set. These are:

- **GS1** Remembering and understanding
- **GS2** Describing and explaining
- **GS3** Comparing and contrasting
- **GS4** Classifying, organising, constructing
- **GS5** Examining, analysing, interpreting
- **GS6** Evaluating, predicting, proposing.

In addition to these broad skills, there is a range of essential practical skills that you will learn, practise and master as you study Geography. The SkillBuilder subtopics found throughout this title will tell you about the skill, show you the skill and let you apply the skill to the topics covered.

The SkillBuilders you will use in Year 10 are listed below.

- Evaluating alternative responses
- Drawing a futures wheel
- Interpreting a complex block diagram
- Writing a fieldwork report as an annotated visual display (AVD)
- Creating a fishbone diagram
- Reading topographic maps at an advanced level
- Comparing aerial photographs to investigate spatial change over time
- Comparing an aerial photograph and a topographic map
- Using geographic information systems (GIS)
- Describing change over time
- Constructing a land use map
- Building a map with geographic information systems (GIS)
- Constructing and interpreting a scattergraph
- Interpreting a cartogram
- Using Excel to construct population profiles
- How to develop a structured and ethical approach to research
- Understanding policies and strategies
- Using multiple data formats
- Debating like a geographer
- Writing a geographical essay

1.3.2 SPICESS

Geographical concepts help you make sense of your world. By using these concepts you can investigate and understand the world you live in, and you can use them to try to imagine a different world. The concepts help you to think geographically. There are seven major concepts: *space*, *place*, *interconnection*, *change*, *environment*, *sustainability* and *scale*. We will explore each of these concepts in detail in the following sections and through the activities and exercises for this subtopic.

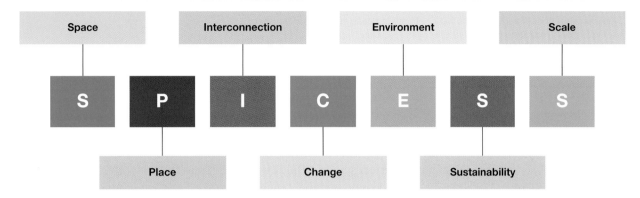

FIGURE 1 A way to remember the seven geographical concepts is to think of the term SPICESS.

| Space | Interconnection | Environment | Scale |

S **P** **I** **C** **E** **S** **S**

| Place | Change | Sustainability |

1.3.3 What is space?

Everything has a location on the space that is the surface of the Earth, and studying the effects of location, the distribution of things across this space, and how it is organised and managed by people, helps us to understand why the world is like it is.

A place can be described by its absolute location; for example latitude and longitude, a grid reference, a street directory reference or an address. Alternatively, a place can be described using a relative location — where it is in relation to another place in terms of distance and direction.

FIGURE 2 A topographic map extract of Narre Warren in 2013, a suburb on the rural–urban fringe of Melbourne

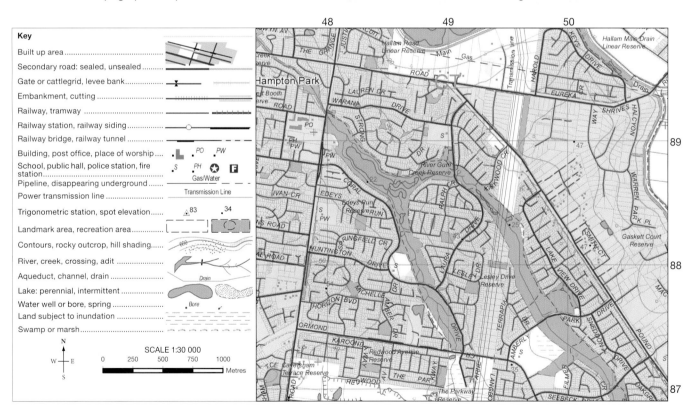

Source: © Vicmap Topographic Mapping Program / Department of Environment and Primary Industries.

Geographers also study how features are distributed across space, the patterns they form and how they interconnect with other characteristics. For example, tropical rainforests are distributed in a broad line across tropical regions of the world, in a similar pattern to the distribution of high rainfall and high temperatures.

┌ Explore more with my**World**Atlas ───

Deepen your understanding of this topic with related case studies and questions.
• Developing Australian Curriculum concepts > **Space**

1.3.4 What is place?

The world is made up of places, so to understand our world we need to understand its places by studying their variety, how they influence our lives and how we create and change them.

Places may be natural (such as an undisturbed wetland) or highly modified (like a large urban conurbation). Places provide us with the services and facilities we need in our everyday life. The physical and human characteristics of places, their location and their environmental quality can influence the quality of life and wellbeing of people living there.

FIGURE 3 The Democratic Republic of the Congo (DRC) has for years been subject to raids by militia groups and the influx of refugees from neighbouring countries. Forests in the country are important places for wildlife habitats and shelter for soldiers. Forests also provide the valuable resources of timber for fuel and building materials for refugees, and cleared land can be planted for food crops.

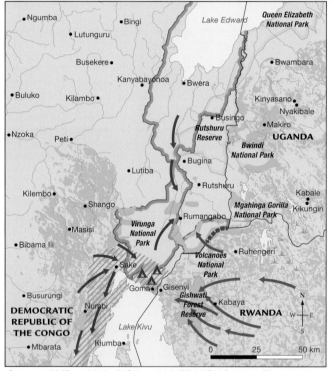
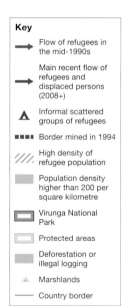

Source: AfriPop 2013. IUCN and UNEP-WCMC 2013, The World Database on Protected Areas (WDPA) [On-line]. Cambridge, UK: UNEP- WCMC. Available at: www.protectedplanet.net [Accessed 30/07/2013]. Made by Spatial Vision.

┌ Explore more with my**World**Atlas ───

Deepen your understanding of this topic with related case studies and questions.
• Developing Australian Curriculum concepts > **Place**

1.3.5 What is interconnection?

People and things are connected to other people and things in their own and other places, and understanding these connections helps us to understand how and why places are changing.

The interconnection between people and environments in one place can lead to changes in another location. The damming of a river upstream can significantly alter the river environment downstream and affect the people who depend on it. Similarly, the economic development of a place can influence its population characteristics; for example, an isolated mining town will tend to attract a large percentage of young males, while a coastal town with a mild climate will attract retirees who will require different services. The economies and populations of places are interconnected.

FIGURE 4 Bangladesh is one of the most flood-prone countries in the world. This is due to a number of factors. Firstly, it is largely the floodplain for three major rivers (the Ganges, Brahmaputra and Meghna), which all carry large volumes of water and silt. Secondly, being a floodplain, the topography therefore is very flat, which allows for large-scale flooding. In addition, the country is located at the head of the Bay of Bengal, which is susceptible to typhoons and storm surges. It is expected that sea level rises associated with global warming will increase the flooding threat even further in the future.

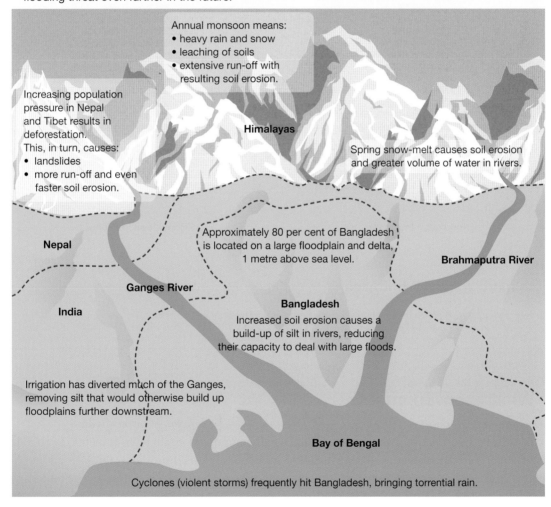

Annual monsoon means:
• heavy rain and snow
• leaching of soils
• extensive run-off with resulting soil erosion.

Increasing population pressure in Nepal and Tibet results in deforestation.
This, in turn, causes:
• landslides
• more run-off and even faster soil erosion.

Himalayas

Spring snow-melt causes soil erosion and greater volume of water in rivers.

Nepal

Approximately 80 per cent of Bangladesh is located on a large floodplain and delta, 1 metre above sea level.

Brahmaputra River

Ganges River

India

Bangladesh

Increased soil erosion causes a build-up of silt in rivers, reducing their capacity to deal with large floods.

Irrigation has diverted much of the Ganges, removing silt that would otherwise build up floodplains further downstream.

Bay of Bengal

Cyclones (violent storms) frequently hit Bangladesh, bringing torrential rain.

Explore more with myWorldAtlas

Deepen your understanding of this topic with related case studies and questions.
• Developing Australian Curriculum concepts > **Interconnection**

1.3.6 What is change?

The concept of change is about using time to better understand a place, an environment, a spatial pattern or a geographical problem.

Topics that are studied in Geography are in a constant process of change over time. The scale of time may be a short time period; for example, the issue of traffic congestion in peak hour or the erosion of a beach during a storm. Other changes can take place over a longer period, such as changes in the population structure of a country, or revegetation of degraded lands.

FIGURE 5 Population pyramids for Kenya, showing the predicted changes from 2012 to 2050. The graphs represent the number of males and females in five-year age groups.

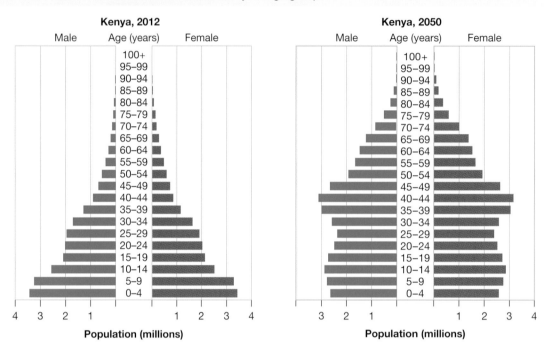

Explore more with **myWorldAtlas**

Deepen your understanding of this topic with related case studies and questions.
• Developing Australian Curriculum concepts > **Change**

1.3.7 What is environment?

People live and depend on the environment, so it has an important influence on our lives.

There is a strong interrelationship between humans and natural and urban environments. People depend on the environment for the source, sink, spiritual and service functions it provides.

Humans significantly alter environments, causing both positive and negative effects. The building of dams to reduce the risk of flooding, the regular supply of fresh water and the development of large-scale urban environments to improve human wellbeing are examples. On the other hand, mismanagement has created many environmental threats such as soil erosion and global warming, which have the potential to have a negative impact on the quality of life for many people.

FIGURE 6 Lake Urmia is the largest lake in the Middle East and one of the largest landlocked saltwater lakes in the world. Since 2005, the lake has lost over 65 per cent of its surface area due to over-extraction of water for domestic and agricultural needs. The lake and its surrounding wetlands are internationally important as a feeding and breeding ground for migratory birds.

(a) 1998

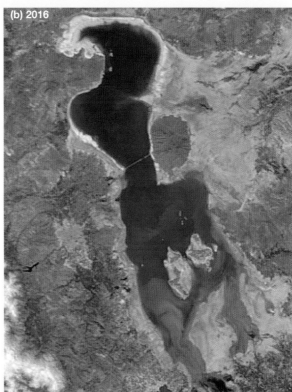

(b) 2016

Explore more with my**World**Atlas

Deepen your understanding of this topic with related case studies and questions.
- Developing Australian Curriculum concepts > **Environment**

1.3.8 What is sustainability?

Sustainability is about maintaining the capacity of the environment to support our lives and the lives of other living creatures.

Sustainability ensures that the source, sink, service and spiritual functions of the environment are maintained and managed carefully to ensure they are available for future generations. There can be variations in how people perceive sustainable use of environments and resources. Some people think that technology will provide solutions, while others believe that sustainable management involves environmental benefits and social justice.

This concept can also be applied to the social and economic sustainability of places and their communities, which may be threatened by changes such as the degradation of the environment. Land degradation in the Sahel region of Africa has often forced people, especially young men, off their land and into cities in search of work.

FIGURE 7 Dust storms are an extreme form of land degradation. Dry, unprotected topsoil is easily picked up and carried large distances by wind before being deposited in other places. Drought, deforestation and poor farming techniques are usually the cause of soil being exposed to the erosional forces of wind and water. It may take thousands of years for a new topsoil layer to form. Therefore, any land practices that lead to a loss of topsoil may be considered unsustainable.

Explore more with my**World**Atlas

Deepen your understanding of this topic with related case studies and questions.
• Developing Australian Curriculum concepts > **Sustainability**

1.3.9 What is scale?

When we examine geographical questions at different spatial levels we are using the concept of scale to find more complete answers.

Scale is a useful tool for examining issues from different perspectives; from the personal to the local, regional, national and global. It is also used to look for explanations or compare outcomes. For example, explaining the changing structure of the population in your local area may require an understanding of migration patterns at a national or even global scale.

FIGURE 8 A map of India showing the distribution of literacy levels (percentage) for 2011

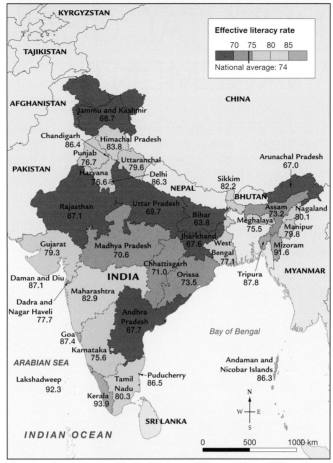

Source: Government of India, Ministry of Home Affairs, Office of Registrar General. (*Note*: Most recent data avaiable)

─ Explore more with myWorldAtlas ────────────────

Deepen your understanding of this topic with related case studies and questions.
* Developing Australian Curriculum concepts > **Scale**

1.4 Review

online only

1.4.1 Key knowledge summary
Use this dot point summary to review the content covered in this topic.

1.4 Exercise 1: Review
Select your learnON format to complete review questions for this topic.

 **Resources**

eWorkbook Crossword (doc-31762)

Interactivity Geographical skills and concepts crossword (int-7668)

UNIT 1
ENVIRONMENTAL CHANGE AND MANAGEMENT

In the twenty-first century, the world faces many environmental challenges. These challenges can range from a local scale — for example, degradation of a nearby creek — through to a global scale — for example, the threat of global warming. Understanding how people and their environments interconnect is vital for explaining environmental changes and helps in planning effective management for a sustainable future.

The future is in our hands.

GEOGRAPHICAL INQUIRY: DEVELOPING AN ENVIRONMENTAL MANAGEMENT PLAN (EMP)

online only

Task

Each class team will research and prepare an EMP that deals with a specific environmental threat and then present it to the class. Decide on an environment and the threat it faces and then devise three key inquiry questions you would like to answer.

Select your learnON format to access:

- an overview of the project task
- details of the inquiry process
- resources to guide your inquiry
- an assessment rubric.

on Resources

💡 **ProjectsPLUS** Geographical inquiry: Developing an environmental management plan (pro-0150)

2 Introducing environmental change and management

2.1 Overview

The Earth is our home and provides us with everything we need to live. What are we doing to it in return?

2.1.1 Introduction

Across the world, humans have caused many environmental changes: pollution, land degradation and damage to aquatic environments. People have different points of view, or world views, on many of these changes. Climate change is a major environmental change as it affects all aspects of the biophysical environment, such as plants and animals; our land; inland water resources; coastal, marine and urban environments. It is vital that we respond intelligently to, and effectively manage, all future environmental changes to minimise negative social and economic impacts.

Human-induced climate change has led to increased severe weather events such as drought. Rivers can dry up, with consequent loss of plant and animal life.

---

on Resources

☑ **eWorkbook**   Customisable worksheets for this topic

▦ **Video eLesson**   What are we doing? (eles-1707)

LEARNING SEQUENCE

2.1 Overview
2.2 Interacting with the environment
2.3 **SkillBuilder:** Evaluating alternative responses
2.4 Is climate change heating the Earth?
2.5 Tackling climate change
2.6 Is Australia's climate changing?
2.7 **SkillBuilder:** Drawing a futures wheel   on line only
2.8 **Thinking Big research project:** Wacky weather presentation   on line only
2.9 **Review**   on line only

To access a pre-test and starter questions and receive immediate, **corrective feedback** and **sample responses** to every question, select your learnON format at www.jacplus.com.au.

2.2 Interacting with the environment

2.2.1 How much space do we need?

If you gathered together all 7.7 billion humans from around the world and gave each person a space of one square metre, the island of Cyprus, which is approximately 8000 square kilometres, would provide standing room for everyone (see **FIGURE 1**). Clearly this would be impractical, and providing services to ensure human wellbeing in an area with a density of almost 1 000 000 per square kilometre would be impossible.

While it is unrealistic to suggest that 0.005 per cent of the total space on Earth is sufficient for humanity, it suggests we need to think about how little personal space we occupy as an individual and, more importantly, how our needs for the Earth's resources can only be satisfied by major modifications to biomes.

2.2.2 Human interaction with the environment

Over 200 years ago, an English scholar named Thomas Malthus proposed that England's population growth would eventually outstrip agricultural production. Malthus's Earth-centred **environmental world view** foretold of problems with supplies of food and warned

FIGURE 1 Humanity crowded onto the island of Cyprus

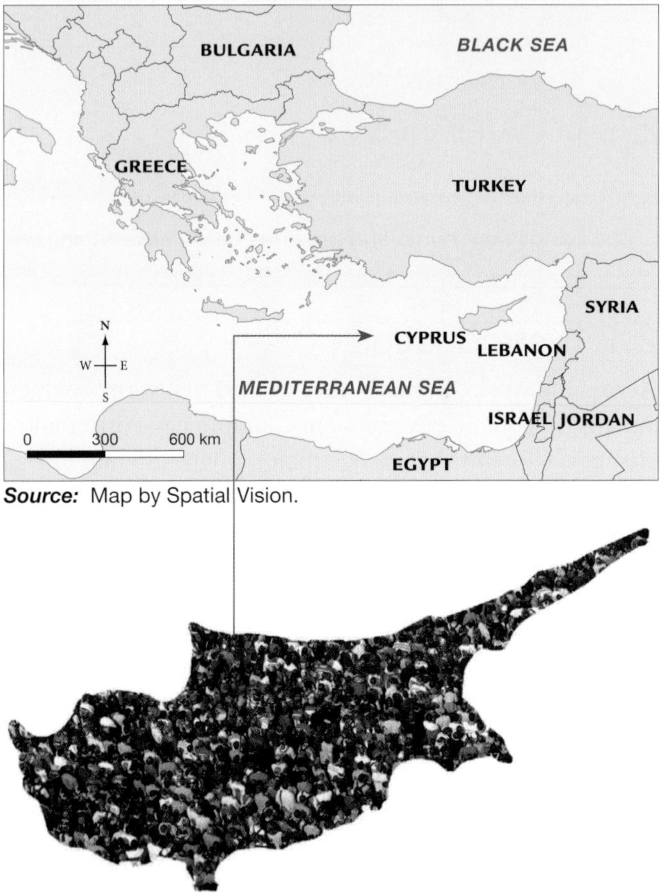

Source: Map by Spatial Vision.

that there would be more deaths because of famine and wars over resources. In 1798 he wrote, 'The power of population is so superior to the power in the earth to produce subsistence for man ...' At the time Malthus wrote his thesis, England was moving into a period known as the Industrial Revolution; a time when the human-centred environmental world view of the government and leaders of industry considered the Earth's resources as limitless and that the development of the economy should take priority over the preservation of the natural world.

Today, many environments have become overloaded with the growing demands for food, land and other resources. This pressure on biomes and ecosystems has led to land degradation, with a consequent loss of habitats and biodiversity. Further consequences of this change are a reduction in human wellbeing and a struggle for social justice as land becomes unproductive because of overuse. Nevertheless, we should remember that change can happen naturally as well as being induced by humans.

Some topics that can help us explore change and the need for careful management include marine environments and coasts, the land, inland waters, and urban or built environments (see **FIGURE 2**).

FIGURE 2 Interaction of environmental change with human wellbeing

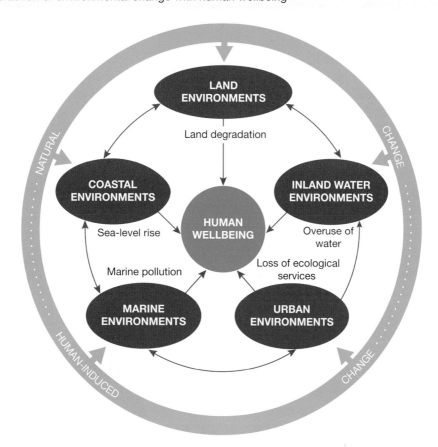

2.2.3 What are ecological services?

A new view of the relationship between the environment and people is one of an **ecological service** or 'what nature provides for humanity'.

Ecological services can be thought of as biological and physical processes that occur in natural or semi-natural ecosystems and maintain the habitability and livelihood of people on the planet. These services are shown in **FIGURE 3**.

Understanding the link (interconnection) between ecological services and human action is important as it can lead to more sustainable practices. The idea of ecological management takes an Earth-centred environmental world view, promoting **stewardship** or custodial management. This view considers caring for the land and the ecological services it provides as paramount. By applying this Earth-centred viewpoint to human uses and management of the environment, future options for human wellbeing will be sustainable. The question is: how do we evaluate human impacts on the environment and what management strategies

FIGURE 3 Ecological services

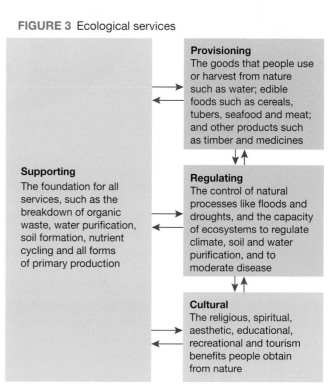

can be implemented to reverse damage and create a sustainable future? As such, we need to consider the costs and benefits, or more simply, the advantages and disadvantages of changes we make to the environment, as there will be consequences in terms of economic viability and social justice.

What is the ecological footprint?

The **ecological footprint** is one means of measuring human demand for ecological services. The footprint takes into account the regenerative capacities of biomes and ecosystems, which are described as the Earth's **biocapacity**. The footprint is given as a number, in hectares of productive land and sea area, by measuring a total of six factors, as shown in **FIGURE 4**. The ecological footprint is a useful indicator of environmental sustainability.

FIGURE 4 Measuring the Earth's ecological footprint

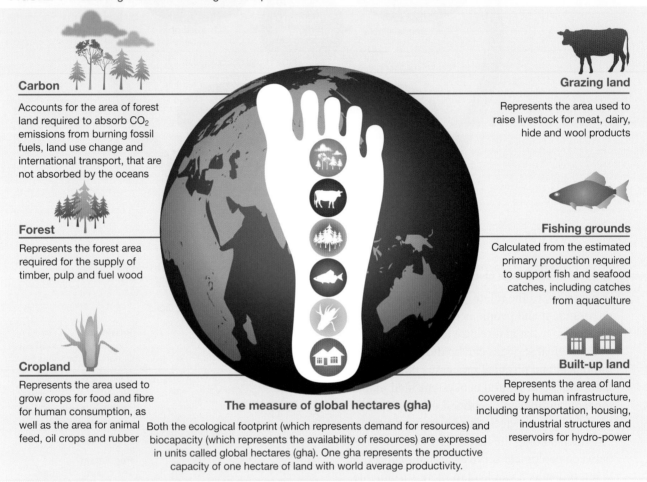

Carbon
Accounts for the area of forest land required to absorb CO_2 emissions from burning fossil fuels, land use change and international transport, that are not absorbed by the oceans

Forest
Represents the forest area required for the supply of timber, pulp and fuel wood

Cropland
Represents the area used to grow crops for food and fibre for human consumption, as well as the area for animal feed, oil crops and rubber

Grazing land
Represents the area used to raise livestock for meat, dairy, hide and wool products

Fishing grounds
Calculated from the estimated primary production required to support fish and seafood catches, including catches from aquaculture

Built-up land
Represents the area of land covered by human infrastructure, including transportation, housing, industrial structures and reservoirs for hydro-power

The measure of global hectares (gha)
Both the ecological footprint (which represents demand for resources) and biocapacity (which represents the availability of resources) are expressed in units called global hectares (gha). One gha represents the productive capacity of one hectare of land with world average productivity.

FIGURE 5 compares the ecological footprint with biocapacity. The elephants represent each region's footprint (per capita) and the balancing balls represent the size of the region's biocapacity (per capita). The dark green background represents the gross footprint of regions that exceed their biocapacity, and the light green background represents those regions that use less than their biocapacity.

In 2018 the total ecological footprint was estimated at 1.7 planet Earths, which means that humanity used ecological services at 1.7 times the biocapacity of the Earth to renew them. The 1.7 ecological footprint figure represents an average for all regions of the Earth. However, the United States, which has an ecological footprint of 8.2, is well above this average. This level of resource use is not sustainable into the future, and raises questions of economic viability, environmental benefit and social justice. **FIGURE 6** shows a map of the Earth's ecological debt. Note that there is a strong relationship between ecological

FIGURE 5 Biocapacity and ecological footprint

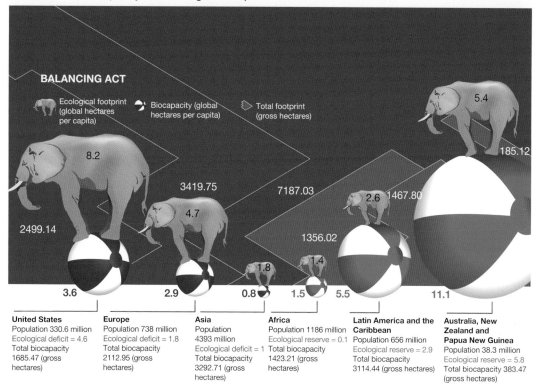

BALANCING ACT

Ecological footprint (global hectares per capita)
Biocapacity (global hectares per capita)
Total footprint (gross hectares)

United States
Population 330.6 million
Ecological deficit = 4.6
Total biocapacity 1685.47 (gross hectares)

Europe
Population 738 million
Ecological deficit = 1.8
Total biocapacity 2112.95 (gross hectares)

Asia
Population 4393 million
Ecological deficit = 1
Total biocapacity 3292.71 (gross hectares)

Africa
Population 1186 million
Ecological reserve = 0.1
Total biocapacity 1423.21 (gross hectares)

Latin America and the Caribbean
Population 656 million
Ecological reserve = 2.9
Total biocapacity 3114.44 (gross hectares)

Australia, New Zealand and Papua New Guinea
Population 38.3 million
Ecological reserve = 5.8
Total biocapacity 383.47 (gross hectares)

FIGURE 6 Ecological debt map

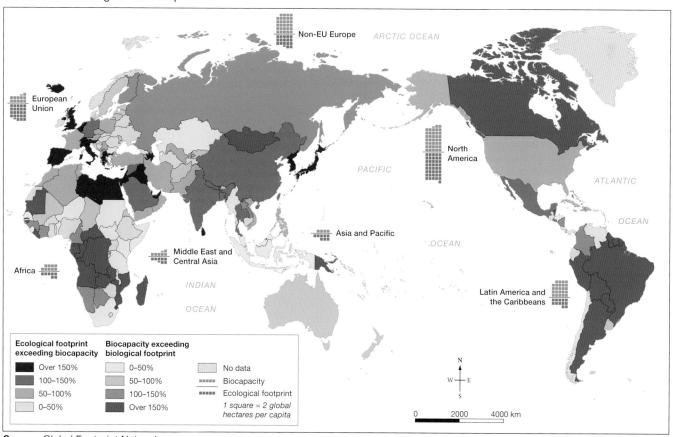

Ecological footprint exceeding biocapacity
Over 150%
100–150%
50–100%
0–50%

Biocapacity exceeding biological footprint
0–50%
50–100%
100–150%
Over 150%

No data
Biocapacity
Ecological footprint
1 square = 2 global hectares per capita

Source: Global Footprint Network.

footprint and a country's wealth and/or population. For example, the United States and much of Europe and Japan are wealthy countries with large ecological footprints and small biocapacities. China and India are highly populated countries with large ecological footprints and small biocapacities. Australia and New Zealand have minimal ecological footprints because they have relatively small populations and high biocapacities.

What is a sustainable world?

A range of indices have been developed in recent years to examine the link between ecological services, human wellbeing and sustainability. These include the Human Development Index (HDI), the Sustainable Society Index (SSI) and the Happy Planet Index (HPI), and each gives a slightly different perspective on human activity and/or sustainability.

The Sustainable Society Index says that sustainable human action must:
- meet the needs of the present generation yet not compromise the ability of future generations to meet their own needs
- ensure that people have the opportunity to develop themselves in a free, well-balanced society that is in harmony with nature.

The Sustainable Society Index gives values to 21 factors across a range of social, political, economic and environmental considerations. It is worthwhile investigating these indices as they put forward many sound ideas about human wellbeing and the sustainability of the ecological services of the natural world.

── Explore more with myWorldAtlas ──────────────────────────

Deepen your understanding of this topic with related case studies and questions.
- Investigating Australian Curriculum topics > Year 10: Environmental change and management > **Global biodiversity**
- Investigating Australian Curriculum topics > Year 10: Environmental change and management > **Indigenous Australians — Caring for Country**

── **on** Resources ──────────────────────────

☑ **eWorkbook** Treading lightly (doc-31761)

🧩 **Interactivity** Nature's bounty (int-3287)

2.2 INQUIRY ACTIVITIES

1. Complete the **Treading lightly** worksheet to explore this subtopic's themes further.
   Examining, analysing, interpreting
2. Indian activist and leader of the movement against British rule in India, Mahatma Gandhi suggested that the Earth provides enough to satisfy everyone's needs, but not everyone's greed. Provide an argument with an Earth-centred viewpoint about this quote and then a counter-argument based on a human-centred viewpoint. Ensure that your arguments are logical, clearly expressed and supported by evidence. Discuss and compare your arguments with a partner. **[Critical and Creative Thinking Capability]**

2.2 EXERCISES

Geographical skills key: GS1 Remembering and understanding **GS2** Describing and explaining **GS3** Comparing and contrasting **GS4** Classifying, organising, constructing **GS5** Examining, analysing, interpreting **GS6** Evaluating, predicting, proposing

2.2 Exercise 1: Check your understanding

1. **GS1** Define 'ecological service'.
2. **GS1** What is an ecological footprint?
3. **GS2** Explain the concept of biocapacity.
4. **GS1** What does one global hectare (gha) represent?
5. **GS2** Outline the four ecological services.

2.2 Exercise 2: Apply your understanding

1. **GS2** Outline six factors that are considered when measuring ecological footprint.
2. **GS2** What are the main criteria that the Sustainable Society Index uses to qualify *sustainable* human action?
3. **GS6** Refer to **FIGURE 5**.
   (a) What reasons can you suggest for the United States having such a large ecological deficit?
   (b) Is Australia in ecological deficit or reserve? How might this be explained?
4. **GS5** Study **FIGURE 6**. Name three countries with biocapacity exceeding ecological footprint by over 150 per cent. What features of these countries might account for this position?
5. **GS5** Refer to **FIGURE 6**. Of the world regions identified on the map, which ones have ecological footprint in excess of biocapacity? Why might this be the case?

Try these questions in learnON for instant, corrective feedback. Go to www.jacplus.com.au.

2.3 SkillBuilder: Evaluating alternative responses

What are alternative responses?

Alternative responses are a range of different ideas/opinions on an issue. Evaluating ideas involves weighing up and interpreting your research to reach a judgement or a decision based on the information.

Select your learnON format to access:

- an overview of the skill and its application in Geography (Tell me)
- a video and a step-by-step process to explain the skill (Show me)
- an activity and interactivity for you to practise the skill (Let me do it)
- questions to consolidate your understanding of the skill.

A special environment

The island's special features include:
- long surf beaches and rocky headlands
- about 40 crystal-clear freshwater lakes. Some of these are perched lakes (that is, they sit, or perch, on an impermeable layer of rock or hardened organic matter lying above the watertable). There are also 'window' and barrage lakes. Window lakes appear when depressions in the land surface dip below the watertable, thus exposing part of it. Barrage lakes form when shifting sand dunes block running water and cause it to pool.
- many streams and creeks
- coloured sand cliffs, some 35 kilometres in length
- salt pans, lagoons, mangrove forests and wetlands
- thick rainforests, some of which are so dense that sunlight does not penetrate the canopy
- offshore seagrass beds to support colonies of dugong
- over 25 species of mammals, including dingoes thought to be the purest strain of the species in Australia
- over 350 species of birds. One of Australia's rarest birds, the endangered ground parrot, is found on the island.
- vast sandblows (that is, tracts of sand moved by the wind) and lofty sand dunes.

Past land uses

Fraser Island once had a sand-mining industry (mining its tracts of mineral-rich black sand). This was stopped in 1976 following a federal government inquiry. There was also a timber industry, disbanded in December 1991 by the Queensland Government after a separate inquiry.

Tourists

Hundreds of thousands of tourists now visit the Fraser coast region every year, injecting some $366 million into the region in 1999. It is estimated that 32 per cent of this visitor expenditure was contributed by tourists to Fraser Island itself. The most obvious risks that tourism brings to the national park have to do with land-clearing, waste, increased traffic and disturbance of the island's flora and fauna.

Ecotourism facilities

The island's Kingfisher Bay Resort and Village has the highest level of accreditation as an ecotourism facility. An environmental impact statement was prepared before the proposed facility was approved for construction.

Dingo management

In the past, many tourists fed the dingoes that roam the island. In April 2001, however, a young boy was tragically killed by dingoes. Tourists are now provided with a 'Dingo Smart' brochure, and are heavily fined if caught feeding a dingo or trying to encourage its attention. Any dingoes known to be a problem are culled.

Managing camping facilities

There are six government-owned camps — at Central Station, Lake Boomanjin (the largest perched lake in the world), Lake McKenzie, Dundubara, Waddy Point and Wathumba — and two that are privately owned. People can also camp on a restricted number of beach areas, but not within 50 metres of a creek, stream or lake. Beach camping areas are temporarily closed sometimes to allow vegetation to regrow or to halt erosion.

Managing four-wheel drive vehicles

Four-wheel drive vehicles are needed to travel around the island. Left unmanaged, these large vehicles could have a significant impact on the island's flora and fauna and on levels of erosion, especially because touring parties tend to drive in the same areas. It is the most attractive parts of the island that are often the most vulnerable.

All vehicles travelling on the island have to display a purchased permit and, more recently, driving and parking on sand dunes have been made illegal. The Environmental Protection Agency has started a campaign to educate four-wheel drivers about the impacts their vehicles have on the island's environment.

Government funding

In 2004, the Australian government's Natural Heritage Trust granted $300 000 to reduce road-related erosion, provide environmentally friendly amenities and better direct pedestrian movement around Fraser Island. Barriers along the sides of the island's roads and better planning have reduced the degree of erosion from run-off. The idea was partly to make pedestrian travel a more attractive option, as well as draw pedestrians away from the island's vulnerable dunes. To do this, boardwalks were built along the banks of Eli Creek (see the photograph on page 26). Stretches of dunes are also regularly closed for rehabilitation. Although the potential threat from tourism-related erosion remains, the stability of the island's sand dunes is improving.

Please tourists, don't pee in the lake

So, we're sure your momma told you not to pee in the swimming pool — but did she also tell you it's bad to pee in a lake? Down in Australia, the beautiful Basin Lake on Fraser Island off Queensland isn't doing well these days, and one of the causes is high levels of urine in the water. The official word is that too many tourists are using the lake as a toilet and that's led the Queensland Parks and Wildlife Service to consider closing one of the access tracks to reduce visitor numbers. Right now 35 000 people visit the lake every year and since there's no in-or-outflow from the lake; whatever goes in, stays in. Our alternative suggestion is to simply stick up a big notice advertising the current urine levels in the lake. We're fairly sure most people would skip the swim.

Source: Fraser Island Travel Guide, 17 October 2008.

Visitor numbers to Fraser Island

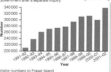

Resources

Video eLesson Evaluating alternative responses (eles-1744)

Interactivity Evaluating alternative responses (int-3362)

2.4 Is climate change heating the Earth?

2.4.1 Climate change and global warming

The world's climate has been changing for millions of years but, more recently, the concentration of greenhouse gases in the atmosphere has increased, leading to **global warming**. It is believed that human particularly burning fossil fuels such as coal and oil, have led to what is known as the **enhanced greenhouse effect**, which is heating the Earth and its atmosphere. Despite ongoing debate about the nature and extent of climate change, the majority of the scientific community agrees that global warming and climate change exist and will result in ongoing changes to world weather patterns and the longer-term climates from the equatorial to polar regions. The wider consequences of global warming will also lead to environmental change across a wide range of biophysical systems (see **FIGURE 1**).

Climate, which can be defined as the long-term weather patterns of a particular area, is highly variable over the Earth's surface. As such, climates in the tropics contrast markedly with climates near the poles. Climate also varies over extensive periods of time, and scientists have described these changes, which date back millions of years, long before the emergence of the human species, as warm periods and ice ages. Currently the Earth is in a warm period, having moved out of ice age conditions as recently as 6000 years ago. Today we realise that human activity is increasing the rate of global warming leading to **climate change**, particularly in the past few hundred years, and this can have serious consequences for the planet (see **FIGURE 2**).

FIGURE 1 Consequences of changes in the global climate

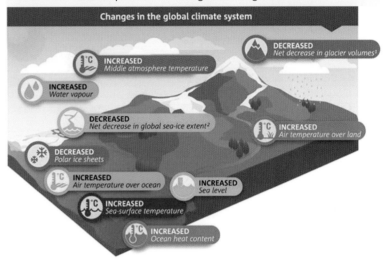

Source: Bureau of Meteorology and CSIRO.

FIGURE 2 Global temperature change to 2017

Global warming relative to 1850–1900 (°C)

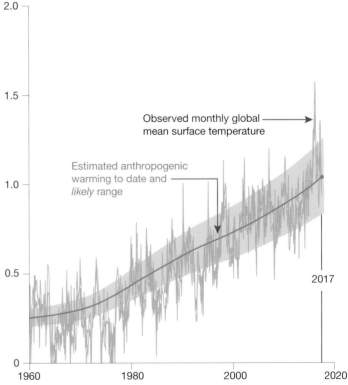

Source: IPCC special report 2018: *Global Warming of 1.5 °C:* Summary for Policymakers, figure a), page 6.

2.4.2 The greenhouse effect

The greenhouse effect is the mechanism where solar energy is trapped by water vapour and gases in the atmosphere, heating the atmosphere and helping to retain this heat, as in a glasshouse (see **FIGURE 3**). The three most important gases responsible for the greenhouse effect are carbon dioxide, nitrous oxide and methane. Without this greenhouse effect, the atmosphere would be much cooler, and ice age conditions would prevail over the planet, making life as we know it impossible.

FIGURE 3 How the greenhouse effect works

Human activity and the enhanced greenhouse effect

Changes in the balance of the greenhouse gases are a natural event, leading to the different climatic conditions on the planet as experienced over geological time. The issue today is how much impact human activity is having on the natural cycle of events, and how this activity is leading to climate change and global warming.

The term 'the enhanced greenhouse effect' has been developed to show that heating of the atmosphere is moving at a rate that is above what could be expected by natural processes of change (see **FIGURE 4**). Recent research by government and non-government organisations has indicated that all parts of the world are vulnerable to the impacts of the enhanced greenhouse effect and associated climate change. Six key risks that have been identified in Australia alone include higher temperatures, sea-level rise, heavier rainfall, greater wildfire risk, less snow cover, reduced run-off over southern and eastern Australia, and more intense tropical cyclones and storm surges along the coast.

FIGURE 4 The enhanced greenhouse effect

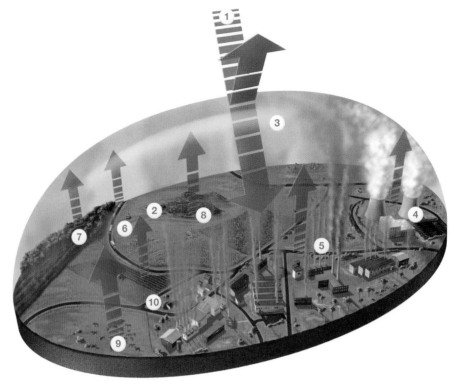

1. Heat from the sun
2. Heat trapped by greenhouse gases
3. Heat radiating back into space
4. Greenhouse gases produced by power stations burning fossil fuels
5. Greenhouse gases produced by industry burning fossil fuels
6. Greenhouse gases produced by transport burning fossil fuels
7. Greenhouse gases released by logging forests and clearing land
8. Methane escaping from waste dumps
9. Methane from ruminant (cud-chewing) livestock, e.g. cattle, sheep
10. Nitrous oxide released from fertilisers and by burning fossil fuels

2.4.3 What can we do?

A switch to renewable energy sources such as solar, wind, water (hydro) and geothermal (heat from inside the Earth's crust) will lead to sustainable energy use in the future, and reduce carbon emissions into the atmosphere, thereby reducing the enhanced greenhouse effect. At the household level, using energy-efficient light bulbs and appliances and installing solar panels to produce hot water and electricity can lead to a significant reduction in greenhouse gas emissions. You could even think of purchasing a new motor vehicle that uses electricity or has a higher fuel efficiency rating.

┌─ Explore more with my**World**Atlas ──────────────────────────────

Deepen your understanding of this topic with related case studies and questions.
- Investigate additional topics > Climate change > **Causes of climate change**
- Investigate additional topics > Climate change > **Larsen Ice Shelf break-up**
- Investigate additional topics > Climate change > **Impacts on polar bears**
- Investigate additional topics > Climate change > **Climate change and Australia**
- Investigate additional topics > Climate change > **Global warming and Antarctica**

2.4 INQUIRY ACTIVITIES

1. In groups, prepare a report that explains how the enhanced greenhouse effect operates, based on the information in **FIGURE 4** in section 2.4.2. You may wish to carry out further research also. Prepare a presentation for the class that includes your suggestions about what we can do to reduce the impacts of the enhanced greenhouse effect. **Examining, analysing, interpreting**
2. Research the scientific debate on climate *change* and global warming, and present cases for both sides of the argument. **Examining, analysing, interpreting**

2.4 EXERCISES

Geographical skills key: GS1 Remembering and understanding **GS2** Describing and explaining **GS3** Comparing and contrasting **GS4** Classifying, organising, constructing **GS5** Examining, analysing, interpreting **GS6** Evaluating, predicting, proposing

2.4 Exercise 1: Check your understanding

1. **GS1** What are the differences between climate *change* and global warming?
2. **GS1** What is the greenhouse effect and what are the three atmospheric gases responsible for this effect?
3. **GS1** What would happen to the Earth if there was no greenhouse effect?
4. **GS2** What *changes* have occurred to the Earth's climate over geological time?
5. **GS2** Why would sea levels be much lower in an ice-age period?

2.4 Exercise 2: Apply your understanding

1. **GS2** What role do trees play in the carbon cycle and in controlling the level of greenhouse gases?
2. **GS6** What impacts will global warming, and in particular higher water temperatures, have on a marine ecosystem such as the Great Barrier Reef?
3. **GS6** Refer to **FIGURE 2**.
   (a) What is the time period shown in the graph?
   (b) What is the total temperature increase shown between the start and end points of the graph?
   (c) What is the general trend shown by the graph?
4. **GS5** Why is the greenhouse effect crucial to maintaining life on Earth?
5. **GS6** In what energy-saving actions does your household participate? Suggest other actions that could be taken into the future.

Try these questions in learnON for instant, corrective feedback. Go to www.jacplus.com.au.

2.5 Tackling climate change

2.5.1 Global action

Climate change is a global phenomenon. The greenhouse gases produced in one country spread through the atmosphere and affect other countries. Action by only a few countries to reduce greenhouse gases will, therefore, have little impact — it requires international cooperation, especially by the largest polluters.

Since the 1990s, countries have met at the United Nations Intergovernmental Panel on Climate Change (IPCC) conferences and agreed to take steps to reduce emissions of greenhouse gases. An early conference developed the **Kyoto Protocol**, an agreement that sets targets to limit greenhouse gas emissions, and 128 countries have agreed to this Protocol. Further conferences in 2009 in Copenhagen, Denmark, 2010 in Cancun, Mexico and in 2015 in Paris led to an important new direction, with all countries agreeing to contain global warming within 2 °C. This means that emissions of CO_2, which were at 395 parts per million (ppm) in 2013, must be kept below 550 ppm to reach this target. If no actions (mitigation measures) are taken, temperatures could increase by 5 °C, as shown in **FIGURE 1**. To date, 192 of the world's 196 countries have signed the Kyoto Protocol, however, close to half have modified their commitment to reach targets for greenhouse emission reductions set for 2020. The United States has signed the Protocol but has not ratified emission targets and Canada has withdrawn from the Protocol.

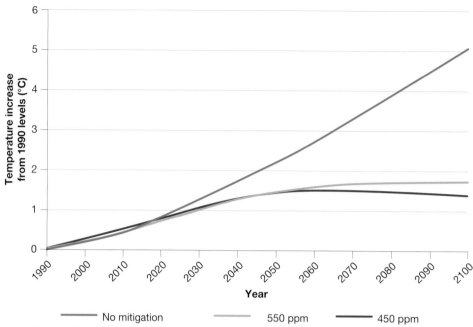

Source: The Garnaut Climate Change Review 2008, p. 88.

To meet the greenhouse gas emissions targets defined by these agreements, countries must make changes that reduce their level of emissions. They can also meet the targets in two other ways:

1. A country can carry out projects in other countries that reduce greenhouse gas emissions and offset these reductions against their own target.
2. Companies can buy and sell the right to emit carbon gases. For example, a major polluter, such as a coal power station, is allowed to emit a certain amount of greenhouse gases. If it is energy efficient, and emits less than its limit, it gains **carbon credits**. It has the right to sell these credits to another company that is having difficulty reducing its emissions. Companies can also gain credits by investing in projects that reduce greenhouse gases (such as renewable energy), improve energy efficiency, or that act as carbon sinks (such as tree planting and underground storage of CO_2).

2.5.2 Australia's action

The Garnaut Report 2011 and the findings of the 2018 IPCC state that it is in Australia's national interest to do its fair share in a global effort to mitigate climate change (see **TABLE 1**). The findings of the 2011 report were confirmed at the IPCC meeting in Paris in 2015. In 2019, the Australian Government announced the Climate Solutions Package, a $3.5 billion investment to deliver on Australia's 2030 Paris climate commitments. The plan will help Australia meet its Kyoto Protocol commitments by:

- providing a $2 billion Climate Solutions Fund to reduce greenhouse gases across the economy through the existing Emissions Reduction Fund, giving farmers, small businesses and Indigenous communities the chance to improve the environment and benefit from new revenue opportunities
- securing our energy future through investments in a high-tech expansion of the Snowy Mountains Scheme and a second interconnector, Marinus Link, between Victoria and Tasmania
- helping households and businesses improve energy efficiency
- implementing a National Electric Vehicle Strategy to ensure a planned and managed transition to new vehicle technology and infrastructure
- helping create green and clean local environments by supporting local communities.

TABLE 1 Potential impacts for each of the three emissions cases by 2100

Emissions case	450 ppm	550 ppm	No action
Likely range of temperature increase from 1990 level	0.8–2.1 °C	1.1–2.7 °C	3–6.6 °C
Percentage of species at risk of extinction	3–13%	4–25%	33–98%
Area of reefs above critical limits for coral bleaching	34%	65%	99%
Likelihood of starting large-scale melt of the Greenland ice sheet	10%	26%	100%
Threshold for starting accelerated disintegration of the West Antarctic ice sheet	No	No	Yes

Source: The Garnaut Climate Change Review 2008, p. 102.

2.5.3 Taking personal action

Australian households produce about one-fifth of Australia's greenhouse gases through their use of transport, household energy and the decay of household waste in landfill. This amounts to about 15 tonnes of CO_2 per household per year. (A tonne of CO_2 would fill one family home.) The Australian Conservation Foundation has suggested a 10-point plan (see **FIGURE 2**) that every Australian household can follow to reduce its level of greenhouse gas pollution.

FIGURE 2 The Australian Conservation Foundation Plan

1. **Switch to green power**
   Choose renewable energy from your electricity retailer and support investment in sustainable, more environmentally friendly energies. Make sure it is accredited GreenPower [electricity produced using renewable resources] — see www.greenpower.gov.au for a list of who qualifies.

2. **Get rid of one car in your household**
   A car produces seven tonnes of greenhouse pollution each year (based on travelling 15 000 km per year). This does not include the energy and water used to build the car — 83 000 litres of water and eight tonnes of greenhouse pollution. So share a car with your family.

3. **Take fewer air flights**
   A return domestic flight in Australia creates about 1.5 tonnes of greenhouse emissions (based on Melbourne to Sydney return). A return international flight creates about 9 tonnes (based on Melbourne to New York return). Holiday closer to home.

4. **Use less power to heat your water**
   A conventional electric household water heater produces about 3.2 tonnes of greenhouse pollution in a year. Using less hot water will reduce your pollution. Using the cold cycle on your washing machine will save 3 kg of greenhouse pollution. Switching off your water heater when you're away will also reduce your energy use.

5. **Eat less meat**
   Meat, particularly beef, has a very high environmental impact, using a lot of water and land to produce it, and creating significant greenhouse pollution. If you reduce your red meat intake by two 150-gram serves a week, you'll save 20 000 litres of water and 600 kg of greenhouse pollution a year.

6. **Heat and cool your home less**
   Insulate your walls and ceilings. This can cut heating and cooling costs by 10 per cent. Each degree change can save 10 per cent of your energy use. A 10 per cent reduction is 310 kg of greenhouse pollution saved.
7. **Replace your old showerhead with a water-efficient alternative**
   This will save about 44 000 litres of water a year and up to 1.5 tonnes of greenhouse pollution from hot water heating (on average).
8. **Turn off standby power**
   Turning appliances off at the wall could reduce your home's greenhouse emissions by up to 700 kg a year.
9. **Cycle, walk or take public transport rather than drive your car**
   Cycling 10 km to work (or school) and back twice a week instead of driving saves about 500 kg of greenhouse pollution each year and saves you about $770. Besides, it's great for your health and fitness!
10. **Make your fridge more efficient**
    Ensure the coils of your fridge are clean and well ventilated — that will save around 150 kg of greenhouse pollution a year. Make sure the door seals properly — this saves another 50 kg. Keep fridges and freezers in a cool, well-ventilated spot to save up to another 100 kg a year. If you have a second fridge, turn it off when not in use.

2.5.4 The role of fossil fuels and renewable energy

Climate authorities have declared that global warming is possibly the most important issue affecting life on Earth now and into the future. The burning of **fossil fuels**, which generate greenhouse gases, is causing the atmosphere to heat up, and it is believed that a sustainable future, in terms of energy use, can be achieved only by reducing the consumption of energy and/or switching to renewable energy forms. While use of fossil fuels is a significant factor in global warming, it should also be realised that there are a number of other human activities that lead to greenhouse gas emissions.

Fossil fuels have been widely used for energy production by human societies since the **Industrial Revolution**. Burning of wood in fires was the earliest use of fuels, and today coal, oil and gas are the fossil fuels of choice. Much of the energy used in society today for transport, domestic use and all forms of industry is from electricity generated by power stations that are fired by fossil fuels (see **FIGURES 3** and **4**).

The environmentally friendly alternative to fossil fuels is renewable energy. This includes hydro-power, solar, wind, wave and tidal, **geothermal**, and bioenergy sources to generate electricity. These sources do not produce greenhouse gases and are replenished in relatively short periods of time (see **FIGURE 5**).

FIGURE 3 World electricity generation by fuel, 2005–30

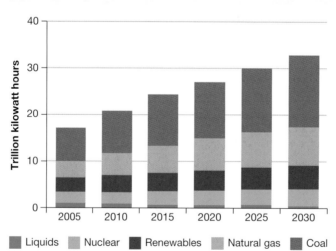

Source: Energy Information Administration (EIA).

FIGURE 4 Australia's primary energy sources, (a) 2016–17 and (b) projected for 2026–27

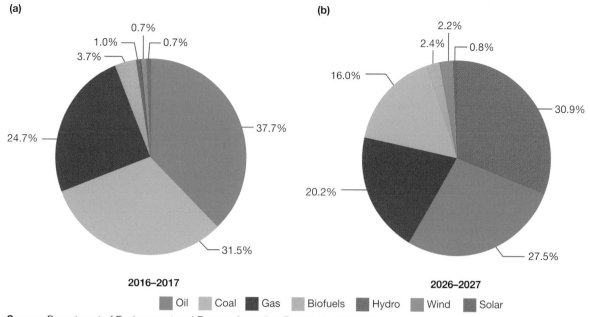

(a)

0.7%
0.7%
1.0%
3.7%
37.7%
24.7%
31.5%

2016–2017

(b)

2.2%
2.4%
0.8%
16.0%
30.9%
20.2%
27.5%

2026–2027

Oil   Coal   Gas   Biofuels   Hydro   Wind   Solar

Source: Department of Environment and Energy, *Australian Energy Update 2018.*

FIGURE 5 Some sources of renewable energy: (a) solar, (b) wind, (c) hydro-electric and (d) geothermal

The movement to environmentally friendly alternative energy fuels will confirm a significant change in thinking, from a human-centred to an Earth-centred world view. This change in thinking will lead to a more sustainable use of energy with a significantly lower impact of greenhouse gas emissions on the environment.

Many countries throughout the world are now using or developing sustainable energy industries. The United States, for example, has established the Clean Energy Plan and currently produces 1.66 per cent of its energy needs from solar power, with renewable energy sources comprising 10 per cent of its total electricity generation. In Europe, Germany has made great progress in harnessing renewable energy sources, which today provide 29.5 per cent of its power generation needs. Wind energy alone provides just over one-third of this amount.

In Australia, with expansive desert regions, there is huge potential to generate solar power. In recent years, the installation of solar panels for domestic households has increased, and this has been supported by a federal government subsidy scheme; however, currently solar energy accounts for only 1.24 per cent of Australia's latest total energy requirements. In other renewable energy fields, wind farms have become more widespread in southern Australia, and there are companies investigating the potential for geothermal energy production.

2.5.5 Future action

In 2015, the IPCC confirmed the 2007 recommendations to reduce greenhouse gas emissions. The recommendations cover a wide range of human activities, with suggestions for management to mitigate global warming (see **TABLE 2**).

For each of the mitigation actions shown in **TABLE 2** there are economic, social and environmental consequences. For example, considering the 'developing safer and cleaner nuclear energy' action, there may be positive economic consequences, such as the creation of energy security and job opportunities, but also negative consequences, such as the cost of waste disposal. Similarly, the social and environmental consequences may be positive, such as reduced air pollution, and negative, such as nuclear accidents.

TABLE 2 Reducing greenhouse gas emissions

	Ways to reduce greenhouse gas emissions
Energy supply	• Increasing use of renewables such as hydro-power, solar, wind, wave and tidal, geothermal and bioenergy • Switching from coal to gas • Carbon capture and storage (CCS) at fossil fuel electricity generating facilities • Developing safer and cleaner nuclear energy (although this is also debated in terms of its sustainability)
Transport	• More fuel-efficient vehicles such as electric, hybrid, clean diesel and biofuels • Changing from road to rail and bus transport systems • Promoting cycling and walking to work
Buildings	• Installing more efficient lighting and day-lighting systems and electrical appliances for heating and cooling, cooking, and washing • Increased use of photovoltaic (PV) solar panels • Improved refrigeration fluids including the recovery and recycling of fluorinated gases
Industry	• More efficient electrical equipment • Heat and power recovery • Material recycling and substitution • Control of gas emissions

(continued)

TABLE 2 Reducing greenhouse gas emissions *(continued)*

	Ways to reduce greenhouse gas emissions
Agriculture	• Improved crop yields and grazing land management • Increased storage of carbon in the soil and reduction of methane gas emissions from livestock manure • Restoration of cultivated soils and degraded lands • Improved nitrogen fertiliser application techniques to reduce nitrous oxide emissions • New bioenergy crops to replace fossil fuels
Forestry/forests	• Planting new forests • Better harvested wood management • Use of forestry products for bioenergy to replace fossil fuel use • Better remote sensing technologies for analysis of vegetation and mapping land-use change
Waste	• Landfill methane recovery • Waste incineration with energy recovery • Composting of organic waste • Controlled waste water treatment • Recycling and waste minimisation

Source: UN IPCC Report 2007.

DISCUSS

Is it the responsibility of the ordinary citizen or the Government to accept the consequences of climate change and do something about it? Discuss your view.   **[Ethical Capability]**

2.5 INQUIRY ACTIVITIES

1. Find what the latest *State of the Climate* report, produced by the Australian Government Bureau of Meteorology and the CSIRO, has to say about the impacts of climate ***change*** on Australia's ***environment***.
   Examining, analysing, interpreting

2. Create a poster to communicate the main points of the Australian Conservation Foundation's 10-point strategy to reduce greenhouse gases. **Classifying, organising, constructing**

3. Use the internet to find out about geothermal energy and its potential as a future energy source.
   Examining, analysing, interpreting

4. Considering nuclear power plant accidents from the past and their impacts on the ***environment***, how might nuclear energy be managed as a safe energy source into the future? Is this a ***sustainable*** option? Conduct research into this topic and write a page outlining your findings and your view.
   Examining, analysing, interpreting

5. Use the internet to access and peruse the IPCC's Climate Change Report. Consider the ***environmental***, social and economic impacts of climate ***change*** mitigation for one of the following: transport, buildings, energy systems or industry. **Examining, analysing, interpreting**

6. Complete the **Tackling climate change** worksheet to explore the concepts in this subtopic further.
   Examining, analysing, interpreting

 Resources

eWorkbook Tackling climate change (doc-31745)

Interactivity Small acts, big changes (int-3288)

2.5 EXERCISES

Geographical skills key: GS1 Remembering and understanding **GS2** Describing and explaining **GS3** Comparing and contrasting **GS4** Classifying, organising, constructing **GS5** Examining, analysing, interpreting **GS6** Evaluating, predicting, proposing

2.5 Exercise 1: Check your understanding

1. **GS1** Which arm of the United Nations is involved in formulating measures to tackle climate *change*?
2. **GS1** What is the Kyoto Protocol?
3. **GS2** Explain why the two basic strategies developed by the Kyoto Protocol can *sustainably* reduce the amount of greenhouse gases in the atmosphere.
4. **GS2** Explain why organisations such as the Australian Conservation Foundation would have different views to companies that produce electricity on the topic of 'climate *change* and global warming'.
5. **GS1** What is meant by the term *fossil fuel*?
6. **GS1** List some major renewable energy sources.
7. **GS2** What would be the negative impacts if all fossil fuels were banned tomorrow?
8. **GS2** What would be the best renewable energy source for the future? Give reasons for your selection.

2.5 Exercise 2: Apply your understanding

1. **GS6** Refer to **FIGURE 1**. How much will temperatures increase by 2070 with no mitigation? Which action will reduce temperature *change* the most by 2100?
2. **GS6** Refer to **TABLE 1** and the 450 ppm case. What is the percentage of species at risk of extinction? What might happen to the Great Barrier Reef under the 550 ppm case?
3. **GS2** What part do international forums play in helping to solve climate *change*?
4. **GS6** Refer to **FIGURE 4**. What percentage of Australia's energy currently comes from renewable sources, and by how much is this projected to *change* by 2026–27?
5. **GS2** Why isn't the use of fossil fuels *sustainable*?
6. **GS6** What would be the *environmental*, social and economic consequences of the different management strategies adopted for renewable energy use in Australia, the United States and Germany, outlined in section 2.5.4?

Try these questions in learnON for instant, corrective feedback. Go to www.jacplus.com.au.

2.6 Is Australia's climate changing?

2.6.1 Impacts of climate change in Australia

Research by government and non-government organisations, such as the Bureau of Meteorology (BOM), the Commonwealth Scientific and Industrial Research Organisation (CSIRO) and the IPCC in 2018, has indicated that Australia is particularly vulnerable to climate change. The consequent changes that will affect all Australian biophysical systems have been identified as eight key risks, which are outlined in **FIGURE 1**.

The challenge for the future is how to manage these risks to minimise negative consequences for the Australian environment, economy and social systems.

What changes are we seeing?

Australia's year-by-year climate statistics can be quite variable against long-term climate records, and floods and droughts have always occurred, breaking records in various regions. These variations in weather patterns are often referred to as climatic anomalies. The concern raised by global warming and climate change is the degree of climate variability and the likelihood of more extreme weather events. For instance, will we experience worse floods and droughts and more bushfires and severe cyclones, tornadoes and the like? Scientific evidence supports the view that there have been more extreme weather events in recent years and that the climate of Australia has undergone significant regional change (see **FIGURE 2**).

FIGURE 1 Key climate and severe weather risks for Australia

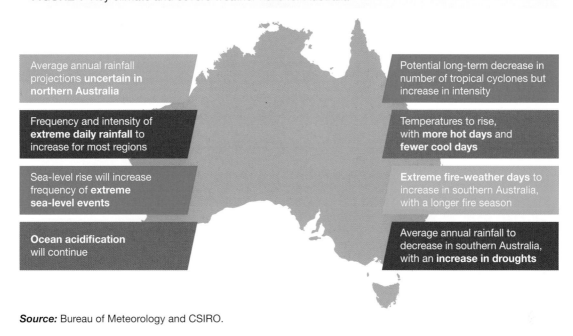

Average annual rainfall projections **uncertain in northern Australia**

Frequency and intensity of **extreme daily rainfall** to increase for most regions

Sea-level rise will increase frequency of **extreme sea-level events**

Ocean acidification will continue

Potential long-term decrease in number of tropical cyclones but increase in intensity

Temperatures to rise, with **more hot days** and **fewer cool days**

Extreme fire-weather days to increase in southern Australia, with a longer fire season

Average annual rainfall to decrease in southern Australia, with an **increase in droughts**

Source: Bureau of Meteorology and CSIRO.

FIGURE 2 The *State of the Climate 2018* report

The *State of the Climate 2018* report, produced by the BOM and CSIRO made the following summary points.
- Australia's climate has warmed just over 1 °C since 1910, leading to an increase in the frequency of extreme heat events.
- Oceans around Australia have warmed by around 1 °C since 1910, contributing to longer and more frequent marine heatwaves.
- Sea levels are rising around Australia, increasing the risk of inundation.
- The oceans around Australia are acidifying (the pH is decreasing).
- April to October rainfall has decreased in the southwest of Australia. Across the same region May–July rainfall has seen the largest decrease, by around 20 per cent since 1970.
- There has been a decline of around 11 per cent in April–October rainfall in the southeast of Australia since the late 1990s.
- Rainfall has increased across parts of northern Australia since the 1970s.
- Streamflow has decreased across southern Australia. Streamflow has increased in northern Australia where rainfall has increased.
- There has been a long-term increase in extreme fire weather, and in the length of the fire season, across large parts of Australia.

Recent severe weather events

In 2018, in their annual report on Extreme Weather, the Climate Council of Australia wrote:

> Climate change is influencing all extreme weather events as they are occurring in a more energetic climate system. Australia is one of the most vulnerable developed countries in the world to the impacts of climate change. Heatwaves are becoming longer, hotter and starting earlier in the year. In the south of the country, where many Australians live and work, dangerous bushfire weather is increasing and cool season rainfall is dropping off, stretching firefighting resources, putting lives at risk and presenting challenges for the agriculture industry and other sectors, such as tourism.

FIGURE 3 outlines records of climatic change for 2018.

FIGURE 3 Indications of climatic change from the Climate Council of Australia

2018/19 ANGRY SUMMER
IN JUST 90 DAYS, OVER 206 RECORDS BROKEN, INCLUDING:
› Record-highest summer temperature: 87 locations
› Record-lowest summer total rainfall: 96 locations
› Record highest summer total rainfall: 15 locations
› Record number of days 35°C or above: 2 locations
› National or state/territory hottest on record: 5 states/territories and (1) Australia.
NORTHERN TERRITORY
› Hottest summer on record (2.67°C above average).
› Rabbit Flat: 34 consecutive days of 40°C or above.
QUEENSLAND
› Cloncurry: 43 consecutive days of 40°C or above (State record).
› Townsville received more than annual average rainfall in 10 days (1,257 mm).
WESTERN AUSTRALIA
› Hottest summer on record (1.73°C above average).
› Marble Bar: 45°C or higher on 32 days during the summer.
NEW SOUTH WALES
› Hottest summer on record (3.41°C above average).
› Bourke: 21 consecutive days above 40°C (State record).
CANBERRA
› Hottest summer on record.
› 35°C or above on 24 days, five times the summer average.
SOUTH AUSTRALIA
› Port Augusta: Hottest temperature this summer - 49.5°C on January 24.
› Adelaide: Hottest temperature for January or any month – 46.6°C on January 24.
VICTORIA
› Hottest summer on record (2.54°C above average).
TASMANIA
› Driest January on record.
› Bushfires burned ~ 200,000 hectares of vegetation.
Note: For all statistics, the average is calculated over the period between 1961 and 1990. Records are for seasonal or monthly mean temperature unless otherwise specified.
CLIMATECOUNCIL.ORG.AU
crowd-funded science information

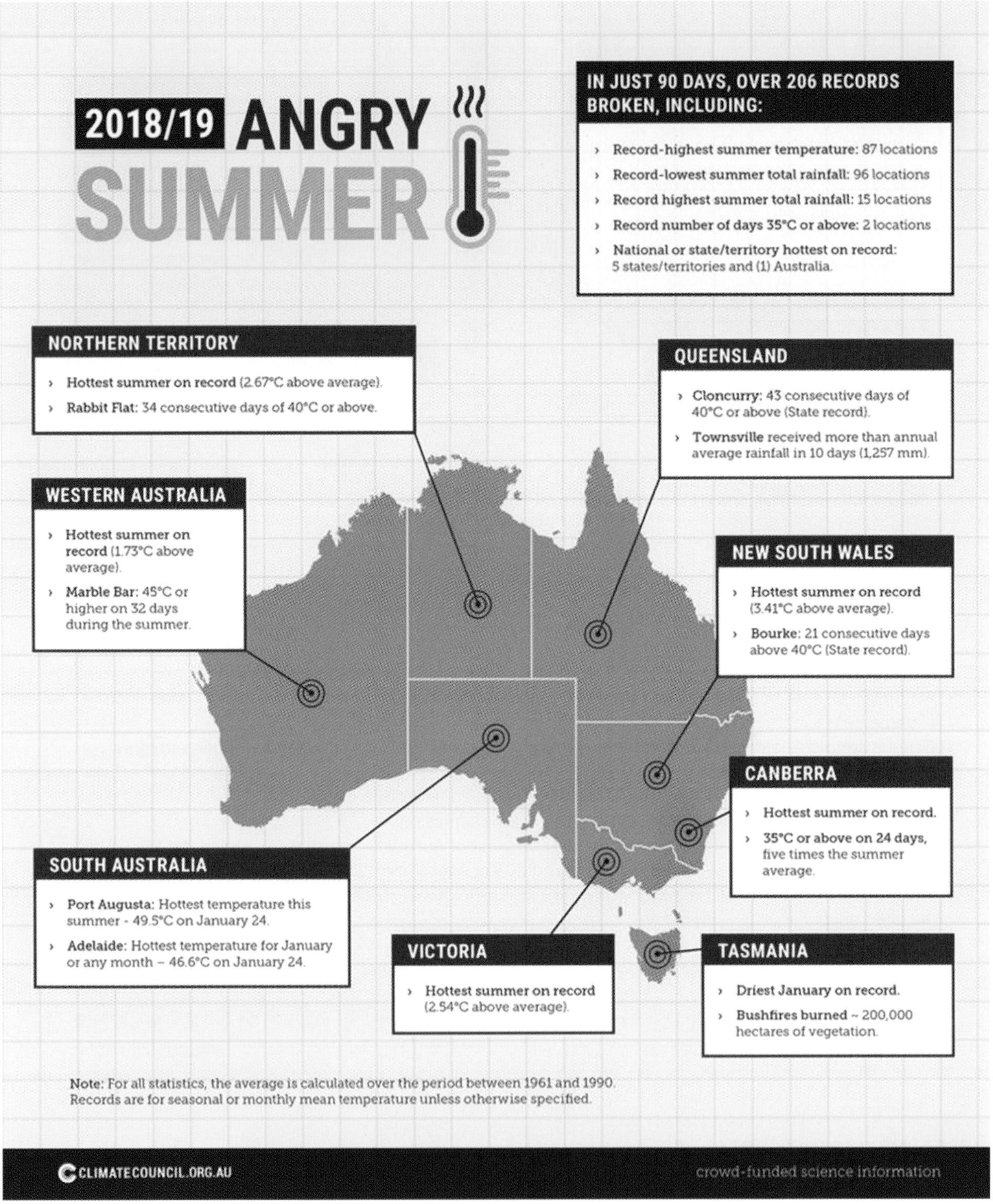

FIGURE 4 Floods in Townsville, February, 2019, due to tropical cyclone Omar

2.6.2 The impact of climate change on the environment, economy and social systems

Some of the impacts of climate change that will require management by governments and communities include:

- impacts on fragile and diverse biomes and ecosystems; for example, the Great Barrier Reef, where warming of 1 °C is expected to have significant impacts on biodiversity, with losses of species and associated coral communities and the potential for up to 97 per cent of the reef to be subject to coral bleaching

FIGURE 5 The fragile Great Barrier Reef ecosystem may be significantly affected by climate change.

- changed temperatures and rainfall regimes affecting the potential of agriculture and forestry to maintain crop yields such as wheat, and timber yields from forests
- reduced river flows in the Murray–Darling Basin with significant impacts on agriculture, industry and urban household use
- more extreme weather events such as heatwave conditions, with an increase in the number of days when the forest fire index rating is very high or extreme
- more severe tropical cyclones, with associated property damage due to strong winds and flooding
- spread of tropical diseases such as dengue fever and malaria to southern regions.

Managing the impact of severe weather

Scientific experts agree that environments will change due to global warming and climate change and this will have a range of economic and social consequences, to which society will need to adapt. Where particular industries such as agriculture and forestry may be affected, there could be a need for governments and other agencies to encourage and facilitate the development of employment opportunities in alternative industries, such as renewable energy.

In dealing with the potential impact of severe weather events, a number of approaches may be taken. The redesign of urban infrastructure to improve storm water drainage is a management strategy to reduce the threat of flooding. If redesign is not able to solve the problem, there may be a need for some people to consider relocating away from the flood-prone coastal and riverine locations in which they currently live.

Successful management strategies in relation to events such as cyclones and bushfires include the development of improved tropical cyclone warning systems, with monitoring conducted and warnings issued by the Bureau of Meteorology, and bushfire warnings, issued by relevant state fire authorities. National and state-based agencies such as Emergency Management Victoria, Emergency New South Wales and the Department of Community Safety in Queensland provide a range of information and resources aimed at minimising the impacts on communities of severe weather events, and assisting with management strategies such as emergency evacuation planning. Improved building design to withstand these severe weather events is another successful form of management strategy.

Government Disaster Relief programs that offer financial and other assistance to individuals and communities to recover after events such as flood, fire and drought are further examples of impact management.

Perhaps most importantly, the root causes of severe weather events as a consequence of global warming and climate change need to be addressed. The Australian Conservation Foundation's 10-point plan (see **FIGURE 2** in section 2.5.3) suggests a range of personal energy-use management strategies that aim to minimise individuals' contribution to greenhouse gas emissions, such as switching to solar energy and other renewables. If adopted by businesses and the general community, these strategies will go a long way towards reducing the environmental impacts of climate change and global warming, thereby mitigating the social and economic impacts.

2.6 INQUIRY ACTIVITY
Use the internet to find out about Pacific Island nations that are threatened by rising sea levels because of climate **change**. Where are they located? Are any plans in place to protect these areas? Discuss your findings with a partner. **Examining, analysing, interpreting**

2.6 EXERCISES
Geographical skills key: GS1 Remembering and understanding **GS2** Describing and explaining **GS3** Comparing and contrasting **GS4** Classifying, organising, constructing **GS5** Examining, analysing, interpreting **GS6** Evaluating, predicting, proposing

2.6 Exercise 1: Check your understanding
1. **GS1** How are Australia's temperatures expected to alter because of climate **change**?
2. **GS1** Name three extreme weather events that are expected to increase in frequency due to climate **change**.
3. **GS2** Study **FIGURE 3**, which outlines climatic records broken in Australia. Describe the general pattern of temperature and rainfall extreme weather events for the 2018 period outlined.
4. **GS2** From **FIGURE 3**, list the types of temperature and rainfall **changes** recently experienced where you live.
5. **GS1** By how much has Australia's climate warmed since 1910?

2.6 Exercise 2: Apply your understanding
1. **GS4** Develop an evacuation plan for a house or town in a bushfire-prone area.
2. **GS6** How might climate **change** affect tourism in the Snowy Mountains region of Australia?
3. **GS6** How will rising sea levels affect Australia's state capital cities that are located on the coast?
4. **GS6** What strategies do you think people who live in tropical cyclone-prone areas could adopt to cope with increased severe weather events?
5. **GS6** Identify three key points from the *State of the Climate* report (**FIGURE 2**) that you consider to be of greatest concern. Justify your choices.

Try these questions in learnON for instant, corrective feedback. Go to www.jacplus.com.au.

2.7 SkillBuilder: Drawing a futures wheel

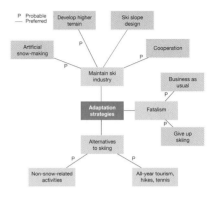

What is a futures wheel?

A futures wheel is a series of bubbles or concentric rings with words written inside each to show the increasing impact of change. It helps show the consequences of change.

Select your learnON format to access:

- an overview of the skill and its application in Geography (Tell me)
- a video and a step-by-step process to explain the skill (Show me)
- an activity and interactivity for you to practise the skill
  (Let me do it)
- questions to consolidate your understanding of the skill.

 Resources

 Video eLesson Drawing a futures wheel (eles-1745)

 Interactivity Drawing a futures wheel (int-3363)

2.8 Thinking Big research project: Wacky weather presentation

SCENARIO

As the regular weather presenter for an evening news program, you have been asked by the producer to compile a segment on the history of extreme weather events in Australia and to outline the link between these events and climate change.

Select your learnON format to access:

- the full project scenario
- details of the project task
- resources to guide your project work
- an assessment rubric.

 Resources

ProjectsPLUS Thinking Big research project: Wacky weather presentation (pro-0211)

2.9 Review

2.9.1 Key knowledge summary

Use this dot-point summary to review the content covered in this topic.

2.9.2 Reflection

Reflect on your learning using the activities and resources provided.

on Resources

eWorkbook Reflection (doc-31763)

Crossword (doc-31764)

Interactivity Introducing environmental change and management crossword (int-7669)

KEY TERMS

biocapacity the capacity of a biome or ecosystem to generate a renewable and ongoing supply of resources and to process or absorb its wastes

carbon credits term for a tradable certificate representing the right of a company to emit one metric tonne of carbon dioxide into the atmosphere

climate change any change in climate over time, whether due to natural processes or human activities

ecological footprint a measure of human demand on the Earth's natural systems in general and ecosystems in particular; the amount of productive land required by each person for food, water, transport, housing, waste management and other purposes

ecological service the benefits to humanity from the resources and processes that are supplied by natural ecosystems

enhanced greenhouse effect increasing concentrations of greenhouse gases in the Earth's atmosphere, contributing to global warming and climate change

environmental world view varying viewpoints, such as environment-centred as opposed to human-centred, in managing ecological services

fossil fuels carbon-based fuels formed over millions of years, which include coal, petroleum and natural gas. They are called non-renewable fuels as reserves are being depleted at a faster rate than the process of formation.

geothermal (power) describes power that is generated from molten magma at the Earth's core and stored in hot rocks under the surface. It is cost-effective, reliable, sustainable and environmentally friendly.

global warming the observable rising trend in the Earth's atmospheric temperatures, generally attributed to the enhanced greenhouse effect

Industrial Revolution the period from the mid 1700s into the 1800s that saw major technological changes in agriculture, manufacturing, mining and transportation, with far-reaching social and economic impacts

Kyoto Protocol an internationally agreed set of rules developed by the United Nations aiming to reduce climate change through the stabilisation of greenhouse gas emissions into the atmosphere

stewardship the caring and ethical approach to sustainable management of habitats for the benefit of all life on Earth

---

3 Land environments under threat

3.1 Overview

From housing to food production, we use land for many different things. What impact are we having on this important resource?

3.1.1 Introduction

Land is one of our most valuable resources. Left alone, it exists in a state of balance, and if managed wisely, will continue to do so. However, the land is under increasing pressure as a direct result of population growth; agriculture, mining and expanding settlements all have the potential to interfere with natural processes and cause environmental damage. Our challenge is to balance the needs of our growing population with sustainable land management practices, to protect this precious resource for future generations.

on Resources

eWorkbook	Customisable worksheets for this topic
Video eLesson	Wasting our land (eles-1708)

LEARNING SEQUENCE

3.1 Overview
3.2 The causes and impacts of land degradation
3.3 **SkillBuilder:** Interpreting a complex block diagram `online only`
3.4 Managing land degradation
3.5 Environmental change and salinity
3.6 Desertification: the drylands are spreading
3.7 Introduced species and land degradation
3.8 Native species and environmental change
3.9 Indigenous communities and sustainable land management
3.10 **SkillBuilder:** Writing a fieldwork report as an annotated visual display (AVD) `online only`
3.11 **Thinking Big research project:** Invasive species *Wanted!* poster `online only`
3.12 **Review** `online only`

To access a pre-test and starter questions and receive immediate, **corrective feedback** and **sample responses** to every question, select your learnON format at www.jacplus.com.au.

3.2 The causes and impacts of land degradation

3.2.1 Explaining land degradation

Land degradation is the process that reduces the land's capacity to produce crops, support natural vegetation and provide fodder for livestock. Land degradation causes physical, chemical and biological changes; the natural environment deteriorates and the landscape undergoes a dramatic change (see **FIGURE 1**). Common causes of land degradation include soil erosion, increased salinity, pollution and desertification.

FIGURE 1 Land degradation causes physical, chemical and biological changes to the natural environment.

BEFORE

AFTER

3.2.2 What causes land degradation?

Land can be degraded in many ways, but most of the causes can be traced back to the influences of human activity on the natural environment. **FIGURE 2** outlines these activities and their impacts.

FACTORS THAT CONTRIBUTE TO LAND DEGRADATION
- Poor management leads to the loss of nutrients vital for plan growth.
- Removal of vegetation makes the land vulnerable to erosion by wind and water.
- When urban development encroaches on agricultural land, vegetation is removed and the waste generated is disposed of in landfill.
- Poor agricultural practices, especially related to irrigation and the use of chemical fertilisers, can lead to the soil becoming saline or acidic.

FIGURE 2 Why land degrades

A   When land is cleared or overgrazed, it becomes vulnerable to erosion by wind and water. The nutrient-rich soil is either washed or blown away, reducing the quality and quantity of crop yields. Dust storms result and sediment is transported to rivers, where it can smother marine species.

B   Introduced species such as rabbits eat grass, shrubs and young trees (saplings) down to the soil, thus exposing it to erosion. Their burrows increase erosion as they destabilise the soil. Rabbits also compete with native animals for food and burrows.

C   Tourism encourages the clearing of sand dunes for high-density housing, and mountain slopes for ski runs, leaving the surface exposed to erosion.

D   Overgrazing leads to nutrient-rich soil being washed or blown away. Animals with hard hoofs such as sheep and cattle trample vegetation and compact the soil, making it increasingly difficult for native species to grow. This leads to increased run-off after heavy rain.

E   Climate change will affect land degradation in the future. Higher sea levels will flood low-lying coastal areas. Expanding cities, removal of vegetation and use of concrete reduces the ability of the land to absorb moisture. This not only increases erosion, but can reduce the amount of rainfall in an area.

F   Urban communities produce large quantities of waste, which is deposited in landfills. Much of the rubbish remains toxic or, in the case of plastic bags, takes hundreds of years to break down. Liquid and solid waste seeps into groundwater and runs off into rivers and eventually into the sea, killing marine species.

G   Introduced plant species such as blackberries and Paterson's Curse (Salvation Jane) choke the landscape and compete with native vegetation. Their dense groundcover prevents light from reaching the soil.

H   Salinity occurs naturally in areas where there is low rainfall and high evaporation and also where the land was below sea level millions of years ago. Salinity is also caused by excess irrigation and clearing natural vegetation. In some cases the watertable rises, bringing salt to the surface.

FIGURES 3 and **4** show that agricultural activities and overgrazing combined account for more than 50 per cent of land degradation in the Asia–Pacific region and globally.

FIGURE 3 Causes of land degradation in the Asia–Pacific region; Australia is ranked fifth in clearing of native vegetation.

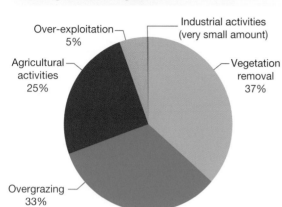

Over-exploitation 5%
Industrial activities (very small amount)
Agricultural activities 25%
Vegetation removal 37%
Overgrazing 33%

FIGURE 4 Main causes of land degradation globally

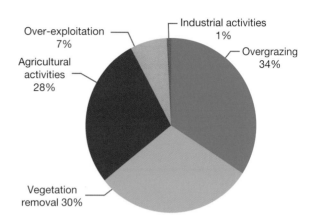

Over-exploitation 7%
Industrial activities 1%
Overgrazing 34%
Agricultural activities 28%
Vegetation removal 30%

3.2.3 The effects of land degradation

Even small changes can have dramatic effects on the land. The shortcut students take from the oval to the classroom can soon reduce a grassy area to dust. Drought can quickly reduce the productivity of an area used for farming. A farmer who neglects the land after one growing season may still be able to raise a good crop the following season, but if the land is neglected year after year it will eventually become unproductive.

The effects of land degradation are far-reaching. Farmland productivity diminishes and yields drop because the soil becomes exhausted through overuse or deforestation (see **FIGURE 5**). Costs increase, as the land requires more treatment with fertiliser, and **topsoil** and nutrients in the soil need to be replaced. Valuable topsoil is often washed away into rivers and out to sea. Nutrients cause foul-smelling blue–green **algal blooms** that choke waterways. These blooms decrease water quality, poison fish and pose a direct threat to other aquatic life.

FIGURE 5 Land clearing and deforestation leave the land vulnerable to erosion. When rain falls on a well-vegetated hillside, the water is absorbed by plant roots and held in the soil. However, if the vegetation is removed, there is nothing to stabilise the soil and hold it together, especially when it rains. Rills and gullies form (see subtopic 3.4) where the unprotected soil is washed away, and landslides may occur.

Forested hillside

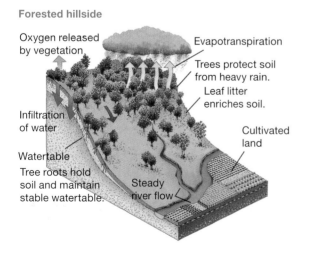

Oxygen released by vegetation
Evapotranspiration
Trees protect soil from heavy rain.
Leaf litter enriches soil.
Infiltration of water
Cultivated land
Watertable
Tree roots hold soil and maintain stable watertable.
Steady river flow

After deforestation

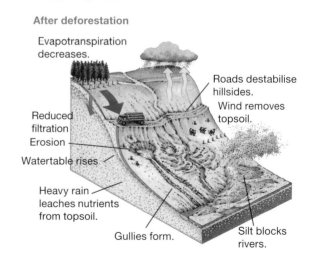

Evapotranspiration decreases.
Roads destabilise hillsides.
Wind removes topsoil.
Reduced filtration
Erosion
Watertable rises
Heavy rain leaches nutrients from topsoil.
Gullies form.
Silt blocks rivers.

3.2.4 How has agriculture degraded the Australian landscape?

Climate, topography, water supply and soil quality are the major physical factors that determine how land can be used. When white settlers first colonised Australia they brought seeds and hoofed animals from Europe with them. They intended to farm here as they had always done at home; they undertook large-scale clearing of trees and shrubs and planted crops and pasture. However, the Australian landscape is much different from what they had left behind. Australia's soils are naturally low in nutrients and have a poor structure. Much of the vegetation is shallow-rooted and easily disturbed when the land is ploughed and made ready for cultivation. Even in areas where the soil is fertile, over-irrigation and deforestation can raise the **watertable** and bring salt to the surface, decreasing soil fertility. Australia also has variable rainfall, and drought can last for years. This leaves the earth dry, parched, barren and unproductive. Floods can wash away a farmer's livelihood and leave the land flooded.

FIGURE 6 A former freshwater lake affected by salinity. The high salt levels have killed the native eucalypts; the smaller plants are more salt tolerant.

Are kangaroos the answer?

Australia's early economic growth and development depended on the success of agriculture. The first settlers knew they had to be self-sufficient, for their own survival and that of the new colony. They had to learn quickly how to farm soil that was often hard, stony and exposed to a variety of climatic extremes. Overgrazing by heavy, hard-hoofed animals such as sheep and cattle increased the rate of land degradation, especially in arid and semi-arid regions. Kangaroo farming has been presented as an alternative sustainable solution to this problem.

Those in favour of kangaroo farming claim it would be more environmentally friendly as kangaroos are not hard-hoofed; issues such as soil compaction and vegetation trampling would be lessened. There could also be human health benefits as kangaroo meat contains less fat and fewer calories than both lamb and beef. Those against the idea argue that because of various species characteristics, kangaroo farming is not commercially viable in the long term (see **FIGURE 7**).

FIGURE 7 Comparing commercial viability of kangaroo farming with sheep farming

- Young dependent on mother for 14 months
- Cannot be sold live
- One-off use (meat and skin)
- 18 months before meat can be harvested
- A 60 kg kangaroo yields 6 kg of prime meat; the rest is suitable only as pet food.
- Can meet only 0.5 per cent of current needs

- Young dependent on mother for a few months
- Can be sold live
- Multiple uses (wool, meat and skin)
- Breed from 12 months; multiple births possible
- Meat can be harvested from 3–6 months
- Yields 20 kg of prime meat
- Easier to herd and care for

3.2.5 Where is the land degrading?

In 1961, globally, there were 0.37 hectares of arable (productive) land available to grow food for every man, woman and child. By 2016, this figure had fallen to 0.19 hectares. **TABLE 1** shows changes to arable land availability over time for Australia and the world. These changes are due to factors such as population growth, urban sprawl, land degradation and climate change. In Australia, approximately two-thirds of the land used for agricultural production is degraded. **FIGURES** 8 and 9 show the severity of land degradation in Australia and globally.

TABLE 1 Arable land per person over time

Year	Arable land per person (hectares)	
	Global	Australia
1961	0.37	2.88
1976	0.29	3.0
1991	0.23	2.64
2016	0.19	1.90

According to the United Nations, around 42 per cent of the world's poorest people live on the most degraded lands. Areas where the land degradation is most rapid are also those where population growth is greatest. In sub-Saharan Africa, for example, the annual rate of population growth is 2.7 per cent annually, significantly higher than the world average of 1.15 per cent. Africa is one of the most vulnerable continents, having lost 65 per cent of its arable lands and 25 per cent of its overall land area to desert. Experts estimate that 10 million hectares of land needs to be rehabilitated each year to reverse the current trends in land degradation.

FIGURE 8 Severity of soil degradation in Australia

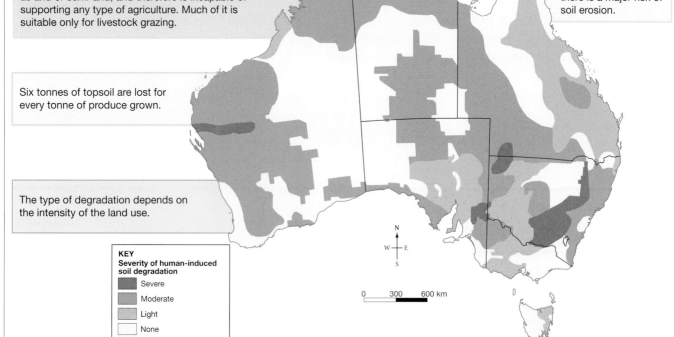

Only about six per cent of the continent is arable without irrigation. Seventy per cent of the landmass is classed as arid or semi-arid, and therefore is incapable of supporting any type of agriculture. Much of it is suitable only for livestock grazing.

When sloping land is cleared of vegetation, there is a major risk of soil erosion.

Six tonnes of topsoil are lost for every tonne of produce grown.

The type of degradation depends on the intensity of the land use.

KEY
Severity of human-induced soil degradation
- Severe
- Moderate
- Light
- None

0   300   600 km

Source: MAPgraphics Pty Ltd Brisbane.

FIGURE 9 Soil degradation is a global problem affecting every permanently inhabited continent.

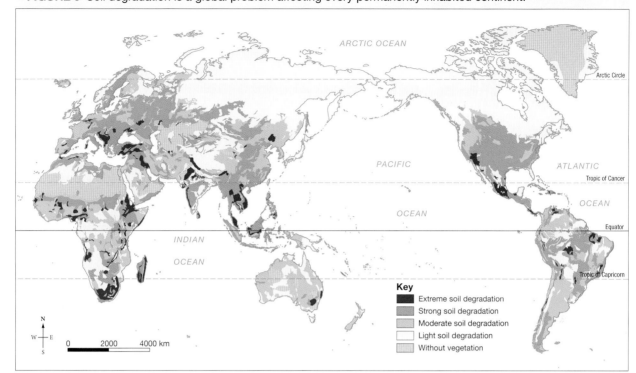

Key
- Extreme soil degradation
- Strong soil degradation
- Moderate soil degradation
- Light soil degradation
- Without vegetation

Source: © Commonwealth of Australia Geoscience Australia 2013. © Commonwealth of Australia Department of Sustainability, Environment, Water, Population and Communities 2013. Map by Spatial Vision.

In 1950, about 65 per cent of the world's population lived in developing nations; however, this figure is expected to rise to 85 per cent by the year 2030. These people are dependent on the most fragile environment for their survival.

3.2.6 Challenges to food production

Today, there are more than three times the number of people living on the Earth than in 1950: over 7.7 billion, compared with 2.5 billion. Our primary energy use is five times higher and our use of fertilisers has increased eightfold. In addition, the amount of nitrogen pumped into our oceans has quadrupled. While global population is increasing, the land upon which food is grown to feed this population is degrading.

Everyone on Earth relies on the land. Apart from providing us with a place to live, the land also provides most of our food and products such as oil and timber. Since 1990, the amount of land irrigated for agriculture has doubled and agricultural production has trebled. With the world's population expected to reach 9.8 billion by 2050 and 11.2 billion by the end of the century, the land and its resources will be placed under even more pressure.

Globally, 75 per cent of the Earth's total land area is classed as degraded, and around 60 per cent of this degraded land is used for agricultural production. In 2017 alone, 24 billion tonnes of fertile soil was lost worldwide, with the worst affected areas being in sub-Saharan Africa. Global food production is already being undermined by land degradation and shortages of both farmland and water resources, making feeding the world's rising population even more daunting.

Land degradation is a global problem. If the current trends continue, our ability to feed a growing world population will be threatened. Although we have a better understanding of factors that contribute to land degradation, the challenge is to manage the land sustainably for the future and reverse the trends.

FIGURE 10 Sustainable land use will be important for ensuring the health of the land while increasing our capacity to feed the Earth's growing population.

 Resources

Interactivity Destroying the land (int-3289)

3.2 INQUIRY ACTIVITIES

1. Investigate a particular type of land degradation and produce an annotated visual display to show the parts of Australia affected by it. Cover major contributing factors and possible management strategies. Add an inset diagram that examines a particular *place*, *scale* and rate of *change* associated with this type of degradation. Include your own recommendations for *sustainable* use of the *environment* to combat the issue you have investigated.   **Classifying, organising, constructing**

2. (a) In small groups, prepare a fold-out educational pamphlet outlining the damage caused by the waste produced by urban communities each year. Make sure you clearly outline the *interconnection* between human activity and *environmental* harm. Devise a strategy to reduce this waste and estimate the difference that this would make to the amount of waste generated. **Classifying, organising, constructing**

   (b) In your group, evaluate your own and others' contributions to this group task, critiquing roles including leadership, and provide useful feedback to your peers. Evaluate your group's task achievement and make recommendations for improvements in relation to team goals.   **[Personal and Social Capability]**

3. Create an overlay theme map. Prepare a base map that shows the extent of land degradation around the world. Prepare an overlay map showing land use. Annotate your overlay with any similarities and differences between the two maps.   **Classifying, organising, constructing**

4. (a) Investigate alternatives to traditional livestock farming of sheep and cattle, such as kangaroos or emus. Use the information presented in this subtopic as a starting point. Present a reasoned argument for or against this type of farming as a *sustainable* alternative.   **Examining, analysing, interpreting**

   (b) Evaluate emotional responses and the management of emotions in terms of this type of farming.
                                                                                   [Personal and Social Capability]

5. Working in pairs, create a presentation showing the different ways people use and manage the land in another country.
   (a) Design a suitable symbol for land degradation and use this to highlight any uses you think might result in land degradation.
   (b) Add annotations to explain how highlighted activities might degrade *environments*, and the *scale* of this *change*.
   (c) Suggest a possible *sustainable* solution for each type of degradation identified.
                                                                            Classifying, organising, constructing

3.2 EXERCISES

Geographical skills key: GS1 Remembering and understanding **GS2** Describing and explaining **GS3** Comparing and contrasting **GS4** Classifying, organising, constructing **GS5** Examining, analysing, interpreting **GS6** Evaluating, predicting, proposing

3.2 Exercise 1: Check your understanding

1. **GS1** List the different ways in which the land can become degraded.
2. **GS1** Outline the impact of land degradation on water resources.
3. **GS2** Explain why land degradation is a current geographical issue.
4. **GS1** Describe in your own words what land degradation is.
5. **GS1** Why were European farming methods unsuitable for the Australian *environment*?

3.2 Exercise 2: Apply your understanding

1. **GS2** Do you think land degradation is happening on a small or large *scale*? Explain.
2. **GS6** Study **FIGURES 1** and **5**.
   (a) In your own words, describe the damage that has occurred to the *environment*.
   (b) Suggest how these *changes* have come about.
   (c) How would you try to restore these *places* and manage their resources in a *sustainable* manner?
3. **GS5** Study **TABLE 1**.
   (a) What is the difference in arable land per person available in 1961 and 2016 globally and in Australia?
   (b) Arable land per person has decreased since 1961. The table shows an anomaly in this trend in Australia in 1976. Explain this anomaly and suggest why this may have occurred.
4. **GS5** Land degradation is often the result of many little actions and events, the effects of which interact and build up over time. Identify some things that you do, consciously or unconsciously, that might be contributing to land degradation where you live. Explain the impact of these actions.
5. **GS6** Describe an area or *place* that is near where you live, that you have visited recently, or that you have heard about in the media, and that you think is degraded. Give reasons for your choice and suggest how and why you think this degradation came about.

Try these questions in learnON for instant, corrective feedback. Go to www.jacplus.com.au.

3.3 SkillBuilder: Interpreting a complex block diagram

What is a complex block diagram?

A complex block diagram is a diagram that is made to appear three-dimensional. It shows a great deal of information about a number of aspects on a topic or location. It shows what is happening at the surface of the land or water, what is happening above the land or water, and what is happening beneath the soil or water at a number of different locations across an area.

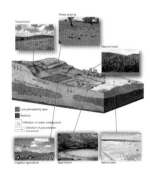

Select your learnON format to access:

- an overview of the skill and its application in Geography (Tell me)
- a video and a step-by-step process to explain the skill (Show me)
- an activity and interactivity for you to practise the skill (Let me do it)
- questions to consolidate your understanding of the skill.

on Resources

Video eLesson Interpreting a complex block diagram (eles-1746)

Interactivity Interpreting a complex block diagram (int-3364)

3.4 Managing land degradation

3.4.1 The development and importance of soil

Soil is a mixture of broken-down rock particles, living organisms and **humus**. Over time, as surface rock breaks down through the process of **weathering** and mixes with organic material, a thin layer of soil develops and plants are able to take root (see **FIGURE 1**). These plants then attract animals and insects and when these die their dead bodies decay, making the soil rich and thick.

Soil formation is a complex process brought about by the combination of time, climate, landscape and the availability of organic material. In some areas it takes hundreds of years to develop, while in others soil can form in a few decades.

The land is one of our most valuable resources. We depend on it for food, shelter, fibres and the oxygen we breathe. Yet the demands of an ever-increasing population place great pressure on it. To meet our needs, swamps and coastal marshes have been drained, vegetation removed and minerals extracted from the ground. Large-scale clearing and poor agricultural practices have left the land vulnerable. While erosion is a natural process, farming, land clearing and the construction of roads and buildings can accelerate the process.

FIGURE 1 Wild flowers taking root in cracks in the rocks

3.4.2 How is soil being lost?

Sheet erosion

Sheet erosion (see **FIGURE 2**) occurs when water flowing as a flat sheet flows smoothly over a surface, removing a large, thin layer of topsoil. Sheet erosion might happen down a bare slope. It occurs when the amount of water is greater than the soil's ability to absorb it.

Strategies to combat this form of erosion include planting slopes with vegetation and adding **mulch** to the exposed soil so that it can absorb greater volumes of water. Another solution is to 'terrace' the landscape — to form the land into a series of steps rather than a steep slope.

FIGURE 2 What evidence of sheet erosion can you observe?

Rill erosion

Rill erosion (see **FIGURE 3**) often accompanies sheet erosion, occurring where rapidly flowing sheets of water start to concentrate in small channels (or rills). These channels, less than 30 centimetres deep, are often seen in open agricultural areas. With successive downpours, rills can become deeper and wider, as fast-flowing water scours out and carries away more soil.

Strategies to combat rill erosion include tilling the soil (turning it over before planting crops) to slow the development of the rills. Building contours in the soil and planting a covering of grass can help slow the flow of water and hold the soil in place.

FIGURE 3 Rill erosion. In which direction do you think the water is flowing? Why?

Gully erosion

Gully erosion (see **FIGURE 4**) often starts as rill erosion. Over time, one or more rills may deepen and widen as successive flows of water carve deeper into the soil. Gully erosion may also start when a small opening in the surface such as a rabbit burrow or a pothole is opened up over time. Soil is often washed into rivers, dams and reservoirs, muddying the water and killing marine species. Large gullies need bridges or ramps to allow vehicles and livestock to cross.

Strategies to combat gully erosion largely involve stopping large water flows reaching the area at risk, through measures such as planting vegetation or crops to soak up the water. Other strategies include building diversion banks to channel the water away from the area, and constructing dams.

Tunnel erosion

Sometimes water will flow under the soil's surface; for example, under dead tree roots or through rabbit burrows, carving out an underground passage or tunnel (see **FIGURE 5**). The roof of the tunnel may be thin and collapse under the weight of livestock or agricultural machinery. When these tunnels collapse they create a pothole or gully.

Strategies to combat tunnel erosion include planting vegetation both to absorb excess water and to break up its flow. Sometimes major earthworks are needed to repack the soil in badly affected areas.

Wind erosion

When the surface of the land is bare of vegetation, the wind can pick up fine soil particles and blow them away (see **FIGURE 6**). It is more common during periods of drought or if the land has been overgrazed. The soil can be transported large distances and deposited in urban areas.

Strategies to combat wind erosion include planting bare areas with vegetation, mulching, planting wind breaks and avoiding overgrazing.

FIGURE 4 Gully erosion. What impact will the falling water have? How could further damage be prevented?

FIGURE 5 Tunnel erosion. What do you notice about the ground around these tunnels? What do you think will happen if water flows through them?

FIGURE 6 Wind erosion results when wind picks up and carries away fine soil particles. Did you know that soil from China has been deposited in the United States?

3.4.3 CASE STUDY: Managing land degradation in Costerfield, Victoria

How has Costerfield changed over time?

Costerfield is around 100 kilometres north of Melbourne (see **FIGURE 7**). The landscape is characterised by gentle slopes with undulating (wavelike) pastures. It has an average annual rainfall of around 575 millimetres, but with relatively high average evaporation rates, the climate is described as semi-arid or **Mediterranean**. Summers are hot and dry, and winters cool and wet. The area is also subject to climate extremes, so heavy rain, drought, frost and dust storms are not uncommon. Costerfield once had a dense covering of trees, predominantly eucalyptus. Native grasses also dominated the area. Soils in the area are generally considered to have a low **carrying capacity** for livestock because of poor fertility levels. Bushfires were a constant threat throughout the nineteenth century.

FIGURE 7 Costerfield lies approximately 100 kilometres north of Melbourne.

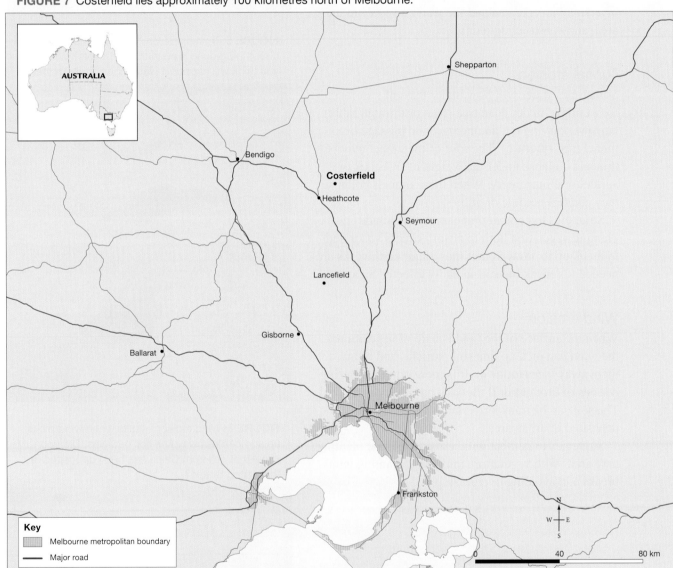

Source: © The State of Victoria, Department of Environment and Primary Industries 2013.

The first **pastoral run** was established in 1835, and the land was extensively cleared. Settlers introduced sheep, cattle, rabbits and foxes soon afterwards. Sheep grazing soon became the dominant activity in the region. The lack of native vegetation cover allowed rainfall to flow across the surface, eroding soil and making the run-off **turbid**. It also allowed rainfall to infiltrate the subsoil, leading to sheet, rill, gully and tunnel erosion (see **FIGURES 8** and **9**). The problem was further exacerbated by the rabbit population. Subdivision plans were drawn up in the early 1850s; however, in 1852 gold was discovered at McIvor Creek, which led to an influx of up to 40 000 gold prospectors in the region, causing the land to become even more degraded.

Following the gold rushes and into the twentieth century, the area around Costerfield was largely used for grazing livestock, predominantly sheep. However, there is evidence that both horses and cattle were raised in the region on a much smaller scale.

It is much easier to prevent gully erosion than control it once it has developed. Without intervention, gullies can continue to become larger and larger. A number of measures were introduced by local Landcare groups to tackle the issues in the Costerfield area.

The gullies were stabilised by constructing banks, gully check dams (see **FIGURE 10**) and terracing, all aimed at reducing and redirecting run-off. Other strategies included:

- re-establishing ground cover, especially plants and grasses that are native to the region (see **FIGURE 11**)
- rabbit eradication programs to control the population and reduce burrowing activities that can create access points for run-off to enter the subsoil and promote development of new gullies. In addition, they protect the newly sown grasses from being eaten by the rabbits
- introduction of chemicals such as lime and gypsum to improve soil structure and pH levels to assist in the revegetation process
- protection of revegetated areas by preventing access, especially by livestock, during the restoration process.

FIGURE 8 A dead tree stump or old fence post can allow water to infiltrate the soil.

FIGURE 9 Notice that tunnel erosion forms where the surface of the land is bare. What do you think is likely to happen next?

What do you think has caused these tunnels?

Turbid tunnel flow

Turbid gully flow

FIGURE 10 Permanent check dams may be constructed using logs or stone. Sometimes they are lined to prevent seepage into the ground so that the water is trapped and can be used for irrigation purposes. They also trap sediment and prevent it being washed into waterways. Additionally, some are designed to trap nutrients and so help maintain water quality. Outlet pipes allow water to be redirected and control the flow of water across the landscape.

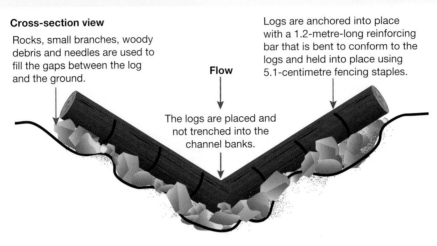

Cross-section view

Rocks, small branches, woody debris and needles are used to fill the gaps between the log and the ground.

Logs are anchored into place with a 1.2-metre-long reinforcing bar that is bent to conform to the logs and held into place using 5.1-centimetre fencing staples.

Flow

The logs are placed and not trenched into the channel banks.

FIGURE 11 Revegetated area near Costerfield

 **Resources**

Interactivity Down in the dirt (int-3290)

3.4 INQUIRY ACTIVITIES

1. With the aid of a flow diagram, show the **interconnection** between sheet, rill and gully erosion. Use the captions and the questions that appear with each image to help you. **Classifying, organising, constructing**
2. Working with a partner, use the internet to investigate an international **environment**, such as the Dust Bowl in the United States or the Yellow River in China, that has been degraded because of soil erosion.
    (a) Annotate a sketch of this **environment** to explain what has happened to the area. Include an inset sketch map that shows the location of this **place**, and describe the **scale** and rate of **change**.
    (b) Swap your eroded **environment** sketch with another pair who will devise a series of management strategies to rehabilitate the **environment** and allow it to be used in a **sustainable** manner. Add these to your annotated sketch. **Examining, analysing, interpreting**
    Classifying, organising, constructing

3.4 EXERCISES

Geographical skills key: GS1 Remembering and understanding **GS2** Describing and explaining **GS3** Comparing and contrasting **GS4** Classifying, organising, constructing **GS5** Examining, analysing, interpreting **GS6** Evaluating, predicting, proposing

3.4 Exercise 1: Check your understanding

1. **GS2** Explain the *interconnection* between the removal of native vegetation and land degradation.
2. **GS2** In your own words, explain how the use of a check dam might reduce the development of gullies.
3. **GS1** Identify the types of soil erosion that occurred in Costerfield.
4. **GS2** Explain what is meant by the phrase 'making the run-off turbid'.
5. **GS2** Explain the term *pastoral run*.

3.4 Exercise 2: Apply your understanding

1. **GS6** Why do you think soil erosion in all its forms is such a significant cause of land degradation?
2. **GS6** How would the arrival of gold prospectors increase land degradation? Explain the type of erosion that would most likely occur and the activities relating to gold mining that might cause these *changes* to develop.
3. **GS2** Explain the *interconnection* between rill erosion and gully erosion.
4. **GS2** An important part of any land management program is the control of introduced species such as rabbits. Explain why rabbits are a problem in areas where the land has been degraded.
5. **GS5** Look at **FIGURE 12**, which depicts the types and *scale* of soil erosion in Victoria.
   (a) In which parts of the state is erosion highest resulting from (i) wind and (ii) water?
   (b) Compare this map with a relief map of Victoria in your atlas. What conclusions can you draw about the *interconnection* between topography and erosion caused by water?
   (c) Use your atlas to find a map showing vegetation in Victoria. Explain why wind erosion is more common in north-west Victoria than south-east Victoria.

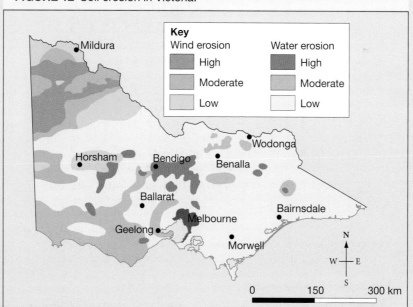

FIGURE 12 Soil erosion in Victoria.

Source: MAPgraphics Pty Ltd Brisbane.

Try these questions in learnON for instant, corrective feedback. Go to www.jacplus.com.au.

3.5 Environmental change and salinity

3.5.1 Where does the salt come from?

Salinity is not a new problem. In fact, it was an environmental issue for the earliest civilisations some 6000 years ago. Historical records indicate that the Sumerians, who farmed the land between the Tigris and Euphrates rivers in the area known as Mesopotamia, ruined their land as a result of their poorly managed irrigation practices.

Salt has become a major contributor to land degradation in Australia. Rising up from below the land surface, it is destroying native vegetation and threatening the livelihood of many Australians. As plants die as a result of salinity, other problems emerge: the soil no longer has a protective cover of vegetation, which means it is more easily blown away or eroded.

Some 140 million years ago, parts of the Australian continent were covered by shallow seas and saltwater lakes. The salt stores from these waters have lain dormant below the surface of the land, much of them in the **groundwater**. In addition, salt continues to be deposited on the land's surface by rain and winds blowing in from the oceans, and by the weathering of mineral-carrying rocks.

Australia's native vegetation had built up some tolerance to the salt levels in the soil. The deep-rooted vegetation also soaked up water in the soil before it could seep down into the groundwater. This meant that the watertable stayed at a fairly constant level, and that the concentrated salt stores stayed where they were. This natural balance changed with the arrival of European settlers. The farming and land-clearing practices they introduced were, and still are, according to many experts, unsuited to Australia's generally harsh, dry climate, as well as to its geological history.

Salt has now become a serious problem. There are two ways in which the soil can become too salty: these are called dryland salinity and irrigation salinity. The areas in Australia affected by salinity are shown in **FIGURE 1**.

FIGURE 1 Salinity distribution

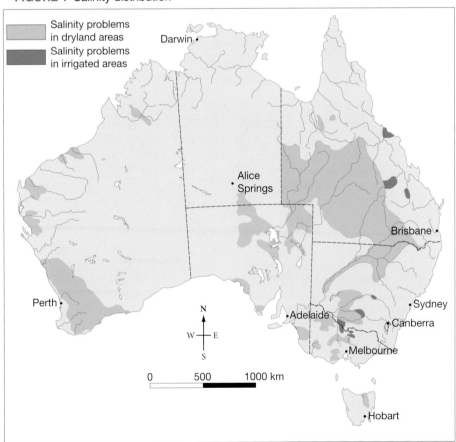

Source: Spatial Vision.

3.5.2 What is dryland salinity?

Dryland salinity occurs in areas that are not irrigated. When settlers cleared the land, they replaced deep-rooted native vegetation with crop and pasture plants. These plants generally have shorter roots and cannot soak up as much rainfall as native vegetation. Excess moisture seeped down into the groundwater, raising the watertable and bringing concentrated saline water into direct contact with plant roots (see **FIGURE 2**). Vegetation, even salt-tolerant plants, started dying as the salt concentrations rose. Once the vegetation dies off, the soil is left bare and is prone to erosion. Often layers of salt, known as **salt scald**, are visible on the surface of the land.

FIGURE 2 The effects of a rising watertable

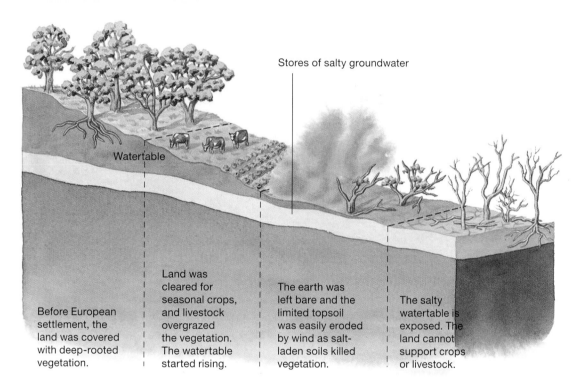

Stores of salty groundwater

Watertable

Before European settlement, the land was covered with deep-rooted vegetation.

Land was cleared for seasonal crops, and livestock overgrazed the vegetation. The watertable started rising.

The earth was left bare and the limited topsoil was easily eroded by wind as salt-laden soils killed vegetation.

The salty watertable is exposed. The land cannot support crops or livestock.

3.5.3 What is irrigation salinity?

One-third of the world's food is produced on irrigated land. Irrigation salinity occurs in irrigated regions and is a direct result of overwatering. When more water is applied to crops or pasture plants than they can soak up, the excess water seeps down through the soil into the groundwater, causing the salty watertable to rise to the surface. Some of this salt is washed into rivers, either as run-off or groundwater seepage, and transported to other places.

HOW MUCH LAND AROUND THE WORLD IS AFFECTED BY SALINITY?
- Africa: 2 per cent of Africa's landmass
- China: 21 per cent of arid lands or around 30 million hectares of land
- Western Europe: 10 per cent of the land area
- United States: Land across 17 states
- South America: Most countries have areas of land affected
- Australia: 2.5 million hectares
- Worldwide: It is estimated that 10 million hectares of arable land succumb to the effects of irrigation-related salinity each year and that, without intervention, the affected area might triple by 2050.

How do we solve the problem?

Many programs are in place to identify and monitor problem areas. Action being taken includes:
- changing irrigation practices to reduce overwatering
- planting deep-rooted native trees and shrubs in open areas
- developing new crops that are more salt tolerant, such as new strains of wheat
- replacing introduced pasture grasses with native vegetation such as saltbush (see **FIGURE 3**)
- using satellite technology to map areas at risk to enable early intervention.

FIGURE 3 Native plants such as saltbush help solve the problem of salinity on Australian grazing lands.

3.5.4 CASE STUDY: Salinity in the Murray–Darling Basin

The Murray–Darling Basin is Australia's largest **drainage area**. Extending across parts of four states and the entire Australian Capital Territory, it contains the country's three longest rivers: the Murray (2508 kilometres), the Darling (2740 kilometres when including its three main tributaries) and the Murrumbidgee (1690 kilometres). It is also one of the country's most significant agricultural regions, producing close to 45 per cent of the nation's food. Because it receives very little rainfall, the area depends heavily on irrigation. In fact, 70 per cent of Australia's irrigation occurs here.

Over time, human activities in the Murray–Darling Basin have increasingly threatened the basin's **ecology** (see **FIGURE 4**). These activities have included introducing non-native plant and animal species, changing the natural flow of the river for irrigation purposes, clearing the land and over-watering crops. It is the last two activities that have particularly contributed to the region's salinity. (Although, in 1829 when explorer Charles Sturt discovered the Darling River during the dry season, he observed that the water was too salty to drink.) It has been estimated that by 2050, 1.3 million hectares (or 93 per cent) of land in the region could be salt affected.

FIGURE 4 The earliest signs of salinity: the watertable has risen, bringing salt to the root zone. Look carefully at the trees in the background. What do you observe?

Tackling the problems

Over the years, a range of strategies have been investigated to better manage the Murray–Darling Basin and reduce salinity problems. Strategies have included the development of action plans such as revegetation programs and educational programs. In 2008, the Commonwealth Government took control of the region to allow for the implementation of a comprehensive management strategy that would provide for the needs of all states and also be environmentally sustainable. The Murray–Darling Basin Authority (MDBA) was created to oversee the entire project.

As part of the measures to control salinity, salt-interception schemes (see **FIGURE 5**) have been established along the Murray River. Collectively, they remove 500 000 tonnes of salt annually from groundwater and drainage basins (see **FIGURE 6**). Prior to these schemes, the Murray River carried huge amounts of salt; for example, 250 tonnes per day and 100 tonnes a day respectively past Woolpunda and Waikerie, between Renmark and Morgan in South Australia. Recent surveys show that salinity levels have decreased to less than 10 tonnes a day in each area.

FIGURE 5 Salt interception in the Murray–Darling Basin

Key

Salt yield of catchments, 1995
Tonnes per km² per annum
- More than 6
- 4 to 6
- 2 to 4
- 1 to 2
- Less than 1
- Salt interception scheme

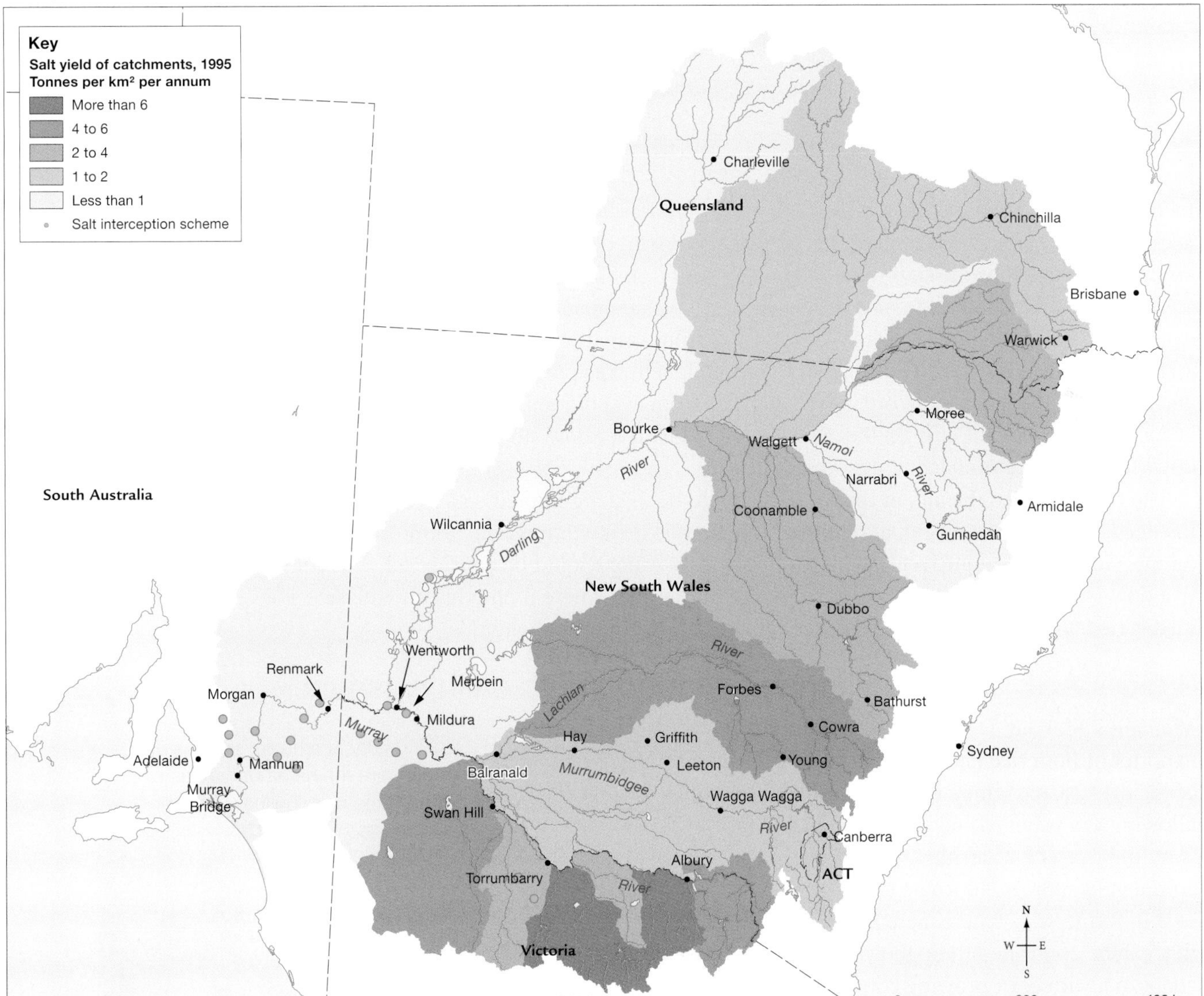

Source: © Commonwealth of Australia Geoscience Australia 2013. Map by Spatial Vision.

FIGURE 6 Salt harvested from evaporation ponds is exported all around the world.

Murray–Darling Basin Royal Commission 2018

In 2018, a report by the ABC program *Four Corners* levelled serious allegations of inadequate management, negligence and water theft in the Murray–Darling Basin. Following this, the South Australian government ordered a **Royal Commission** into the management of the Basin.

The Royal Commission found that the existing management strategies were in need of a complete overhaul to ensure that more water was diverted from irrigation and back into the environment.

The report was also highly critical of the Menindee Lakes project (in NSW) which involved shrinking and emptying the lakes more often to save water from evaporation — an action that failed to take into account the impact not only on the environment when river flow stopped, but also on those further downstream in Victoria and South Australia.

The report argued the aim of the MDBA should be to share limited water resources and ensure that the needs of the environment, agriculture, Indigenous communities and 2.6 million people who depend on the Murray–Darling River system to supply their water should be considered — especially those who live downstream, and all the way to the mouth of the river where it empties into the sea. The sustainable management of this vital water resource remains an ongoing challenge.

3.5.5 CASE STUDY: Vietnam — adapting to salinity issues

Vietnam's Mekong Delta region (see **FIGURE 8**) is a major exporter of both rice and shrimp. Drought and the early arrival of the dry season, which is being attributed to climate change, is allowing sea water to encroach on valuable farming land.

While rice is a water-thirsty crop, it does not like to be completely submerged for its entire growing season. Rice grows best in fertile soils where there is an abundant supply of water that can be controlled throughout the growing season, but it can adapt to a variety of growing conditions.

Scientists are now developing strains of rice that are not only salt-resistant but can also withstand being submerged in water for almost three weeks, whereas traditional strains die within a week of being flooded and fully submerged.

FIGURE 7 Dry rice husks suitable only for poultry are the result of brackish water from the sea flowing inland.

Additionally, in some regions, farmers are making use of the **brackish** water that results from periods of saltwater intrusion. While brackish water is not suited to rice farming, it is ideal for cultivating shrimp. With the onset of the monsoon season, the farmers rely on the heavy rains to flush out the salt water and allow them to plant their rice crops.

FIGURE 8 The Mekong Delta

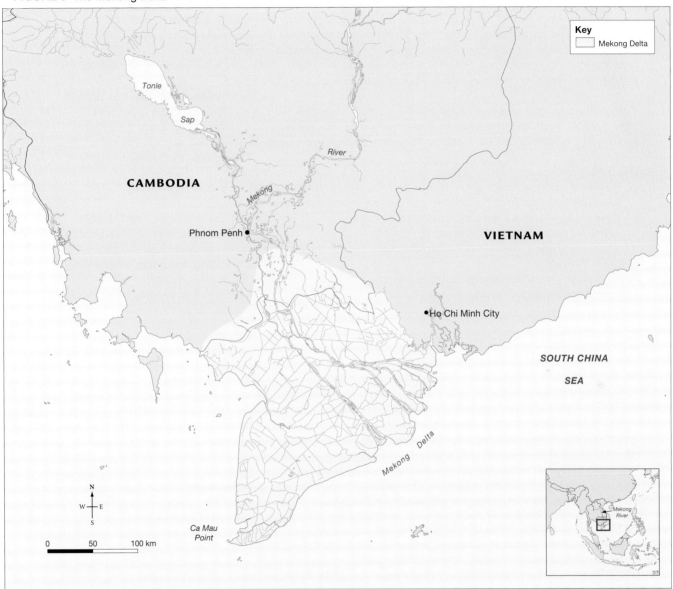

Source: Vector Map Level 0 Digital Chart of the World. Map by Spatial Vision.

─Explore more with myWorldAtlas─────────────────

Deepen your understanding of this topic with related case studies and questions.
- Investigating Australian Curriculum topics > Year 10: Environmental change and management > **Salinity in the Murray–Darling Basin**

3.5 INQUIRY ACTIVITIES

1. Investigate the history of agriculture in an ancient civilisation, such as Mesopotamia.
   (a) Include a sketch map of the area. Annotate this map to show how the region was affected and why.
   (b) What lessons might modern farmers learn from ancient practices?　　**Examining, analysing, interpreting**
2. In groups, investigate a method of combating salinity and **sustainable** practices that will improve the productivity of agricultural land. Before you begin, decide as a class which groups will cover dryland salinity and which will focus on irrigation salinity. Present your findings as a news report.
   Examining, analysing, interpreting
3. Find out the total land area of Australia and the world. If areas affected by irrigation salinity are expected to triple by 2050, estimate the proportion of land that will be affected on a national and global **scale**. Use your findings as the basis for writing a letter to the Editor, urging governments to take action and halt this trend.
   Examining, analysing, interpreting
4. In groups, investigate the Mekong River, its delta and its importance to Vietnam.
   (a) Prepare a report on how the river is used and the issue of land degradation. In your report make reference to the **scale** of the problem and the rate at which **change** is occurring.
   (b) What strategies have been suggested or used to deal with this issue? Are these strategies a **sustainable** option for caring for the **environment**? Why/why not?　　**Examining, analysing, interpreting**
5. Investigate land use in the Murray–Darling Basin and explain the **interconnection** between land use and salinity in this region.　　**Examining, analysing, interpreting**
6. With a partner, investigate how saltbush may help to reduce salinity. Create an annotated diagram of a saltbush plant and its root system to explain this role.　　**Examining, analysing, interpreting**

3.5 EXERCISES

Geographical skills key: GS1 Remembering and understanding **GS2** Describing and explaining **GS3** Comparing and contrasting **GS4** Classifying, organising, constructing **GS5** Examining, analysing, interpreting **GS6** Evaluating, predicting, proposing

3.5 Exercise 1: Check your understanding

1. **GS2** Explain the **interconnection** between soil salinity and land degradation.
2. **GS2** Why would planting deep-rooted trees help solve the problem of salinity?
3. **GS2** What actions could an irrigation farmer take to reduce the risk of salinity?
4. **GS2** Why is the Murray–Darling Basin a significant part of the Australian **environment**?
5. **GS2** Refer to **FIGURE 1**. Describe the distribution of dryland salinity areas in Australia.

3.5 Exercise 2: Apply your understanding

1. **GS2** Which general area of the Murray–Darling Basin is most affected by salinity: the north-east, the western region or the south-east? Describe the **scale** of the problem.
2. **GS2** With the aid of a diagram, explain what a delta is and why it is important.
3. **GS5** Refer to **FIGURE 5**.
   (a) What is the aim of such salt-interception schemes? Explain.
   (b) Do you think this is a **sustainable** management strategy? Explain.
   (c) Discuss the impact of this scheme on river ecosystems.
   (d) What do you think happens to the salt that is extracted?
   (e) Do you think a similar scheme could be developed for the Mekong Delta? Justify your point of view.
4. **GS2** With the aid of a Venn diagram, compare salinity issues that exist in the Murray–Darling Basin and Vietnam's Mekong Delta. Include references to the **scale** of the issue and the rate of **change**.
5. **GS2** What factors have contributed to the degradation of the Murray–Darling Basin's land and water resources since the arrival of Europeans?

Try these questions in learnON for instant, corrective feedback. Go to www.jacplus.com.au.

3.6 Desertification: the drylands are spreading

3.6.1 Which regions are most at risk from desertification?

As the Earth's population increases, more pressure is placed on the land to provide both food and shelter. In many parts of the world this has meant that land has been overused and become exhausted. This is especially true in **dryland** regions. Many of these areas have become so degraded that they are at risk of being turned into desert, placing the survival and livelihood of the people who depend on them in jeopardy.

The United Nations estimates that approximately 41 per cent of the Earth's land surface is at risk of turning into desert. This is a process known as **desertification**, an extreme form of land degradation that affects arid, semi-arid and dry sub-humid areas of the Earth. These dryland regions often border existing deserts, but unlike deserts, they support population and agriculture. Drylands are fragile environments that degrade rapidly when the land is not carefully managed.

The areas affected are home to more than two billion people in 168 countries. **FIGURE 1** shows the Earth's desert areas and those places most at risk from desertification.

FIGURE 1 The drylands of the Earth are spreading. It is estimated that 12 million hectares of productive land (an area three times larger than Switzerland) is lost annually due to desertification.

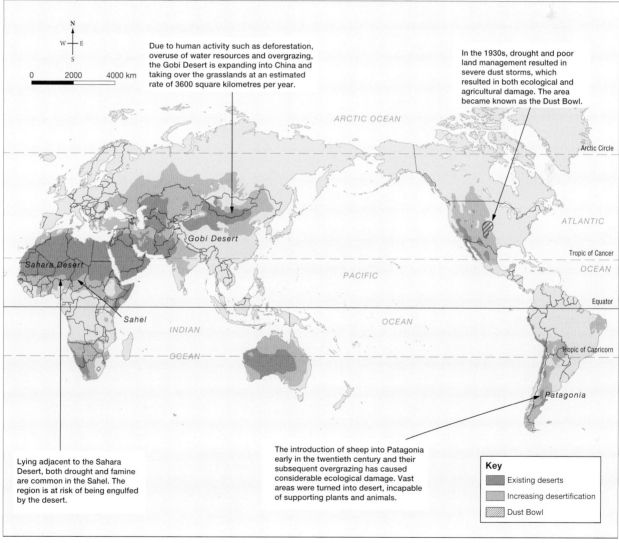

Due to human activity such as deforestation, overuse of water resources and overgrazing, the Gobi Desert is expanding into China and taking over the grasslands at an estimated rate of 3600 square kilometres per year.

In the 1930s, drought and poor land management resulted in severe dust storms, which resulted in both ecological and agricultural damage. The area became known as the Dust Bowl.

The introduction of sheep into Patagonia early in the twentieth century and their subsequent overgrazing has caused considerable ecological damage. Vast areas were turned into desert, incapable of supporting plants and animals.

Lying adjacent to the Sahara Desert, both drought and famine are common in the Sahel. The region is at risk of being engulfed by the desert.

Key
- Existing deserts
- Increasing desertification
- Dust Bowl

Source: UNEP World Conservation Monitoring Centre.

Estimates predict that by 2025, without intervention, two-thirds of the arable lands in Africa will be lost, along with one-third in Asia and one-fifth in China. Based on current trends, Bangladesh will have no fertile soil available in 50 years. Desertification is a global issue as it is present on all continents in both developed and developing economies.

3.6.2 What causes desertification?

Desertification is largely a result of human-induced environmental change, caused by the complex interconnection of environmental, political, cultural and economic factors. It generally arises from the poor management of dryland environments. Increasing populations, the demand for more agricultural production and overuse of the soil degrades the land to the extent that once productive places turn into wastelands (see **FIGURE 2**).

FIGURE 2 Factors contributing to desertification

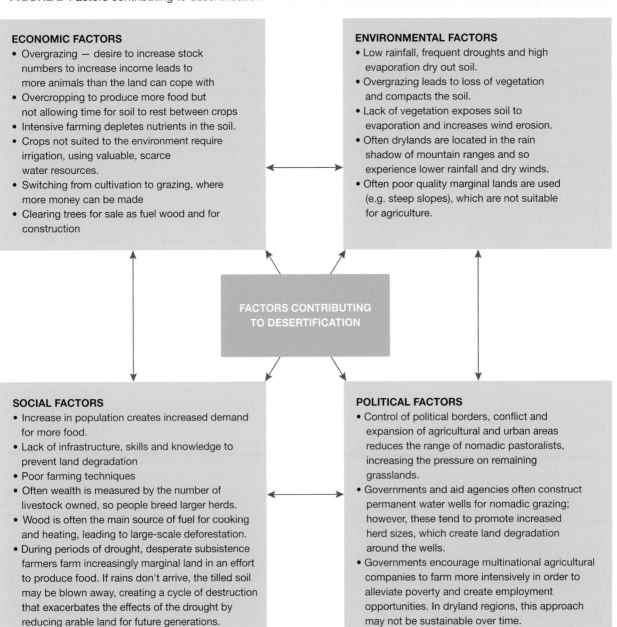

ECONOMIC FACTORS
- Overgrazing — desire to increase stock numbers to increase income leads to more animals than the land can cope with
- Overcropping to produce more food but not allowing time for soil to rest between crops
- Intensive farming depletes nutrients in the soil.
- Crops not suited to the environment require irrigation, using valuable, scarce water resources.
- Switching from cultivation to grazing, where more money can be made
- Clearing trees for sale as fuel wood and for construction

ENVIRONMENTAL FACTORS
- Low rainfall, frequent droughts and high evaporation dry out soil.
- Overgrazing leads to loss of vegetation and compacts the soil.
- Lack of vegetation exposes soil to evaporation and increases wind erosion.
- Often drylands are located in the rain shadow of mountain ranges and so experience lower rainfall and dry winds.
- Often poor quality marginal lands are used (e.g. steep slopes), which are not suitable for agriculture.

FACTORS CONTRIBUTING TO DESERTIFICATION

SOCIAL FACTORS
- Increase in population creates increased demand for more food.
- Lack of infrastructure, skills and knowledge to prevent land degradation
- Poor farming techniques
- Often wealth is measured by the number of livestock owned, so people breed larger herds.
- Wood is often the main source of fuel for cooking and heating, leading to large-scale deforestation.
- During periods of drought, desperate subsistence farmers farm increasingly marginal land in an effort to produce food. If rains don't arrive, the tilled soil may be blown away, creating a cycle of destruction that exacerbates the effects of the drought by reducing arable land for future generations.

POLITICAL FACTORS
- Control of political borders, conflict and expansion of agricultural and urban areas reduces the range of nomadic pastoralists, increasing the pressure on remaining grasslands.
- Governments and aid agencies often construct permanent water wells for nomadic grazing; however, these tend to promote increased herd sizes, which create land degradation around the wells.
- Governments encourage multinational agricultural companies to farm more intensively in order to alleviate poverty and create employment opportunities. In dryland regions, this approach may not be sustainable over time.

3.6.3 The impacts of desertification

Currently, the world loses approximately 12 million hectares of land annually, an area almost three times the size of Switzerland, enough to grow 20 million tonnes of grain. The cost to global economies is estimated to be $490 billion per annum.

Desertification brings about environmental change as the loss of topsoil and protective vegetation enables desert sand dunes to migrate and smother former farmland (see **FIGURE 3**).

FIGURE 3 Fence drowned by a huge sand dune in the United Arab Emirates

Desertification also affects the wellbeing of over one billion people in the world. While poverty can contribute to desertification, it is also a consequence of it, as poverty forces people to over-exploit the land, which can then accelerate land degradation. It can also increase the risk of food insecurity as food production decreases. As the land fails, social and cultural networks become lost as whole villages can effectively be abandoned as people leave farming in search of employment in urban areas.

3.6.4 Tackling the problem

Desertification, climate change and the loss of biodiversity were identified as the greatest challenges to sustainable development during the 1992 Rio Earth Summit. As a result of this, the United Nations developed the United Nations Convention to Combat Desertification (UNCCD), an agreement supported by 193 countries with the aims of:
- improving living conditions for people living in drylands
- maintaining and restoring land and soil productivity
- reducing the impacts of drought.

This worldview encourages cooperation in exchanging knowledge and technology between developed and developing countries and promotes the idea of a 'bottom-up' approach to a problem. This means encouraging and supporting people to develop their own solutions rather than a government-led 'top-down' approach.

3.6.5 CASE STUDY: Combating desertification in China

China is one of the countries most severely impacted by desertification, which affects about 25 per cent of its total land area (approximately 3327 million km^2) and negatively affects 400 million people (see **FIGURE 4**).

FIGURE 4 The predicted scale of desertification in China

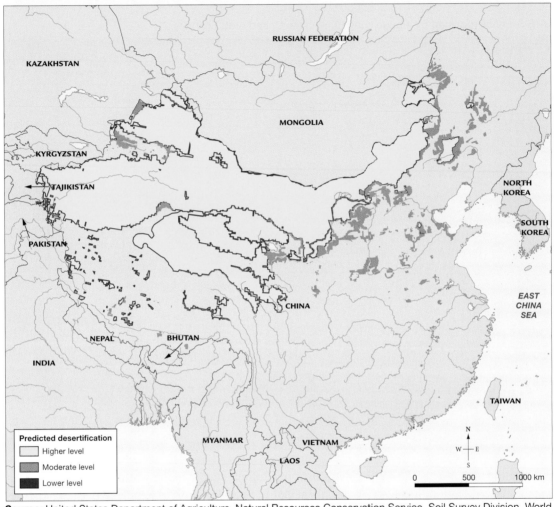

Source: United States Department of Agriculture, Natural Resources Conservation Service, Soil Survey Division, World Soil Resources; Paul Reich, Geographer. 1998. Global Desertification Vulnerability Map. Washington, D.C.

In the wake of rapid population growth (from 550 million in 1950 to 1.4 billion in 2019), the demand for food, fuel, construction timber and livestock feed surged. With a viewpoint of 'growth at all costs', more farmland was opened up on desert fringes and the number of livestock increased exponentially. This expansion was done without any consideration of the environmental impacts. Thus, human activities in the form of inappropriate land use have magnified the problem of desertification (see **FIGURE 5**).

FIGURE 5 Causes of desertification in China

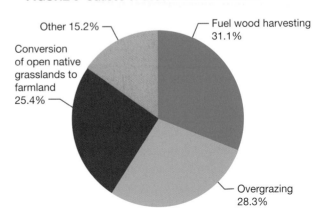

Other 15.2%

Conversion of open native grasslands to farmland 25.4%

Fuel wood harvesting 31.1%

Overgrazing 28.3%

Each year, the Gobi Desert in Mongolia swallows 360 000 hectares of grasslands, and dust storms remove 2000 square kilometres of topsoil. Sand and dust from the desert regions of China is carried eastwards by the prevailing winds, choking the city of Beijing: destroying crops, closing airports and creating a surge in respiratory ailments. The sand storms, or 'Yellow Dragon' as they were traditionally called, continue their journey and affect international communities in South Korea, Japan, Russia and even the United States of America (see **FIGURE 6**).

The Chinese government has been working relentlessly on the problem, implementing a range of schemes, two of which are described in section 3.6.6.

China currently spends $5 billion each year on combating desertification. The aim is to reclaim all the treatable land area by 2050. Already China has slowed the rate of desertification by more than one-third of 1999 levels, making it a world leader in this field.

3.6.6 The Great Green Wall of China

To halt the spread of deserts and reduce the impacts of climate change, the Chinese government

FIGURE 6 A dust storm from China affecting neighbouring countries

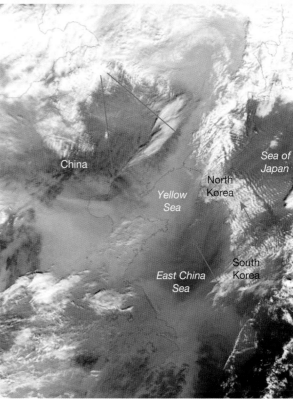

Source: Image courtesy Jacques Descloitres, MODIS Rapid Response Team at NASA GSFC.

embarked on a plan to create the Great Green Wall of China. Green walls are ambitious initiatives designed to act as a barrier against desert winds and prevent desertification. Both the **Sahel** region in Africa and China have embarked on massive replanting projects that are expected to reduce erosion, enhance biodiversity, provide new grazing lands, boost agriculture and provide employment (see **FIGURE 7**).

FIGURE 7 The Great Green Wall of China: an ambitious attempt to stop the advance of desert sands from the Gobi Desert

The Chinese government envisaged a 4800-kilometre series of forest strips spanning the country from east to west, and stretching 1500 kilometres from the southern edge of the Gobi Desert, to protect valuable farmland and waterways against wind erosion. To make this target a reality, every citizen over the age of 11 was expected to plant at least three saplings each year. Since the start of the millennium, Chinese citizens have planted over 66 billion trees. By 2016, the Chinese State Forestry Commission had succeeded in creating almost 30 million hectares of forest, with the goal of reaching a national forest cover of 42 per cent by 2050. However, an early study done by geographers at the University of Alabama noted that 'the reforestation efforts have done little to abate China's great yellow dust storms' (see **TABLE 1**).

TABLE 1 Impacts of planting green walls

Environmental benefits	Environmental drawbacks
<ul><li>Mass planting of fast-growing species (known as **monoculture**) helps to slow desertification.</li><li>Trees act as windbreak and reduce erosion</li><li>Long-term possibility of harvesting trees as a commercial wood crop or for pulp and paper</li><li>Growth of trees acts as a carbon store, reducing greenhouse gases.</li></ul>	<ul><li>Monoculture reduces biodiversity and provides poor habitat for endangered native animal and bird species.</li><li>Monoculture is highly susceptible to disease. A pest can wipe out an entire plantation, ruining decades of work.</li><li>Many tree species chosen were not native and after initial growth soon died. In some places up to 85 per cent of the plantings failed.</li><li>Initial rapid growth of trees used a lot of soil moisture and lowered watertables.</li><li>Trees out-competed native grasses, which have a more extensive root system for holding soil.</li><li>Plantations generate less leaf litter than native forests, so less nutrients are entering the soil.</li></ul>

While China reports an overall increase in forested areas, from 5 per cent to 12 per cent, Greenpeace reports that only 2 per cent of China's original vegetation remains. Many of the trees planted have a lifespan of only 40 years. Nevertheless, in areas where the local community is prepared to care for the newly planted trees, the spread of the desert appears to have been halted, with the area of land affected by desertification shrinking by almost 2000 square kilometres annually. Sandstorms are now also reported to have decreased by 20 per cent.

As one Chinese ecologist, Jian Gaoming, has stated, there is a need for 'nurturing the land by the land itself'. This is an earth-centred approach to the problem of desertification. His research in Inner Mongolia noted that native grasslands will restore themselves in as little as two years, if protected from grazing animals and human activities.

Restoring grasslands

It is estimated that 80 per cent of China's natural grasslands (42 per cent of its land area) are degraded as a result of overgrazing. A wide range of rehabilitation programs are being introduced. These include:

- *Moving people:* In places especially at risk of desertification, people are being resettled to prevent further damage and halt the spread of the desert. Relentless sandstorms threaten the traditional lifestyles and farming practices of nomads in both Tibet and Mongolia. Failing crops and a lack of pasture for grazing livestock is forcing them to join other climate refugees and move into new settlements (see **FIGURE 8**).
- *Changing land use:* Land use is converted from grazing to tree crops and forests, with farmers receiving compensation for the loss of stock and income.
- *Total grazing bans:* Over the years 2005 to 2010, a total ban was placed on animal grazing on seven million hectares of land (an area twice the size of Germany). This was part of a larger plan to restore more than 660 million hectares of grasslands at an estimated cost of approximately AU$4 billion. This has meant that more than 20 million animals had to be farmed indoors and hand fed. In test projects, after three years of grazing bans the vegetation rate increased from 20 per cent to over 60 per cent, and local sand storms have reduced. The grazing ban has since been extended, with farmers paid a subsidy

to safeguard their livelihood. Additional bans were also put in place banning hunting and declaring some areas national parks. Money received by the traditional nomadic herders has been used to leave the land.

FIGURE 8 Nomadic grazing on grasslands in Mongolia. Why would this area be prone to desertification?

─Explore more with myWorldAtlas──────────────────────

Deepen your understanding of this topic with related case studies and questions.
• Investigating Australian Curriculum topics > Year 10: Environmental change and management > **Desertification in Mauritania**

─ Resources ──────────────────────

🔗 **Weblinks** Great Green Wall of China (1)
                 Great Green Wall of China (2)

3.6 INQUIRY ACTIVITIES

1. Investigate one of the causes of desertification outlined in **FIGURE 2** and write a news report that explains the *interconnection* between this factor and *environmental change*. In your report include the following:
   (a) a description of the impact of this factor over *space*
   (b) an example of a *place* that has been *changed* as a result of this factor
   (c) the *scale* of this *change*
   (d) a strategy for the *sustainable* management of the *environment* to combat this factor.
   Classifying, organising, constructing
2. Use the **Great Green Wall of China** weblinks in the Resources tab to find out more about the Great Green Wall of China. With a partner, discuss the actions taken. Do you think the plan will succeed? Why or why not? **Examining, analysing, interpreting**

3.6 EXERCISES

Geographical skills key: GS1 Remembering and understanding **GS2** Describing and explaining **GS3** Comparing and contrasting **GS4** Classifying, organising, constructing **GS5** Examining, analysing, interpreting **GS6** Evaluating, predicting, proposing

3.6 Exercise 1: Check your understanding

1. **GS1** What is the difference between deserts and drylands?
2. **GS1** Why are drylands especially vulnerable to desertification?

▶

3. **GS1** Refer to **FIGURE 1**. Describe the distribution of those *places* in the world most at risk of desertification.
4. **GS1** Identify an economic, social and *environmental* impact of desertification in China.
5. **GS2** In your own words, explain what is meant by *desertification* and why it is a global issue.

3.6 Exercise 2: Apply your understanding

1. **GS6** How would you envisage the issue of desertification in China in the year 2050? Give reasons for your answer.
2. **GS6** The Chinese ecologist Jian Gaoming's viewpoint on managing desertification is 'nurturing the land by the land itself'. How does this earth-centred viewpoint compare to the green wall scheme, which is a human-centred viewpoint? Which do you think is the more *sustainable* approach? Outline and justify your views.
3. **GS2** Explain the difference between a 'top-down' and a 'bottom-up' approach to resolving a problem.
4. **GS6** How effective do you think a top-down approach can be in combating desertification in China? How effective do you think a bottom-up approach might be? Explain your view.
5. **GS2** Evaluate, according to *environmental*, social and economic impacts, the effectiveness of:
    (a) green walls
    (b) grazing bans for combating desertification in China.

Try these questions in learnON for instant, corrective feedback. Go to www.jacplus.com.au.

3.7 Introduced species and land degradation

3.7.1 What is an invasive species?

An invasive species (sometimes referred to as an **exotic species**) is any plant or animal species that colonises areas outside its normal range and becomes a pest. Such species take over the environment at the expense of those that occur naturally in the region, generally causing damage to native habitats and degrading the landscape. Invasive species are a major cause of land degradation. Often introduced for a specific reason, they can soon take over the environment, threatening indigenous plant and animal species and taking over what was once valuable farming land.

Many of the most damaging invasive species were introduced either for sport (rabbits and foxes), as pets (cats) or as livestock (goats) and pack animals (camels and horses). Some, such as the cane toad and mosquito fish, were introduced to control other species (the cane beetle and mosquitoes), and instead became pests themselves. Others, including rats and mice, arrived accidentally as stowaways on ships. Similarly, invasive plants were introduced in a variety of ways: as crops, pasture or garden plants, or to prevent erosion. However, some spread into the bush where they continued to thrive, causing immense damage to the environment.

3.7.2 Animal pests — goats, foxes and rabbits

Goats

Goats were introduced into Australia with the arrival of the First Fleet in 1788 as a source of both milk and meat. Some breeds were later introduced for their hair. During the nineteenth century, sailors released goats onto some of the offshore islands and mainland areas as an emergency food source. Over time, however, domestic goats escaped, were abandoned or were deliberately released and became feral, posing a threat to inland pastoral areas and native forests.

FIGURE 1 Goats were a food source for early settlers and sailors alike. Their size, hardiness and ability to eat a range of plants made them an ideal source of both meat and milk.

It is now estimated that at least 2.6 million feral goats occupy approximately 28 per cent of Australia (see **FIGURE 2**), in concentrations of up to 40 animals per square kilometre. They are found in all states and territories and on offshore islands, but are most common in semi-arid regions. The absence of predators and the establishment of a water supply for sheep grazing has created ideal conditions in which goats can thrive. Their numbers have been adversely affected by drought and eradication programs; however, high fertility levels have meant they are difficult to control.

FIGURE 2 Spread of feral goats in Australia

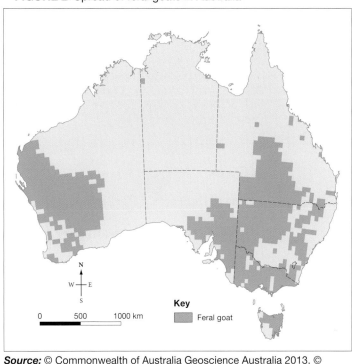

Key
Feral goat

Source: © Commonwealth of Australia Geoscience Australia 2013. © Commonwealth of Australia Department of Sustainability, Environment, Water, Population and Communities 2013. Map by Spatial Vision.

Feral goats cause widespread damage to native vegetation. They damage the soil and overgraze native grasses, herbs, trees and shrubs, causing erosion and preventing plant regeneration. They introduce weeds through seeds contained in their dung, and pollute water courses. They dramatically increase the rate of erosion on steep hillsides, where widespread gullying can quickly develop. By focusing on a favoured food source and preventing its regrowth, goats can totally remove some species of vegetation from an area, allowing more invasive plant species to take over.

During times of drought they also compete with native wildlife and domestic livestock for food, water and shelter, creating an additional imbalance in the food chain.

The impact of feral goats is worse in regions where rabbits are also out of control; together they can reduce to bedrock what was once a well-vegetated environment, leaving it open to erosion by both wind and water.

Foxes and rabbits

Both foxes and rabbits were introduced into Australia by the early settlers. With no natural predators, each species spread rapidly.

Left unchecked, foxes pose a significant threat to agriculture and native fauna. Fox predation accounts for one-third of new lamb deaths, and native animals such as the bandicoot are easy prey. Foxes carry a wide range of diseases and parasites such as hepatitis, distemper, mange and rabies.

Rapid-breeding feral rabbits place significant pressure on the environment, competing with native wildlife for food and damaging vegetation. Rabbits eat plant roots as well as foliage. They **ringbark** trees and eat seeds and seedlings, so plants cannot regenerate.

FIGURE 3 Distribution of red foxes and feral European rabbits in Australia

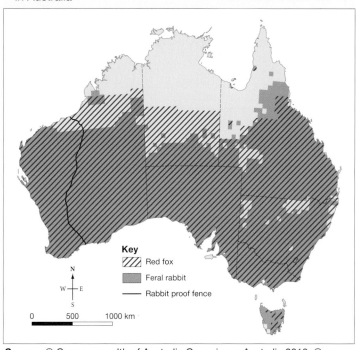

Key
- Red fox
- Feral rabbit
- Rabbit proof fence

N
W—E
S

0   500   1000 km

Source: © Commonwealth of Australia Geoscience Australia 2013. © Commonwealth of Australia Department of Sustainability, Environment, Water, Population and Communities 2013. Map by Spatial Vision.

3.7.3 The problem of introduced plants

Invasive plant species are often referred to as **weeds**. Many were introduced as garden plants, but soon spread to other areas and now pose a significant threat to both the natural environment and agricultural industries. Two of the most troublesome introduced plant species are Paterson's Curse and Viper's Bugloss (see **FIGURE 4**), both of which were introduced in the 1850s as garden plants because of their attractive flowers.

Paterson's Curse and Viper's Bugloss look similar, and are often found in similar regions, if not together. As their seeds germinate earlier than native plants, they are able to establish extensive root systems and spreading leaves, which crowd out other plant species. Their seeds can be spread in the fur of livestock or by water in areas where erosion is already present and run-off levels are high. They thrive in areas where

FIGURE 4 (a) Paterson's Curse has two long stamens protruding from the flower plus two shorter ones, and its flowers are more purple. (b) Viper's Bugloss has four long stamens protruding from the flower. Its flowers are more blue, and prickles are visible on the stem.

rainfall is high in winter and have adapted to cope with dry summers. The nutritional value of Paterson's Curse is low, and if livestock eat it in significant quantities it can be toxic, especially horses and small animals. Their stomachs cannot fully process the plant, and this leads to liver damage, loss of condition and, in extreme cases, death.

Contact with these plants can also cause skin irritations and other allergic reactions in humans and livestock. Once Paterson's Curse colonises an area, soil fertility is reduced. It has been estimated that 33 million hectares of land is infested across the nation, and that the cost to Australia's grazing industry is in excess of $250 million annually.

3.7.4 Controlling invasive species

Can one problem be part of the solution for another?

While we might want to remove invasive species from Australia and other parts of the world, in many cases this is not possible. Total eradication may be feasible in island communities where the risk of re-infestation is limited; however, on a large scale such as mainland Australia, control appears to be the best option.

Introduced species pose a serious threat to the productivity of land and diversity of natural environments. In Western Australia, trials have discovered that goats, which themselves pose a threat to both native vegetation and pasture land, can be used to control a wide variety of **invasive plant species**, such as saffron thistle (see **FIGURE 5**).

While we know that some weeds are spread passing through the digestive systems of animals, this is not the case with goats. Less than 1 per cent of the saffron thistle seeds were found in the goat dung, and these would not germinate. Similar results were found in test sites for the control of blackberries (see **FIGURE 6**). Within 12 months of goats being allowed to feed on both weed types, there was a notable reduction in their spread. Goats can also be used to control hundreds of different invasive plant species such as English Ivy, Paterson's Curse and Viper's Bugloss, which are toxic to grazing livestock.

Goats have the added advantage of being an environmentally friendly method of weed control. They eliminate the need for using herbicides and fertilisers. Soil quality is improved naturally by goat droppings. Fossil-fuel-burning machinery is not needed to remove the weeds, and goats can be used in environments where other control methods are not viable, such as on steep slopes. In recent times, they have been used as a method of weed control in plantation forests and in limited numbers on large pastoral runs. They are also still kept as livestock.

FIGURE 5 In Western Australian trials, goats almost completely eradicated saffron thistle within three years. Researchers will monitor the situation carefully, however, as seeds can lie dormant for up to 10 years.

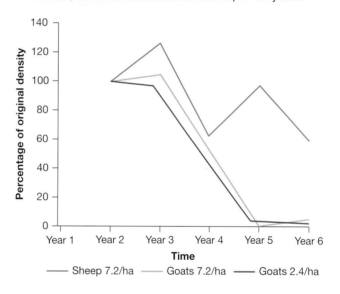

FIGURE 6 Goats have been used successfully in the Tolt River Dam region in Seattle, Washington. There, a herd of 200 goats is used to control the spread of blackberries on ground that is too steep and uneven for mowing by machinery.

Can we control foxes and rabbits?

Both foxes and rabbits have proven difficult to control and pose the same risks today as they did in the past. Foxes are the only natural predator of rabbits. Currently, rabbits are controlled through the following methods:

- biological — introduction of viruses, such as myxomatosis and calicivirus
- chemical — laying poison baits or fumigating warrens
- mechanical — destroying warrens, shooting and laying traps

Rabbits are highly destructive to the environment, but once they are eradicated from an area, the environment can regenerate (see **FIGURE 7**).

FIGURE 7 (a) Rabbits reduced Phillip Island (near Norfolk Island, off the east coast of Australia) to a wasteland. (b) After the rabbits were eradicated, the island's recovery was spectacular.

Hunting, baiting (poisoning) and shooting reduce adult fox populations in the short term; however, their populations soon recover. Scientists are now trialling biological controls and working to develop some form of virus or birth control that will interfere with the reproductive system of the fox, making them infertile and incapable of breeding.

Middle Island Maremma Project

Middle Island, a small rocky island about 2 hectares in size off the Victorian coast near Warrnambool, is home to a Little Penguin colony. At low tide, less than 12 centimetres of water separates the island from the mainland, providing easy access for predators such as foxes.

In 1999, Middle Island had a thriving colony of Little Penguins, comprised of about 600 birds. By 2005, foxes had reduced the population to fewer than 10, with only two breeding pairs remaining. In 2006 an ambitious experiment was launched using Maremma dogs to guard and protect the remaining penguins. This breed of dog has long been used to guard livestock, including chickens, with reports that once the dogs were on duty, fox kills stopped.

After some initial teething problems, the program has proved highly successful. Within a short time, there was evidence of the penguins breeding. By 2013 the Little Penguin population had rebounded to 180. In a cruel twist of fate, however, 70 penguins were killed in a fox attack in 2017, reducing the breeding population to just 14. The dogs had been taken off the island for the winter season due to high tides and bad weather. However, numbers appear to have again rebounded with 60 breeding penguins in residence for the 2018 breeding season.

FIGURE 8 Using Maremma dogs to guard the penguins on Middle Island from fox attacks proved to be a highly successful strategy. The island's penguin population, previously close to extinction, continues to grow under the protection of the dogs.

Explore more with my**World**Atlas

Deepen your understanding of this topic with related case studies and questions.
- Investigate additional topics > Endangered and introduced species > **Introduced species in Australia**

on Resources

🔗 **Weblink** Weed species

3.7 INQUIRY ACTIVITIES

1. Visit a local river or creek near your school and make a field sketch of the area. Survey the area around the creek and annotate your sketch to show the location of areas where there are invasive plant species. Add additional annotations to suggest a *sustainable* solution to this problem.

Classifying, organising, constructing

2. (a) Using information in this subtopic and your own general knowledge, copy and complete the table below. List as many species (both plant and animal) that were introduced into Australia as you can, and state why they might have been introduced. Use your atlas and the **Weed species** weblink in the Resources tab as other sources of information.

Introduced species	Reasons for introduction

(b) Refer to the table you completed in part (a). Find an image of one exotic plant and one exotic animal species in your table. Annotate your images with reasons for their introduction, and their impact on the *environment*. Compare your findings with those of other members of the class.

Classifying, organising, constructing

3. Working in teams, devise your own *sustainable* and *environmentally* friendly strategy for controlling an invasive species.

Classifying, organising, constructing

4. Use the **Weed species** weblink in the Resources tab to prepare an educational leaflet that will assist people in recognising one of these plant species.

Classifying, organising, constructing

3.7 EXERCISES

Geographical skills key: GS1 Remembering and understanding **GS2** Describing and explaining **GS3** Comparing and contrasting **GS4** Classifying, organising, constructing **GS5** Examining, analysing, interpreting **GS6** Evaluating, predicting, proposing

3.7 Exercise 1: Check your understanding

1. **GS1** Why are goats an effective method of controlling invasive plant species?
2. **GS1** Why do people say that rabbits cause the demise of native plant and animal species?
3. **GS1** Describe the distribution over *space* of foxes and rabbits in Australia.
4. **GS2** Explain why goats would be considered an *environmentally* friendly method of controlling invasive plant species.
5. **GS2** Explain why it is easier to eradicate invasive species from island communities than from mainland Australia.

3.7 Exercise 2: Apply your understanding

1. **GS6** The invasive animal species described in this subtopic have proven more difficult to control than the plant species. Suggest a reason for this.
2. **GS6** Why do you think that foxes are not found in Australia's tropical region?
3. **CS3** Consider the three methods of rabbit control by completing the following tasks.
    (a) Copy the table below. In the second column, write your own definition for each of the control methods.
    (b) Complete the remaining columns, detailing the advantages and disadvantages of the three methods.
    (c) Compare the advantages and disadvantages of the three main methods of rabbit control. Which method do you think is the most effective? Give reasons for your answer.

Method	Definition of method	Advantage	Disadvantage
Biological			
Chemical			
Mechanical			

4. **GS2** Examine **FIGURE 7**. Describe the appearance of the *environment* in each image. Do these images represent the same *place*? Suggest reasons for the *changes* that have occurred in this *environment*.
5. **GS6** Devise your own *sustainable* and *environmentally* friendly strategy for controlling an invasive species. Outline the steps in your plan and why you think it would be successful.

Try these questions in learnON for instant, corrective feedback. Go to www.jacplus.com.au.

3.8 Native species and environmental change

3.8.1 Protecting biodiversity

To protect native species, governments have established **national parks**. The world's first national park was Yellowstone National Park in the United States of America established in 1874. The second, the Royal National Park, was established in Sydney, Australia five years later. These parks are intended to provide safe habitats for native plant and animal species and thus help safeguard **biodiversity**. However, biodiversity is not just under threat from introduced species; left unchecked, native species can also cause widespread damage to the landscape.

3.8.2 Should we cull iconic Australian natives?

Many native Australian species have been threatened by the spread of urban settlements. As human populations expand, the natural range of animals such as koalas, kangaroos and wallabies are diminished. Despite the intricate pattern of National Parks that exist, native species can be found not only on the fringe of urban areas, but also taking refuge in our backyards where they face increasing risk from vehicles, domestic pets and poisons.

In some protected regions, natural increase can put entire colonies at risk. Koalas are preferential feeders and have only one source of food — eucalyptus leaves. In times of drought and following extended favourable breeding conditions, there is simply not enough food to sustain the entire population. In the Otway Ranges and on French Island, for example, the situation has at times become so dire that koalas have faced starvation.

In 2013 and 2014, 700 koalas were killed in what the Victorian government described as humane euthanasia to prevent the animals from starving to death. Opponents of the move described it as a secret **cull**.

FIGURE 1 A koala drinking from a dog's water bowl in a suburban backyard

In late 2015, researchers called for koalas to be culled in parts of Victoria, New South Wales and Queensland, in areas where the local populations were infected with the disease **chlamydia**. In koalas, chlamydia can lead to a range of issues such as conjunctivitis, urinary tract infections, reproductive tract infections and pneumonia. Culling, the researchers suggested, would prevent further spread of the disease to healthy animals and allow the population to rebound over the next 10 years. Others have argued that the disease, in its early stages, can be treated with antibiotics.

In 2019, the koala was listed as vulnerable in Queensland, New South Wales and the ACT, with loss of habitat cited as the major area of concern. Even within remaining habitats, a lack of diversity exists among eucalyptus species. Preferred control methods are sterilisation and relocation. Sterilisation prevents a population outstripping its food source; however, it does not allow the population to rebound after natural disasters such as bushfires. Relocation is only a short-term fix, as shown on Kangaroo Island — koalas relocated there have flourished, mainly because of their proximity to a commercial blue gum plantation; however, the population will once again be decimated once the trees are harvested.

DISCUSS

Conduct a class debate on the issue of culling as a method of controlling different species, both native and introduced.   **[Ethical Capability]**

3.8.3 What about the African elephant?

In 1930, some 10 million elephants roamed across 37 countries in Africa. By 1979, almost 90 per cent of the wild elephant population had been wiped out, with their numbers estimated at 1.3 million. By 2007, the estimated range of elephant numbers had declined even further (see **FIGURE 2**). In 2018, the population had continued to decline with numbers down to around 415 000. At this rate, it is possible that they could become extinct in the wild by 2050. Some estimates put the number of animals killed at up to 100 per day, as poachers seek to make their fortune selling meat, ivory and body parts to the lucrative Asian market. From 2011 to 2014 the price of ivory alone tripled in China, peaking at $2100 per kilogram. In 2017, a ban imposed by the Chinese government on ivory saw the price fall to $750 per kilogram.

FIGURE 2 African elephant range and estimated population

African elephant range and estimated population

AFRICA

Mali
Niger
Burkina Faso
Benin
Chad
Central African Republic
Cameroon
Ethiopia
Uganda
Kenya
Democratic Republic of the Congo
Tanzania
Angola
Zambia
Malawi
Zimbabwe
Botswana
Mozambique
South Africa

W-Arly-Pendjari Complex of protected areas — largest remaining elephant population

At risk of local extinction

Greatest decline since 2007

African elephant range and estimated population

Range in 1979
1.3 million elephants

Range in 2007
472 000–690 000

Range in 2016
350 000–400 000

0   1000   2000 km

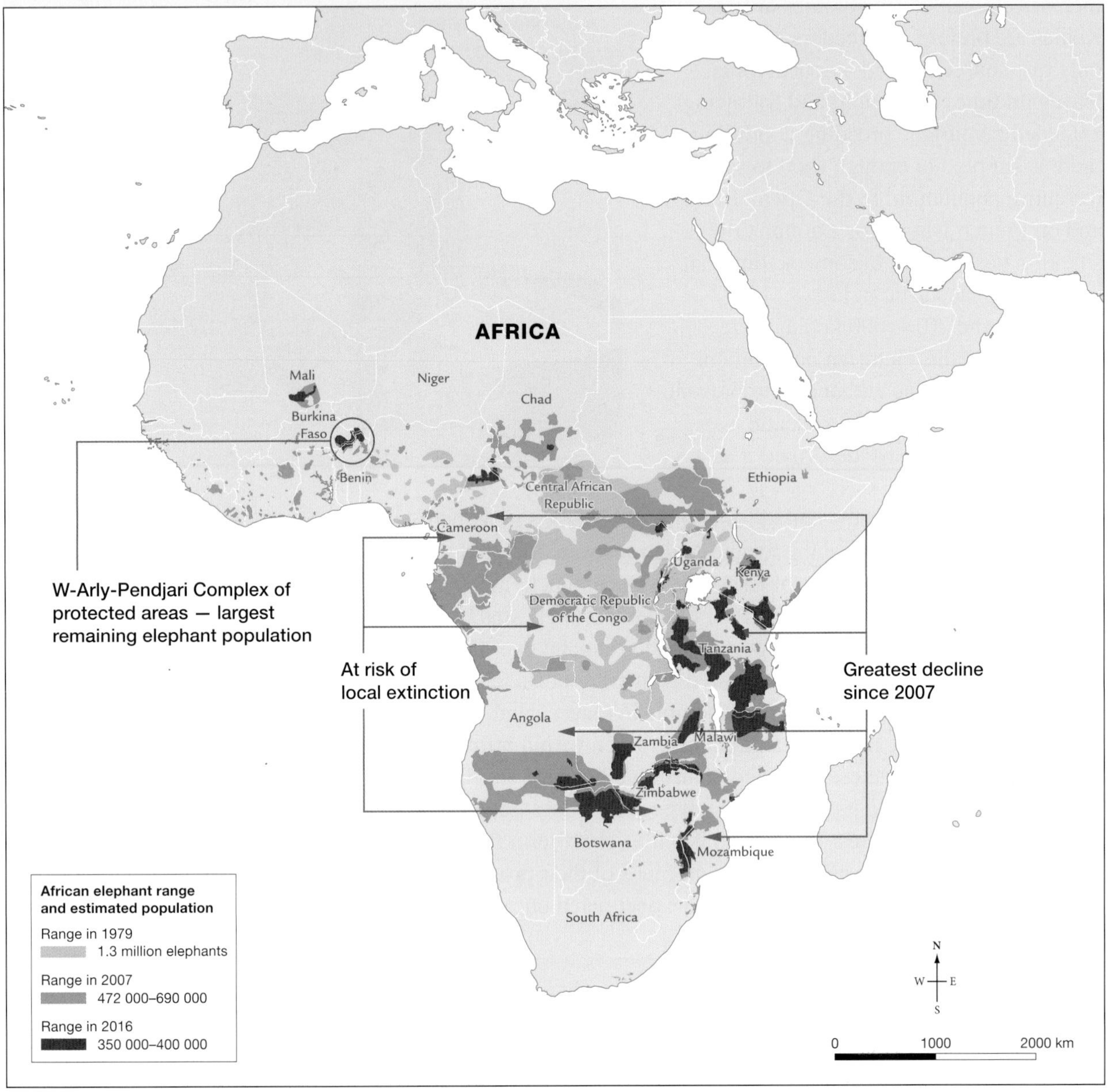

Source: National Geographic and IUCN Red List

Elephants are also under threat from expanding human populations. As the number of people increases, so too does their need for land to grow crops and raise livestock, and people encroach further and further into the elephants' rangelands. Struggling farmers can also earn more from a single elephant kill than from a year of toiling on the land. When elephants enter these newly created farmlands and damage crops, the temptation to kill them intensifies.

Elephants are a key ecological species, sometimes referred to as the caretakers of the environment. They create and maintain their ecosystem and in the process create the habitat for a wide range of plant and animal species with which they coexist. Up to 30 per cent of native African tree species, for example, are dependent on elephants to assist with dispersal and germination of their seeds. The loss of elephants poses a significant threat to local ecologies.

However, the practice of confining large animals such as elephants to national parks free from predation and with an abundant water supply can result in a population explosion. An adult elephant consumes up to

136 kilograms of food in a single day. The search for food can see them cover vast distances, in the process stripping bark from, ripping branches off, and pushing over trees. As the process of confining animals to reserves continues, traditional migration paths are interrupted. The population flourishes and the landscape becomes degraded as the rate of change speeds up.

Africa's first national park was established in 1926. Covering almost 20 000 square kilometres, it is one of the largest reserves in Africa. Culling of elephants within the national park was banned in 1994. While elephant numbers are declining elsewhere, within Kruger National Park they are increasing. This has prompted some to suggest that culling as a means of population control should be reinstated to limit numbers to a sustainable level of 7000 to 8000, rather than the current 17 000, which places both the elephants and their habitat in jeopardy.

3.8.4 Is it a pest?

In Australia, the possum is a protected species that can cause considerable damage in urban areas. Because they are protected, control measures are largely centred on possum-proofing homes and gardens. There are strict regulations relating to the trapping of possums, and they generally must be released on the same property on which they were captured, within 50 metres of their capture site. The only exception is in the state of Tasmania, where they can be removed in order to protect crops and trapped for commercial trade in meat and pelts.

In New Zealand, however, possums are considered 'public enemy number one'. Originally introduced in 1837 to establish a fur trade, with no natural predators they spread rapidly and today some 30 million possums occupy 90 per cent of the landmass. The damage they cause to native forests is unmistakable, laying large expanses of new forest growth bare. They compete with native birds for habitat and food and have been observed raiding nests. Additionally, they are known to spread bovine tuberculosis, thus posing a significant threat to dairy and deer farmers (see **FIGURE 3**). The main methods of possum control in New Zealand include trapping, baiting and shooting.

FIGURE 3 The spread of possums in New Zealand

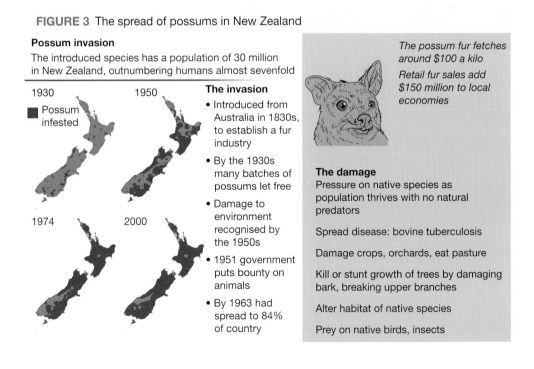

Possum invasion

The introduced species has a population of 30 million in New Zealand, outnumbering humans almost sevenfold

1930   1950

■ Possum
  infested

1974   2000

The invasion
- Introduced from Australia in 1830s, to establish a fur industry
- By the 1930s many batches of possums let free
- Damage to environment recognised by the 1950s
- 1951 government puts bounty on animals
- By 1963 had spread to 84% of country

The possum fur fetches around $100 a kilo

Retail fur sales add $150 million to local economies

The damage
Pressure on native species as population thrives with no natural predators

Spread disease: bovine tuberculosis

Damage crops, orchards, eat pasture

Kill or stunt growth of trees by damaging bark, breaking upper branches

Alter habitat of native species

Prey on native birds, insects

3.8 EXERCISES

Geographical skills key: GS1 Remembering and understanding **GS2** Describing and explaining **GS3** Comparing and contrasting **GS4** Classifying, organising, constructing **GS5** Examining, analysing, interpreting **GS6** Evaluating, predicting, proposing

3.8 Exercise 1: Check your understanding

1. **GS1** Describe the spread of possums over time in New Zealand.
2. **GS1** Why do you think possums are considered a pest in New Zealand but are a protected species in Australia?
3. **GS2** Explain how native species can cause the land to be degraded.
4. **GS2** Explain how a national park can act as a safeguard for biodiversity.
5. **GS2** Describe the distribution of elephants in Africa.

3.8 Exercise 2: Apply your understanding

1. **GS2** Explain how human population growth leads to the fragmentation of koala habitats.
2. **GS6** Predict what might happen to the koala population, over time, in the Otway Ranges or on French Island if no action was taken to control the koala population.
3. **GS2** Explain the threats to the African elephant population.
4. **GS5** Referring to the threats to African elephants, what do you consider to be the greatest threat to the long-term survival of this species? Justify your response.
5. **GS6** Do you think that culling is a viable solution to ensure the long-term survival of elephants or koalas? Give reasons for your answer.

Try these questions in learnON for instant, corrective feedback. Go to www.jacplus.com.au.

3.9 Indigenous communities and sustainable land management

3.9.1 Traditional land management

Before the arrival of European settlers, Indigenous communities had their own system of land management. They maintained grasslands through the use of fire, which encouraged plant regrowth and attracted a variety of animals. Indigenous peoples' life was governed by the seasons, with each change dictating a change in the use of the land and its management.

Environmental change is not new. Indigenous communities around the world, including the Australian Aborigines, have had to manage their environments carefully. In **FIGURE 1** you can see how fire was used to manage the landscape. It is interesting to note that the use of fire in this way over the past 50 000 years had a significant impact on the species of plants that thrive in Australia today. Those plants that adapted to the use of fire thrived; for example, eucalypts. Using fire was just one of a variety of strategies employed by Indigenous Australians to ensure the land was used in a sustainable manner.

FIGURE 1 Indigenous Australians' traditional land management practices and connection to the land

Indigenous people have adapted to environmental change over the last 50 000 years.

Aboriginal land practices involve working with the land and its elements rather than seeking to make dramatic changes.

Indigenous Australians took only what they needed and little was wasted.

Habitat loss, soil erosion and weed infestation were unknown until the time of European settlement.

Evidence also shows that fuel reduction (back burning) was used to prevent bushfires. This practice prevented large bushfires that could burn for months and permanently damage the landscape.

They take collective ownership of the land.

In some Indigenous communities some native species such as the kangaroo and platypus are considered sacred.

Their technology was simple. They used spears and fire sticks, designed to minimise environmental impact.

Fire is used to control plant growth and maintain a grassland environment. Many native seeds need fire in order to germinate.

Although nomadic, there was a pattern to their movement across the land, designed to coincide with the seasons.

Every aspect of their life is governed by the land and the seasons. The land provided all their needs — they had no need to grow crops or raise livestock.

Their spiritual and cultural connections to the land, the health of the land and its water are central to their own wellbeing.

3.9.2 What happens when?

FIGURE 2 is an example of an Aboriginal seasons calendar. It is for the Yolngu people who live in north-east Arnhem Land. The calendar relates the months of the year to aspects of the environment, although a traditional Indigenous community had no use for the months as we know them. Look carefully and you will see that this particular calendar includes information about the weather and the plants and animals that thrive across the year. Traditional communities were made up of hunters and gatherers. They hunted and fished for particular species and gathered bulbs, fruits and other edible vegetation at different times of the year. The calendar varied from place to place, but whatever the location, it enabled Indigenous people to predict seasonal events; for instance, the arrival of march flies signalled the time to collect crocodile eggs and bush honey.

FIGURE 2 The hunting and gathering activities of Indigenous communities were defined by observable changes in the seasons.

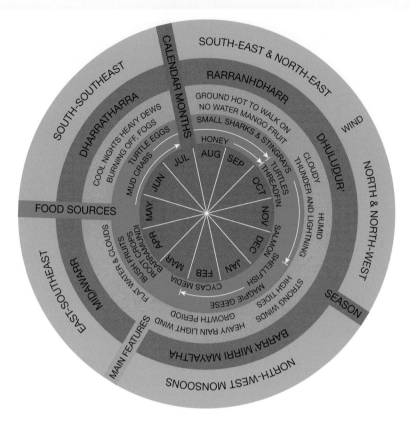

3.9.3 CASE STUDY: Managing Kakadu wetlands

Kakadu is a kaleidoscope of both cultural and ecological biodiversity. The landscape varies from savannah and woodlands to escarpments and ridges as well as wetland, flood plains and tidal flats. The region includes more than 2000 plant species that have provided food, medicine and weaving materials for the Indigenous communities that have inhabited the region for some 50 000 years. Over this time they have defined six distinct seasons, all signalled by subtle changes in the weather patterns that mark the transition from one season to the next. They managed and maintained the landscape through the use of fire.

The arrival of European settlers saw a massive change in the region. Buffalo were introduced in the early- to mid-1800s to serve as a food supply for new settlers. However once these new settlements were abandoned in the mid-1900s the buffalo population expanded from a modest population of less than 100 animals to more than 350 000. The impact on local habitats was extreme.

Natural habitats were devastated. The now feral buffalo took over wetland areas, disturbed native vegetation, caused significant soil erosion and changed the characteristics of the region's floodplains. Saltwater intrusion of freshwater wetlands caused the region to become further degraded, leading to a rapid decline in the flora and fauna including waterbirds that had sustained Indigenous communities.

In the late twentieth century, a massive culling program was commenced to remove the feral buffalo and allow the region to regenerate. However, an invasive native plant species that had once been the main food source of the buffalo spread unchecked. It choked the wetlands and prevented waterbirds from feeding and recolonising the region.

The CSIRO undertook extensive research into the sustainable practices of the region's traditional landowners. A joint management initiative was introduced into the area. At the heart of the initiative was the traditional method of fire management.

The results have been dramatic, and the wetlands are once again home to a rich assortment of flora and fauna. The project provides an internationally recognised example of sustainable land management using the practices carried out by Indigenous communities across multiple generations.

FIGURE 3 Wetlands after removal of buffalo and before burning

FIGURE 4 Wetlands after burning

DISCUSS

There is much to learn from Indigenous Australians' traditional land management practices. Identify and analyse the challenges as well as the benefits of doing this. **[Intercultural Capability]**

Explore more with myWorldAtlas

Deepen your understanding of this topic with related case studies and questions.
- Investigating Australian Curriculum topics > Year 10: Environmental change and management > **Indigenous Australians — Caring for Country**

3.9 INQUIRY ACTIVITIES

1. Land degradation such as salinity was not an **environmental** issue in Australia when Indigenous people were its sole inhabitants. With the aid of diagrams, explain how land-use practices have **changed** over time. Make sure you include references to Indigenous practices that promoted **sustainable** use of the **environment**. Include links to how these **changes** would have resulted in salinity and degraded the **environment**. **Classifying, organising, constructing**
2. Consider the Aboriginal calendar in **FIGURE 2**. Develop your own calendar that reflects the **interconnection** between the seasons and **changes** in your life. **Classifying, organising, constructing**

3.9 EXERCISES

Geographical skills key: GS1 Remembering and understanding **GS2** Describing and explaining **GS3** Comparing and contrasting **GS4** Classifying, organising, constructing **GS5** Examining, analysing, interpreting **GS6** Evaluating, predicting, proposing

3.9 Exercise 1: Check your understanding

1. **GS1** Why wasn't land degradation an issue before European settlers arrived?
2. **GS2** Explain how Indigenous people managed the land before the arrival of European settlers.
3. **GS2** How does the Aboriginal calendar demonstrate an **interconnection** between Indigenous peoples' connection with the land and **sustainable** management of the **environment**?
4. **GS2** Explain why the use of fire was an important component of managing the land.
5. **GS2** Explain why water buffalo are considered a significant threat to the Kakadu wetlands.

3.9 Exercise 2: Apply your understanding

1. **GS6** The Aboriginal calendar (**FIGURE 2**) demonstrates an intricate understanding of the ***environment***.
   (a) How long do you think it would have taken the Aboriginal people to have developed this understanding?
   (b) How do you think this knowledge would have been passed from generation to generation?
2. **GS6** It has been suggested that the four seasons currently used in Australia do not adequately reflect the changing nature of our seasons.
   (a) Do you agree or disagree with this suggestion? Give reasons for your opinion based upon the area you live in.
   (b) Do you think the Aboriginal seasons calendar should be adopted and used as an additional strategy for the ***sustainable*** management of ***environmental*** issues? Justify your point of view.
3. **GS3** Study **FIGURES 3** and **4**. Describe the ***environment*** shown in **FIGURE 3** and the ***changes*** that have occurred in **FIGURE 4**.
4. **GS5** Describe the steps that might have been taken to turn the landscape shown in **FIGURE 3** to the one shown in **FIGURE 4**.
5. **GS6** Do you think the Indigenous land management practices used in Kakadu could be used in other parts of Australia, including Victoria? Give reasons for your answer.

Try these questions in learnON for instant, corrective feedback. Go to www.jacplus.com.au.

3.10 SkillBuilder: Writing a fieldwork report as an annotated visual display (AVD)

What is a fieldwork report?

A fieldwork report helps you process all the information that you have gathered during fieldwork. You sort your data, create tables and graphs, and select images. You interpret the data as text or annotated images to convey your ideas. To convey your ideas, you synthesise, or pull together, all the data in a logical presentation.

Select your learnON format to access:

- an overview of the skill and its application in Geography (Tell me)
- a video and a step-by-step process to explain the skill (Show me)
- an activity and interactivity for you to practise the skill (Let me do it)
- questions to consolidate your understanding of the skill.

 **Resources**

Video eLesson Writing a fieldwork report as an annotated visual display (AVD) (eles-1747)

Interactivity Writing a fieldwork report as an annotated visual display (AVD) (int-3365)

3.11 Thinking Big research project: Invasive species *Wanted!* poster

SCENARIO

The time has come to take action and eradicate invasive species before they cause further damage to the environment! You will research an invasive species and create a *Wanted!* poster, to be featured on the Department of Environment's website, to raise community awareness and educate people about the damage this invader causes.

Select your learnON format to access:

- the full project scenario
- details of the project task
- resources to guide your project work
- an assessment rubric.

 Resources

💡 **ProjectsPLUS** Thinking Big research project: Invasive species *Wanted!* poster (pro-0212)

3.12 Review

3.12.1 Key knowledge summary

Use this dot-point summary to review the content covered in this topic.

3.12.2 Reflection

Reflect on your learning using the activities and resources provided.

 Resources

 eWorkbook   Reflection (doc-31765)

Crossword (doc-31766)

 Interactivity   Land environments under threat crossword (int-7670)

KEY TERMS

algal bloom rapid growth of algae caused by high levels of nutrients (particularly phosphates and nitrates) in water

biodiversity the variety of plant and animal life within an area

brackish (water) water that contains more salt than fresh water but not as much as sea water

carrying capacity the ability of the land to support livestock

chlamydia a sexually transmitted disease infecting koalas

cull selective reduction of a species by killing a number of animals

desertification the transformation of land once suitable for agriculture into desert by processes such as climate change or human practices such as deforestation and overgrazing

drainage area (or basin) an area drained by a river and its tributaries

▶

dryland ecosystems characterised by a lack of water. They include cultivated lands, scrublands, shrublands, grasslands, savannas and semi-deserts. The lack of water constrains the production of crops, wood and other ecosystem services.

ecology the environment as it relates to living organisms

exotic species species introduced from a foreign country

groundwater water held underground within water-bearing rocks or aquifers

humus decaying organic matter that is rich in nutrients needed for plant growth

invasive plant species commonly referred to as weeds; any plant species that dominates an area outside its normal region and requires action to control its spread

mediterranean (climate) characterised by hot, dry summers and cool, wet winters

monoculture cultivating a single crop or plant species over a wide area over a prolonged period of time

mulch organic matter such as grass clippings

national park an area set aside for the purpose of conservation

pastoral run an area or tract of land for grazing livestock

ringbark remove the bark from a tree in a ring that goes all the way around the trunk. The tree usually dies because the nutrient-carrying layer is destroyed in the process.

Royal Commission a public judicial inquiry into an important issue, with powers to make recommendations to government

Sahel a semi-arid region in sub-Saharan Africa. It is a transition zone between the Sahara Desert to the north and the wetter tropical regions to the south. It stretches across the continent, west from Senegal to Ethiopia in the east, crossing 11 borders.

salinity an excess of salt in soil or water, making it less useful for agriculture

salt scald the visible presence of salt crystals on the surface of the land, giving it a crust-like appearance

topsoil the top layers of soil that contain the nutrients necessary for healthy plant growth

turbid water that contains sediment and is cloudy rather than clear

watertable upper level of groundwater; the level below which the earth is saturated with water

weathering the breaking down of rocks

weed any plant species that dominates an area outside its normal region and requires action to control its spread

4 Inland water — dammed, diverted and drained

4.1 Overview

Humans would find life very hard without healthy inland water sources. Are we being careful with how we use and change them?

4.1.1 Introduction

Water makes life on Earth possible — rivers are like blood running through the veins of a body. Our inland waters are important sources of water for both environments and people.

Over time we have dammed, diverted and drained our water sources, and this has brought about significant environmental change. Careful stewardship of our water resources will help ensure a sustainable future.

---

on Resources

- ✓ **eWorkbook**   Customisable worksheets for this topic
- **Video eLesson** Drained away (eles-1709)

To access a pre-test and starter questions and receive immediate, **corrective feedback** and **sample responses** to every question, select your learnON format at www.jacplus.com.au.

4.2 Wet and wonderful — inland water

4.2.1 What is inland water?

Have you ever stopped to think that the water flowing down a river or rippling across a lake is our life-support system? The rivers, lakes and wetlands that make up our inland water then supply our domestic, agricultural, industrial and recreational water use. They also provide important habitats for a wide range of aquatic and terrestrial life.

Inland water systems cover a wide range of landforms and environments, such as lakes, rivers, floodplains and wetlands. The water systems may be **perennial** or **ephemeral**, standing (such as lakes), or flowing (such as rivers — see **FIGURE 1**). There are interconnections between surface water and **groundwater**, and between inland and coastal waters. Inland water is an important link in the water cycle, as water evaporates from its surface into the atmosphere. In return, rainfall can be stored in rivers and lakes, or soak through the soil layers to become groundwater.

FIGURE 1 The Parana River floodplain in northern Argentina shows a variety of different types of inland water.

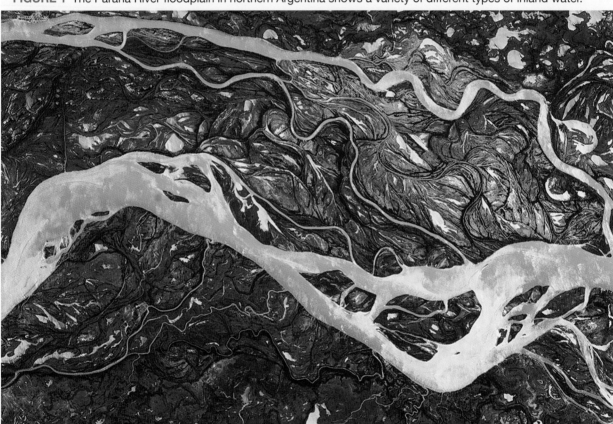

4.2.2 Why is inland water important?

Inland water provides both the environment and people with fresh water, food and habitats. It provides environmental services; for example, it can filter pollutants, store floodwater and even reduce the impacts of climate change. The economic value of these services cannot easily be measured. Their importance, however, can be taken for granted and not appreciated until the services are lost or degraded.

4.2.3 What are the threats to inland water?

Inland water is extremely vulnerable to change. It has been estimated that in the last century, North America, Europe and Australia have lost over 50 per cent of their inland water (excluding lakes and rivers). Those systems remaining are often shrunken and polluted. This loss is largely a result of human-induced environmental changes. **TABLE 1** illustrates some of the reasons for changes to inland water systems, and their possible impacts on the environment and people. As water is such a valuable resource, much of our inland waterways have been dammed, diverted or drained to meet the needs of people.

TABLE 1 Threats to inland water

Cause of change to inland water systems	Environmental functions threatened	Impacts of change
• Increasing population and increasing demand for water across space	• Most services (e.g. fresh water, food and biodiversity) • Regulatory features such as recharging groundwater and filtering pollutants	• Increased withdrawal of water for human and agricultural use • Large-scale draining of wetlands to create farmland
• Construction of infrastructure including dams, weirs and levee banks, diverting water to other drainage basins	• Services supporting the quality and quantity of water • Biodiversity, habitat, river flow and river landforms	• Changes to the amount and timing of river flow. The transport of sediment can be blocked and dams can restrict fish movements.
• Changing land use (e.g. draining of wetlands, urban development on floodplains)	• Holding back floodwaters and filtering pollutants • Habitats and biodiversity	• Alters run-off and infiltration patterns • Increased risk of erosion and flood
• Excessive water removal for irrigation	• Reduced water quantity and quality • Less water available for groundwater supply	• Reduced water and food security • Loss of habitat and biodiversity in water bodies
• Discharge of pollutants into water or onto land	• Change in water quality, habitat Pollution of groundwater	• Decline in water quality for domestic and agricultural use • Changes ecology of water systems

FIGURE 2 Wetlands are an example of inland water systems that are vulnerable to human-induced damage.

4.2 INQUIRY ACTIVITY

Make a simplified sketch of **FIGURE 1** and clearly label an example of each of the following features: *main channel, tributary, anabranch, meander, oxbow lake (or billabong), floodplain*. A dictionary may help you define the terms.

Classifying, organising, constructing

4.2 EXERCISES

Geographical skills key: GS1 Remembering and understanding **GS2** Describing and explaining **GS3** Comparing and contrasting **GS4** Classifying, organising, constructing **GS5** Examining, analysing, interpreting **GS6** Evaluating, predicting, proposing

4.2 Exercise 1: Check your understanding

1. **GS2** Match the following terms with their correct definition in the table below: *main channel, tributary, anabranch, meander, oxbow lake (or billabong), floodplain*

Term	Definition
	A smaller stream that flows into a larger stream
	A bend in the river
	An area of relatively flat, fertile land on either side of a river
	A main river
	A cut-off meander bend
	Where a river branches off and joins back into itself

2. **GS4** Make a list of as many inland water storage features as you can think of, and classify them according to whether they are surface or underground, natural or man-made. (Note that some can be both natural and man-made.)
3. **GS2** Explain how groundwater is part of the water cycle.
4. **GS1** What is the difference between a perennial and an ephemeral river?
5. **GS2** Suggest two reasons why a wetland, such as that shown in **FIGURE 2**, might be drained.

4.2 Exercise 2: Apply your understanding

1. **GS5** Refer to **TABLE 1**. Indicate whether each of the following statements is true or false.
   (a) Large-*scale* draining of wetlands will not affect groundwater.
   (b) The spread of settlement over a floodplain will alter the amount of water available to replenish groundwater.
   (c) Habitat destruction can occur with both draining of wetlands and construction of dams.
   (d) Water that is diverted from one drainage basin to another is lost to the water cycle.
2. **GS6** Suggest two short-term and two long-term examples of human-induced *changes* that could have an impact on the wetland in **FIGURE 2**.
3. **GS5** The Parana River is 4880 kilometres long, making it the second longest river in South America. The river flows from the south-east central plateau of Brazil south to Argentina. **FIGURE 1** is a small section of this river. What evidence is there to suggest that this river frequently floods?
4. **GS5** Refer to **FIGURE 1**. The brown shading visible in the water and on the land represents the river's muddy sediment. This is material such as sand and silt carried and deposited by a river.
   (a) Where has this sediment come from?
   (b) How does the sediment get onto the floodplain?
   (c) If the river is dammed upstream, what *changes* are likely to happen to the sediment carried and to the floodplain?
5. **GS6** Refer to **FIGURE 1**. Imagine that the Parana River flooded, and the floodwaters have now subsided. Would the floodplain look the same as it does in this image? Explain your answer.

Try these questions in learnON for instant, corrective feedback. Go to www.jacplus.com.au.

4.3 Damming rivers — the pros and cons

4.3.1 Why dam rivers?

Are dams marvelous feats of modern engineering or environmental nightmares? Without them, we would not have a dependable supply of water or electricity, nor would we feel relatively safe from floods. For many decades, dams have been seen as symbols of a country's progress and economic development. But increasingly, the social, economic and environmental costs are emerging.

A reliable water supply has always been critical for human survival and settlement. The global demand for water has increased by 600 per cent in the last century — more than twice the rate of population growth. If this rate continues, global water demand will exceed supply by 2030. Water is also unevenly distributed across the globe. Some places suffer from regular droughts, while others experience massive floods. As a result, people have learned to store, release and transfer water to meet their water, energy and transport needs. This could be in the form of a small-scale farm dam or a large-scale multi-purpose project such as the Snowy River Scheme. Constructing dams is one of the most important contributors to environmental change in river basins. Globally, over 60 per cent of the world's major rivers are controlled by dams.

FIGURE 1 shows the degree of **river fragmentation**, or interruption, in the world's major drainage basins. River fragmentation is an indicator of the degree to which rivers have been modified by humans. Highly affected rivers have less than 25 per cent of their main channel remaining without dams, and/or the annual flow pattern has changed substantially. Unaffected rivers may have dams only on tributaries but not the main channel, and their discharge has changed by less than 2 per cent. Today, 48 per cent of the world's river volume is moderately to severely affected by dams.

FIGURE 1 Degree of river fragmentation in the world's major drainage basins

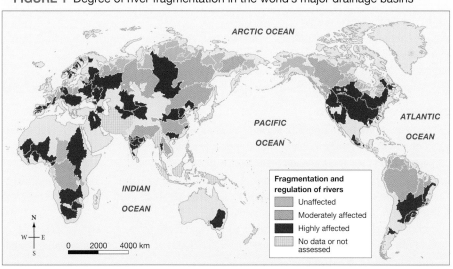

Source: Made with Natural Earth. University of New Hampshire UNH/Global Runoff Data Centre GRDC http://www.grdc.sr.unh.edu/. Map by Spatial Vision.

Dams, **reservoirs** and **weirs** have been constructed to improve human wellbeing by providing reliable water sources for agricultural, domestic and industrial use. Dams can also provide flood protection and generate electricity.

However, while there are many benefits, large-scale or mega dams bring significant changes to the environment and surrounding communities, both positive and negative, as shown in **FIGURE 2**.

FIGURE 2 The advantages and disadvantages of large-scale dams

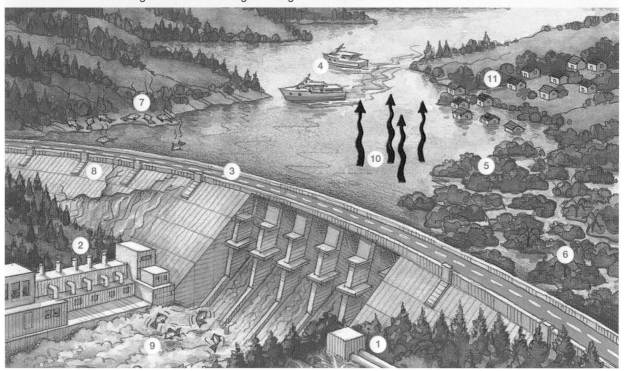

Positive changes

1. A regular water supply allows for irrigation farming. Only 20 per cent of the world's arable land is irrigated, but it produces over 40 per cent of crop output.
2. Released water can generate hydro-electricity, which accounts for 16 per cent of the world's total electricity and 71 per cent of all renewable energy.
3. Dams can hold back water to reduce flooding and even out seasonal changes in river flow.
4. Income can be generated from tourism, recreation and the sale of electricity, water and agricultural products.

Negative changes

5. Large areas of fertile land upstream become flooded or inundated as water backs up behind the dam wall. Alluvium or silt is deposited in the calm water that previously would have enriched floodplains.
6. Initially, flooded vegetation rots and releases greenhouse gases.
7. The release of cold water from dams creates thermal pollution. Originally the Colorado River had a seasonal fluctuation in temperature of 27 °C. Today, temperatures average 8 °C all year. The water is too cold for native fish reproduction, but is ideal for some introduced species.
8. Some dams are constructed in tectonically unstable areas, which are prone to earthquakes.
9. Dams block the natural migration of fish upstream. Since 1970, the world's freshwater fish population has declined by 80 per cent.
10. Over 7 per cent of the world's fresh water is lost through evaporation from water storages.
11. A conservative estimate has stated that dams have negatively affected 472 million people worldwide. Tens of millions have been relocated from dam sites while other communities both upstream and downstream have lost their livelihoods or had their land flooded.

4.3.2 Why should a river flow?

Traditionally, water flowing out to sea was seen as a waste. If it could be stored, then it could be used. Little thought was given to the health of the river and the importance of keeping water in a stream. Governments around the world have favoured damming rivers to make use of water resources. But is this the only solution to our growing water needs?

Mega dams have been linked to economic development and improvement in living standards. Only in recent times have people questioned the real cost of these schemes — environmentally, economically and socially. **FIGURE 3** shows the number of downstream communities in each country that have the potential to be affected by the construction of mega dams.

FIGURE 3 Distribution of downstream communities affected by large dams

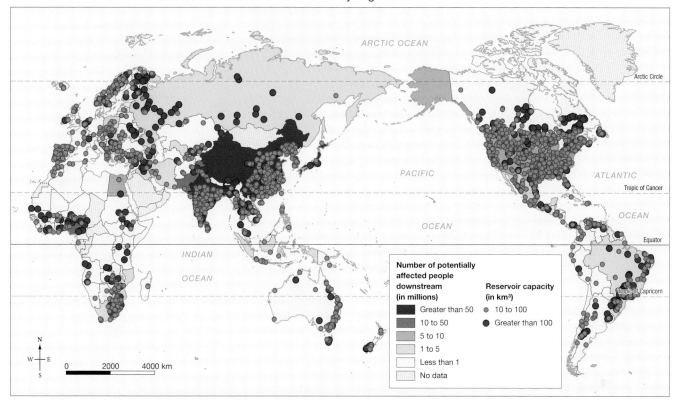

Source: Lehner et al.: High resolution mapping of the world's reservoirs and dams for sustainable river flow management. Frontiers in Ecology and the Environment. GWSP Digital Water Atlas (2008). Map 81: GRanD Database (V1.0). Available online at http://atlas.gwsp.org.

There is also the concern that multi-purpose dams have conflicting aims. To generate hydro-electricity you need to release a large volume of stored water. To provide **flood mitigation** you need to keep water levels low. To use water for irrigation you need a large store. So, what do you do?

More than one billion people worldwide lack access to a decent water supply, yet it has been estimated that only 1 per cent of current water use could supply 40 litres of water per person per day, if the water was properly managed. The problem is not so much the quantity or distribution of water resources but the mismanagement of it. During the twentieth century, over $2 trillion was spent on constructing more than 50 000 dams. The emphasis now is to switch from *controlling* river flow to *adapting to* river flow. In other words, shifting from a human-centred to an earth-centred approach. This means building small-scale projects that promote social and environmental sustainability (see subtopic 4.4). In many regions of the world, there are ongoing community protests against the need for mega dams in preference to smaller schemes that benefit local people directly.

4.3.3 People versus power?

Across the globe, from Africa to Asia to South America, there has been a growing movement of community and environmental groups challenging the construction of mega dams in terms of location, sustainability and the potential social, economic and environmental impacts. Organisations such as International Rivers work with local groups to help restore justice to dam-affected communities, find better alternatives and promote the restoration of rivers through better dam management.

4.3.4 CASE STUDY: Belo Monte Dam, Brazil

For over 20 years, there has been an ongoing protest over the construction of the Belo Monte Dam in Brazil. The original design called for five huge dams on the Xingu River, but after large-scale local and international protests by indigenous groups and environmentalists, the scheme was scaled back to one large dam — the world's third largest (see **FIGURE 4**).

FIGURE 4 (a) Location of the Belo Monte Dam and (b) the related environmental changes

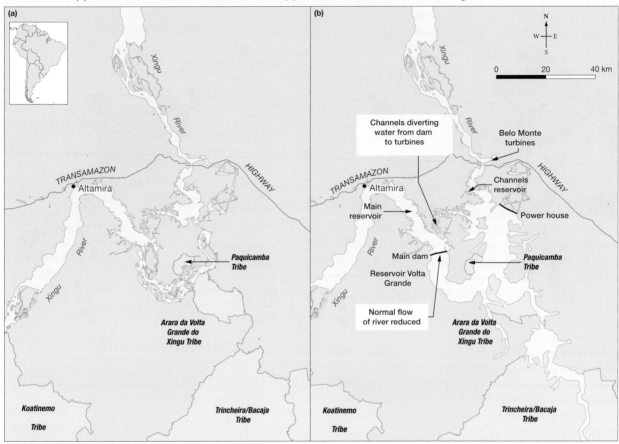

Source: Spatial Vision.

Belo Monte was designed to divert more than 80 per cent of the flow of the Xingu River, drying out over 100 kilometres of river, known as the Big Bend (see **FIGURE 5**). As a result, over 516 km^2 of rainforest was flooded, and between 20 000 and 40 000 indigenous people were displaced from their homelands.

Construction was delayed and battles fought in court over the legality of the **environmental impact assessment**, which was done after work had already started on the project. For the indigenous people, diverting water from the river channel meant a reduction in fish populations. Additionally, because there were few roads in the region, river trading was essential, but has now been reduced. The loss of

FIGURE 5 First stages of the construction of the Belo Monte Dam

rainforest, lowering of watertables and drying out of soils are further predicted impacts. Traditional livelihoods and cultures based on small-scale fishing, floodplain farming and forest management have been threatened.

While the government has claimed the dam will provide **green energy**, the amount of greenhouse gases released from drowned and rotting vegetation behind the dam wall will contribute to global warming for some years. River flow in the region is seasonal, so hydro-electricity can be generated at peak flow for only a few months of the year. During the dry months, only 1000 MW of a potential 11 000 MW will be generated. There is a distinct possibility that another dam will need to be built upstream to supply a more even and continual flow of water for power generation.

Downstream, the small town of Altamira (shown in **FIGURE 4**) rapidly expanded during the three-year construction period, when 60 000 labourers flooded in looking for construction work. Land prices skyrocketed, the cost of living rose, and crime rates soared to create the most dangerous town in the country. Once construction was halted, workers left as jobs disappeared.

In early 2016 the project was suspended and the owners, Norte Energia were fined $317 000 for failing to provide promised protection for local communities. A two-year 'emergency program' established in 2011 was designed to compensate people for the schools and clinics that were promised but not supplied. Each village was allocated 30 000 reais (around $12 500) per month

FIGURE 6 Protesters at the dam site cut a channel through earthworks to restore flow in the Xingu River. The wording translates as 'Stop Belo Monte'.

for two years. After centuries of living a subsistence life, local tribes were introduced to the modern world. Fishing and hunting was replaced with supermarket fast food, alcohol and sweets. Motorbikes and outboard motors replaced canoes, and the role of the tribal Elders was pushed aside by younger people who could speak Portuguese with the construction workers. Traditional social activities were replaced by televisions, and plastic and other garbage accumulated as no-one knew what to do with it.

For nearly three decades the Jurana tribe have fought the dam construction and much of their traditional lifestyle activities have been replaced by meetings with government and company officials, environmental activists and journalists. Attitudes towards the native communities have changed. As one dam employee noted, 'In the old days you just gave the Indians a mirror and they were happy. Now they want iPads and four-wheel drives'.

Scientists are now questioning whether large-scale infrastructure projects can balance economic benefits with environmental and social costs. With the increasing threat of climate change and recent drought, which has reduced flow along the Xingu River, the Belo Monte dam may never meet its promised economic or energy-producing goals.

In 2018, the Brazilian government announced that they would cease constructing mega dams in the Amazon. Brazil has the potential to generate 50 gigawatts of energy by 2050 if they built all the dams under design, but 77 per cent of these would to some extent impact indigenous land or federally protected areas. It appears that the ongoing resistance of indigenous peoples and environmentalists, combined with other political and economic influences have led to a hard-won change in policy.

DISCUSS

Does a large company such as Norte Energia have obligations to the people dislocated by such a large-scale scheme? Before deciding, carefully consider the consequences of the company being deemed responsible or not responsible. **[Ethical Capability]**

4.3 INQUIRY ACTIVITIES

1. Complete the **Controversial dams** worksheet to explore this topic further.

Examining, analysing, interpreting

2. Research one other controversial dam site around the world and compare it with the Belo Monte dam in terms of (a) size, (b) purpose and (c) impacts.   **Examining, analysing, interpreting**

4.3 EXERCISES

Geographical skills key: GS1 Remembering and understanding **GS2** Describing and explaining **GS3** Comparing and contrasting **GS4** Classifying, organising, constructing **GS5** Examining, analysing, interpreting **GS6** Evaluating, predicting, proposing

4.3 Exercise 1: Check your understanding

1. **GS1** What human activities are responsible for *changing* or fragmenting rivers?
2. **GS2** Using **FIGURE 1**, describe the location of *places* with rivers that are largely unaffected by river fragmentation.
3. **GS1** Traditionally, why has water flowing out to sea been considered a waste? Is this a human-centred or Earth-centred viewpoint?
4. **GS1** What is the primary aim of an *environmental* impact assessment?
5. **GS2** Refer to **FIGURES 4 (a)** and **(b)** to describe the *environmental changes* brought to the Xingu River by the dam.

4.3 Exercise 2: Apply your understanding

1. **GS4** Using information from **FIGURE 2** and the text in this subtopic, construct a table with the following headings to classify the impacts of dam building.

Positive effects for people	Negative effects for people

Positive effects for the environment	Negative effects for the environment

2. **GS5** Refer to **FIGURE 3**.
   (a) Which countries in the world have the greatest number of people affected by large dams? Suggest a reason why.
   (b) Where are the world's largest (over 100 km^3) dams?
   (c) Would people and *environments* upstream of large dams be affected by the dams? Explain.
3. **GS6** Suggest reasons why large-*scale* dam projects have been seen as indicators of development and progress.
4. **GS6** Suggest reasons that make a *place* suitable for a large dam. Consider landforms, climate, soil and rock type.
5. **GS6** Is there a *sustainable* future for mega dam projects such as the Belo Monte Dam? Justify your answer.
6. **GS6** 'The positive impacts of large dam-building projects on people outweigh the negative impacts on the *environment*.' Do you agree or disagree with this statement? Give reasons for your point of view.

Try these questions in learnON for instant, corrective feedback. Go to www.jacplus.com.au.

4.4 Alternatives to damming

4.4.1 How can water be saved?

Traditionally, managing water has been focused on exploiting resources rather than conserving them. However, there are viable alternatives to dams that are often cheaper and have fewer social and environmental impacts. The focus has to be, first, on reducing demand for water and, second, on using existing water more efficiently.

Agriculture

Globally, more than 70 per cent of fresh water is used for agriculture. In many expanding economies; for example, India, farming uses more than 85 per cent of available water. Often governments subsidise and encourage farmers to grow water-thirsty crops, such as cotton, in semi-arid regions. Irrigation is often very inefficient, with over half of the water applied not actually reaching the plants. High rates of evaporation and leaking **infrastructure** waste water. Poorly designed and managed irrigation schemes can become unsustainable if they develop waterlogging and salinity problems, but there is often little financial incentive to improve efficiency. For example, a country may have industrial and domestic water users paying up to $3 per cubic metre, while agricultural users pay only $0.10. Such low costs do little to encourage change.

We could save vast amounts of water by improving irrigation methods, switching to less water-consuming crops and taking poor quality land out of production. Pakistan; for example, has one of the most wasteful water systems in the world. With the same quantity of water and an efficient system, Israel produces 70 per cent more food. Globally, if the amount of water consumed by irrigation was reduced by 10 per cent, water available for domestic use could double.

Urban use

Researchers state that 30 per cent of all clean drinking water is said to be lost through leaking pipes. The United States loses eight trillion litres of water each year through deteriorating infrastructure. Countries could make savings by:

- reducing leaking pipes and improving water delivery infrastructure
- encouraging the use of water- and energy-efficient appliances and fixtures
- changing the pricing of water to a 'the more you use, the more you pay' system
- offering incentives to industry to reduce water waste and recycle
- harvesting rainwater, collecting rainwater off roofs, recycling domestic wastewater and other efficiency schemes. For example, 40 per cent of Singapore's water needs are met using treated wastewater.

Small-scale solutions

Currently, researchers estimate it will cost US$114 billion per year to meet the United Nations' goal of achieving universal access to clean water and adequate sanitation. Hence the growing awareness of investing in small-scale technologies. Rather than one large, expensive dam, smaller projects that benefit local communities can be more desirable (see **FIGURE 1**). These are often constructed and maintained by people who benefit directly from control over their own resources, at a minimal cost.

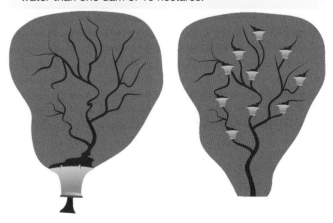

FIGURE 1 Research in India has shown that 10 micro dams with one-hectare catchments will store more water than one dam of 10 hectares.

4.4.2 How can we reduce the need for dams?

As many countries are actually running out of suitable places to locate large dams, we need to find alternatives. **Rainwater harvesting** schemes, as illustrated in **FIGURE 2 (a)** and **(b)**, can be used for storing water. **Micro hydro-dams** (see **FIGURE 3**) can be used for generating electricity. Both of these schemes are easier and cheaper to build than large dams, and have lower environmental impacts.

FIGURE 2 Two methods for water harvesting: (a) rainwater tank and (b) groundwater recharging

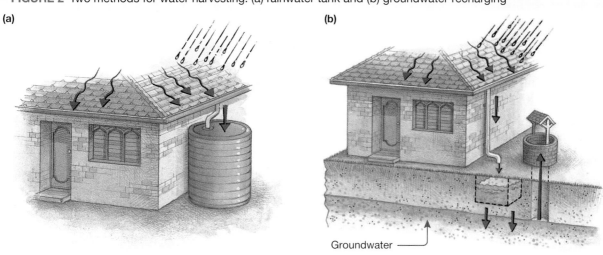

(a)

(b)

Groundwater

FIGURE 3 Water collected from a stream uphill rushes down the pipe and drives a small turbine in the hut to generate electricity for a local community in the Philippines.

4.4.3 CASE STUDY: Traditional water harvesting in India

The state of Rajasthan is located in the arid north-west of India (see **FIGURE 4**). The region has only 1 per cent of the country's surface water and a total **fertility rate** of 2.4, compared to the national average of 2.2 (and compared to Australia's 1.74). Rajasthan adds more than one million people to its population every year; by comparison, Victoria added 120 000 people to its population between 2017 and 2018. The largest state in India faces both water scarcity and frequent droughts. Continual pumping of groundwater has seen underground water supplies dropping.

Traditionally, forests, grasslands and animals were considered property to be shared by all, local communities managed them carefully with a strict set of rules. Resources were used sustainably to ensure regeneration of plants and trees to enable farming to continue each year. However, by the mid twentieth century, government initiatives had taken control of local resources and promoted excessive mining and logging in the area. Large-scale deforestation resulted in severe land degradation, which increased the frequency of flash floods and droughts. There was little motivation for villages to maintain traditional water systems, or *johads*, and so there was a gradual decline in people's economic and social wellbeing.

In 1985, aid agency Tarun Bharat Sangh (TBS) set about trying to re-establish traditional water management practices. The basic principle is to capture, hold and store rainfall whenever it occurs, to then be used during dry periods. TBS focused its attention on constructing and repairing some 10 000 johads in over 1000 villages. Johads are often small, dirt embankments that collect rainwater and allow it to soak into the soil and recharge groundwater **aquifers** (see **FIGURE 5**).

Another johad design features small concrete dams across gullies that would seasonally flood, trapping the water and allowing it to infiltrate. Water, stored in aquifers, can later be withdrawn when needed via wells. The benefits have been remarkable, with the estimated average cost around only US$2 or 100 rupees per person. This is compared to over 10 000 rupees per head for water from the Narmada River Dam Project.

FIGURE 4 Distribution of rainfall in India. The state of Rajasthan is outlined.

Source: World Climate – http://www.worldclim.org/ Made with Natural Earth. Map by Spatial Vision.

FIGURE 5 A *johad* or traditional small water harvesting dam in India

What have been the benefits?

Environmental benefits

- Groundwater has risen from depths ranging 10–120 metres up to 3–13 metres below the surface.
- Five rivers that flowed only after the monsoon season now flow all year (fed by **base flow**).
- Revegetation and agroforestry schemes have increased forest cover by from 7 to 40 per cent, which helps improve the soil's ability to hold water and reduce evaporation and erosion.
  - The area under single cropping (one crop grown per year) has increased from 11 to 70 per cent and the area under double cropping (two crops per year) has increased from 3 to 50 per cent.
  - The water is shared among the villagers, and farmers are not allowed to use it to grow water-thirsty crops.

Social benefits

- More than 700 000 people across Rajasthan have benefited from improved access to water for household and farming use.
- There has been a revival of traditional cultural practices in constructing and maintaining johads.
- The role of the village council (Gram Sabha) is promoted for encouraging community participation and social justice.
- With a more reliable water supply, communities became more economically viable.

Resources

☑ **eWorkbook** Water harvesting (doc-31712)

4.4 INQUIRY ACTIVITIES

1. Complete the **Water harvesting** worksheet to learn more about traditional water harvesting schemes.
   Examining, analysing, interpreting
2. Investigate the different methods of irrigating crops, such as flood, furrow and drip irrigation.
   (a) What are the advantages and disadvantages of each in terms of water use and waste?
   (b) Which irrigation method would:
      i. be the most economically viable
      ii. have the most *environmental* benefit?
      Examining, analysing, interpreting

4.4 EXERCISES

Geographical skills key: GS1 Remembering and understanding **GS2** Describing and explaining **GS3** Comparing and contrasting **GS4** Classifying, organising, constructing **GS5** Examining, analysing, interpreting **GS6** Evaluating, predicting, proposing

4.4 Exercise 1: Check your understanding

1. **GS2** Explain how water can be wasted through poor farming methods.
2. **GS2** List two advantages and two disadvantages of micro hydro-dams (refer to **FIGURE 3**).
3. **GS2** Refer to **FIGURE 4**. Describe the distribution of rainfall in Rajasthan.
4. **GS1** Suggest one *environmental*, one social and one political factor that have contributed to the decline in water availability in Rajasthan.
5. **GS1** What have been the benefits of revegetation schemes around the villages restoring johads?

4.4 Exercise 2: Apply your understanding

1. **GS6** The two goals of *sustainable* water management are first, to reduce the demand for water, and second, to use existing water more efficiently. Propose two methods that your family could use to meet these goals.
2. **GS5** Study the information in **FIGURE 4**. Explain why Rajasthan has water issues. Use data in your answer.

---

3. **GS6** Some *places* in India can receive up to 2500 mm of rainfall per year, but this can all fall in 100 hours. Suggest possible repercussions of this for local communities.
4. **GS6** Have small-*scale* water management schemes in Rajasthan been successful? Why or why not?
5. **GS6** Do you think the johad method of water harvesting could be used in other *places* around the world? Give reasons for your answer.

Try these questions in learnON for instant, corrective feedback. Go to www.jacplus.com.au.

4.5 SkillBuilder: Creating a fishbone diagram

What is a fishbone diagram?

Fishbone diagrams are useful to help visualise a problem or effect, and to show the causes of that problem. Bones above and below the central line are used to identify causes, while the 'head' of the diagram gives the problem or effect. Each major category of cause then flows to other causes and even sub-causes. These are all linked to convey the interconnection of ideas.

Select your learnON format to access:

- an overview of the skill and its application in Geography (Tell me)
- a video and a step-by-step process to explain the skill (Show me)
- an activity and interactivity for you to practise the skill (Let me do it)
- questions to consolidate your understanding of the skill.

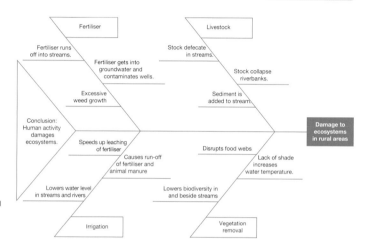

Resources

Video eLesson Creating a fishbone diagram (eles-1748)

Interactivity Creating a fishbone diagram (int-3366)

4.6 Using our groundwater reserves

4.6.1 What is groundwater?

Of all the fresh water in the world not locked up in ice sheets and glaciers, less than 1 per cent is available for us to use — and most of that is groundwater. More than two billion people use groundwater, making it the single most used natural resource in the world. It is also the most reliable of all water sources. Fresh water stored deep underground is essential for life on Earth.

Groundwater is one of the invisible parts of the water cycle, as it lies beneath our feet. Rainfall that does not run off the surface or fill rivers, lakes and oceans will gradually seep into the ground. **FIGURE 1** shows where groundwater is stored in porous rock layers called aquifers. Water is able to move through these aquifers and can be stored for thousands of years. Unlike most other natural resources, groundwater is found everywhere throughout the world.

FIGURE 1 Diagram showing groundwater

4.6.2 What are the advantages of using groundwater?

Since the mid-twentieth century, advances in drilling and pumping technology have provided people with an alternative to surface water for meeting increasing water demands. Groundwater has many advantages:

- It can be cleaner than surface water.
- It is less subject to seasonal variation and there is less waste through evaporation.
- It requires less and cheaper infrastructure for pumping as opposed to dam construction.
- It has enabled large-scale irrigated farming to take place.
- In arid and semi-arid places, groundwater has become a more reliable water supply, which has led to improved water and food security.

If groundwater is removed unsustainably, that is, at a rate that is greater than is being replenished naturally by rainfall, run-off or underground flow, then **watertables** drop and it becomes harder and more expensive to pump. In areas of low rainfall there is very little **recharge** of groundwater so it may take thousands of years to replace. Over-extraction of groundwater can result in wells running dry, less water seeping into rivers and even land **subsidence** or sinking. **FIGURE 2** identifies those places in the world most at risk of groundwater depletion. Many of these are important food bowls for the world.

FIGURE 2 The world's use of groundwater

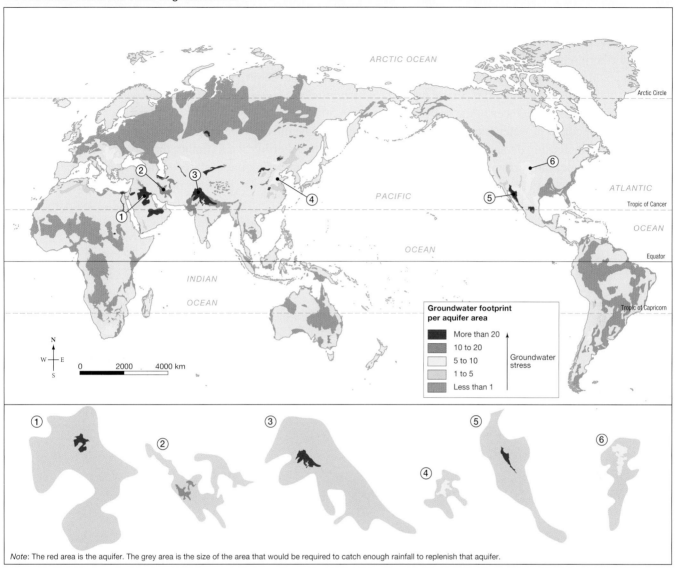

Note: The red area is the aquifer. The grey area is the size of the area that would be required to catch enough rainfall to replenish that aquifer.

Source: BGR & UNESCO 2008: Groundwater Resources of the World 1 : 25 000 000. Hannover, Paris. Map by Spatial Vision.

4.6.3 Can we improve our use of groundwater?

In the past, we had limited knowledge of the interconnection between groundwater and surface water. As agriculture is the biggest user of groundwater, any improved efficiencies in water use can reduce the demand to pump more water. Improved irrigation methods and the reuse of treated effluent water are all methods that could reduce our unsustainable use of groundwater. Many countries share aquifers, so pumping in one place can affect water supplies in another. There is a need for more international cooperation and management of the aquifer as a single shared resource.

4.6.4 CASE STUDY: Why is China drying up?

What is happening to groundwater in China?

What do you do if you don't have access to a reliable water source? You do what hundreds of millions of people do around the world every day. You dig for it. Beneath our feet lie vast quantities of fresh water that may have taken thousands of years to slowly work its way deep into rock layers. Since ancient times, people have used groundwater to provide for their water needs.

Rapid growth in both population and irrigated agriculture, combined with increasing demand for water, has seen the increased pumping of groundwater in northern China. As a consequence, the watertable around Beijing has been dropping by 5 metres per year. Groundwater supplies more than 70 per cent of the water needs for over 100 million people living on the North China Plain (see **FIGURE 3**).

The northern regions of China receive only 20 per cent of the country's rainfall. The southern regions, home to about half the population, receive the other 80 per cent. Eleven provinces in the north have less than 100 cubic metres of water per person per year, which officially classifies these places as being 'water stressed'. Eight other provinces in the north have less than 500 cubic metres of water per person per year. To date, management of water resources has been poor and unsustainable. The emphasis has always been on meeting an increasing demand for water using large-scale engineering 'fixes', rather than looking at ways of using water more efficiently, slowing down demand by increasing the cost, reducing irrigation wastage and improving the catchment areas to recharge the aquifers.

FIGURE 3 Decline in the aquifer under the North China Plain

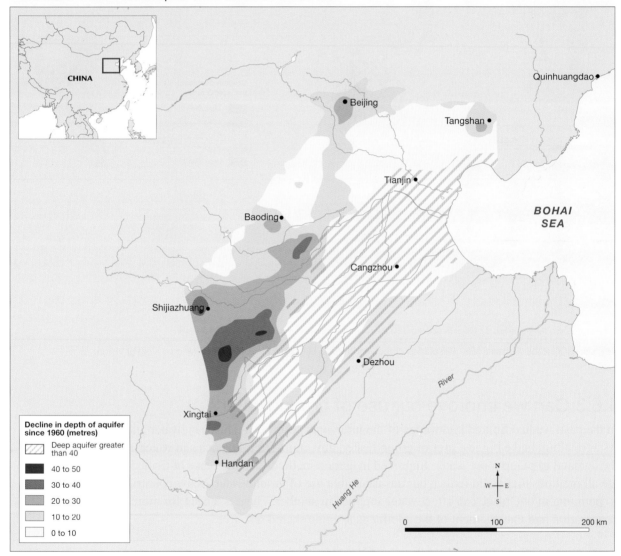

Source: UNEP Global Environmental Alert Service GEAS. Map by Spatial Vision.

After extensive flooding in the 1960s, the government set about building dams and canals to reduce flood impacts and provide water to rapidly growing cities. Farmers were encouraged to increase grain production by drawing on groundwater to irrigate a second crop each year. As cities continued to expand, they too began to pump unsustainable amounts of groundwater for domestic and industrial use. Scientists now also believe that climate change has reduced rainfall in the region, which will only make the situation worse.

The South–North Water Transfer Project

In 2002, an ambitious 50-year project was started to effectively 're-plumb' the country: the South–North Water Transfer Project. At an estimated cost of US$62 billion, over 44.8 billion cubic metres of water per year will be diverted north from the Yangtze River via three canals into the Huang He River Basin in the north of the country (see **FIGURE 4**). Before the transfer project development, the Yangtze River, on average, released 960 billion cubic litres of fresh water into the sea each year. Construction of the central and eastern sections has been completed, but the western route is still being planned. The completed central section now supplies 73 per cent of the Beijing's tap water, to provide for its population of 21.5 million people. The water transfer has reduced the exploitation of groundwater by 800 million cubic metres. As the extra surface water filters into the ground, the watertable has started to rise, with levels increasing by around half a metre.

FIGURE 4 South–North Water Transfer Project for transfer of water from the Yangzte River in the south to the Huang He River in the north

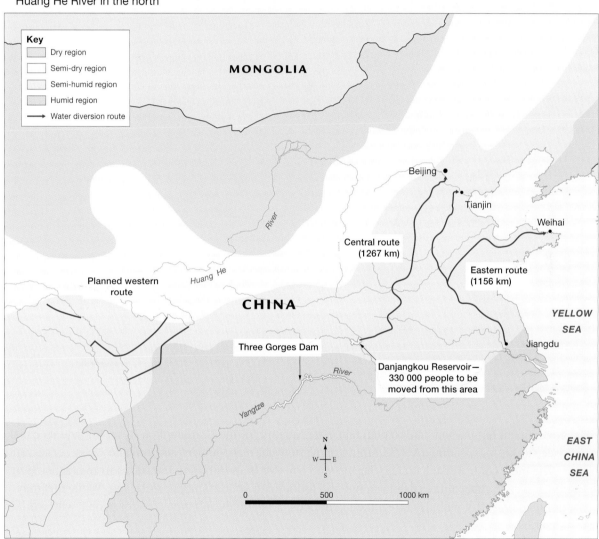

Source: BBC News, http://news.bbc.co.uk/2/hi/8545321.stm

The project's water will largely go to expanding industries and cities such as Beijing and Tianjin. Little of the water will be directed towards food production. Irrespective of the cost and relocation of hundreds of thousands of people, the biggest ongoing concerns of the scheme will be about water quality. The Huang He River collects over 4.29 billion tonnes of waste and sewage each year (see **FIGURE 5**), and over 40 per cent of China's total waste water is dumped in the Yangtze River (see **FIGURE 6**). These figures are likely to increase as more and more industry moves close to new water sources. With less water to flow downstream, there will be less water available to dilute the polluted water. This will affect river environments and it is possible that the water reaching the north will be too contaminated for human or even agricultural use.

FIGURE 5 Levels of pollution along the Huang He River

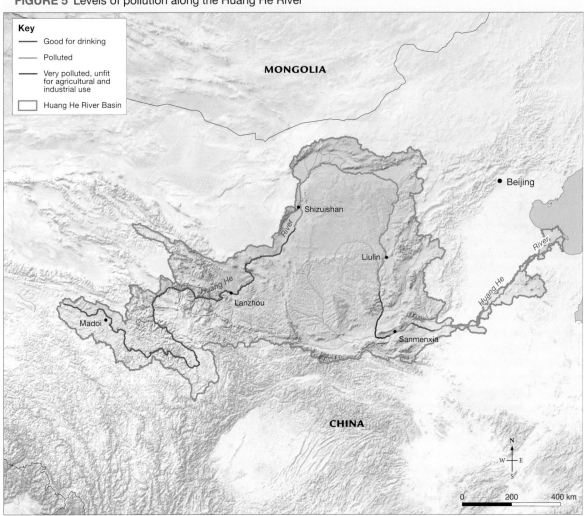

Source: Map by Spatial Vision.

The government has pushed hard to maintain water quality for the Danjiangkou Reservoir and its canal system (see **FIGURE 4**), spending over $3 billion on wastewater management and soil and water conservation systems. The most controversial strategy has been to ban two high-polluting industries from practising in the catchment: cage aquaculture (fish farming) and turmeric (a yellow spice) processing. While this may reduce polluted run-off it will also affect the livelihoods of hundreds and thousands of people who work in these industries.

The South–North Water Transfer Project is by far the most ambitious water transfer project in the world, in all taking 50 years and potentially well over the initial $62 billion estimated cost to construct a network of pipes, canals and tunnels that would stretch in a straight line from Melbourne to Fiji. Is it sustainable? Beijing consumes more than 3.6 billion cubic metres of water each year, supplied partially by its own traditional surface and groundwater sources and, increasingly, by the transfer scheme. Predictions are that its needs will soon outstrip the new scheme's capacity to supply. Scientists are already questioning the 'big scheme' approach rather than the use of water recycling, desalination and harvesting more rainwater as more environmentally friendly and sustainable methods of supplying water.

FIGURE 6 Polluted water flows into the Yangtze River.

  Resources

☑ **eWorkbook** Water transfer (doc-31792)

🔧 **Interactivity** That sinking feeling (int-3293)

4.6 INQUIRY ACTIVITIES

1. Using an atlas, find a map of world food production and compare this with any three *places* from **FIGURE 2** in section 4.6.2. Conduct research to determine the following:
   (a) What types of food are produced in those regions of the world where watertables are severely depleted?
   (b) What are the future implications for *sustainable* food production in these regions?
   Examining, analysing, interpreting
2. Complete the **Water transfer** worksheet to learn more about the South–North Water Transfer Project.
   Examining, analysing, interpreting

4.6 EXERCISES

Geographical skills key: GS1 Remembering and understanding **GS2** Describing and explaining **GS3** Comparing and contrasting **GS4** Classifying, organising, constructing **GS5** Examining, analysing, interpreting **GS6** Evaluating, predicting, proposing

4.6 Exercise 1: Check your understanding

1. **GS2** What are the advantages and disadvantages of using groundwater for domestic and agricultural purposes?
2. **GS2** Refer to **FIGURE 1**.
   (a) What is the difference between groundwater and the watertable?
   (b) Describe how water can move vertically and horizontally through the ground.
   (c) What is the *interconnection* between atmospheric, surface and groundwater?

▶

3. **GS2** Refer to **FIGURE 2**. Describe the location of *places* in the world that have the highest groundwater stress. (You may wish to refer to your atlas.)
4. **GS3** Looking at **FIGURE 2**, compare the *scale* of the selected aquifers with the *scale* of the area needed to recharge them.
5. **GS1** Using **FIGURE 3**, describe the location of the North China Plain. Use distance, direction and place names in your answer.

4.6 Exercise 2: Apply your understanding

1. **GS1** Refer to **FIGURE 4**. Suggest a reason why northern China uses groundwater to supply over 70 per cent of its water needs.
2. **GS6** Suggest possible sources for the pollution you can see entering the river in **FIGURE 6**.
3. **GS6** Is the South–North Water Transfer Project nothing but a pipe dream? Is a human-centred rather than earth-centred viewpoint the best option for water management in northern China? Write a paragraph outlining your views.
4. **GS4** There is often talk about transferring water from the wetter regions of northern Australia to the water-hungry regions further south. What would you need to know before planning a project similar to the one in China? Thinking geographically, write a list of 10 questions to consider before designing such a project.
5. **GS6** Who owns groundwater? How can we manage the resource *sustainably*? Write a paragraph expressing your viewpoint.

Try these questions in learnON for instant, corrective feedback. Go to www.jacplus.com.au.

4.7 The impacts of drainage and diversion

4.7.1 What are wetlands?

Often referred to as the area where 'earth and water meet', **wetlands** are one of the most important and valuable biomes in the world. Wetlands are areas that are covered by water permanently, seasonally or ephemerally and can include fresh, salty and brackish waters. They include such things as ponds, bogs, swamps, marshes, rice paddies and coastal lagoons. Wetlands are intricately connected to other elements in the landscape, especially rivers and floodplains as water, nutrients and sediments move between them (see **FIGURE 1**).

FIGURE 1 Interconnections between the river and wetlands on the floodplain

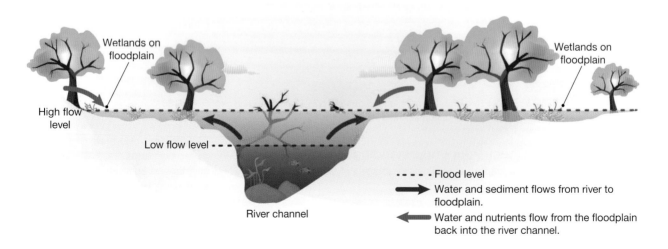

The importance of wetlands

Wetlands perform many important functions. They purify water; for example, much of Melbourne's sewage water is filtered through a series of lagoons and wetlands at the Western Treatment Plant in Werribee, producing high-quality recycled water. Wetlands located along river floodplains reduce the impact and speed of floods by holding vast quantities of flood water and then slowly releasing it back into the river system. Water in wetlands also infiltrates the soil, recharging groundwater reserves. In addition, wetlands provide habitat and breeding grounds for 40 per cent of the world's species, such as aquatic fish, insects, reptiles and birds. Globally, more than one billion people rely on wetlands for a living, for their water and food supply, and for tourism and recreation (see **FIGURE 2**).

FIGURE 2 A wetland in Queensland. What features in this image are typical of a wetland?

4.7.2 What are the threats to wetlands?

The degradation and loss of wetlands and the species that inhabit them, have been more rapid than any other ecosystem, in fact, they are disappearing three times faster than forests. It has been estimated that between 1970 and 2015, the world lost 35 per cent of its wetlands. Competition from other land uses and increasing populations have contributed to the decline.

- Agricultural expansion is the largest contributor to wetland loss and degradation globally. Farming often requires the draining of wetlands to create more land, reducing biodiversity. In addition, the water run-off from agriculture is often polluted with fertilisers and pesticides and increased pumping from aquifers depletes groundwater resources.
- Dams alter seasonal floods and block supply of sediment and nutrients onto the floodplain and deltas. Often, damming means little water and sediment reaches the mouths and deltas of large rivers.
- Loss of wetlands affects populations and the migratory patterns of birds and fish. The introduction of invasive species results in changed ecosystems and loss of biodiversity. For example, 70 per cent of amphibian species are affected by habitat loss.
- Clearing for urban growth, industry, roads and other land uses replaces wetlands with hard **impervious** surfaces, which reduces infiltration and leads to polluted run-off and increased impacts of flooding.

- While wetlands can naturally filter many pollutants, excessive amounts of fertilisers and sewage causes algal blooms and **eutrophication**, depriving aquatic plants and animals of light and oxygen.
- Climate change is expected to increase the rate of wetland degradation and loss.

Wetlands, like all other water resources are prone to over-exploitation and need to be managed carefully to ensure sustainable use.

4.7.3 Why is water diverted?

Because populations and water sources are distributed unevenly, we often need to transfer or divert large amounts of water. This means piping or pumping water from one drainage basin to another; for example, in Australia, water from the Snowy River is diverted into the Murray and Murrumbidgee rivers. Diverting water can alleviate water shortages and allows for the development of irrigation and the production of hydro-electricity. Diversions, however, are not always the most sustainable use of water resources.

Many of the world's greatest lakes are shrinking, and large rivers such as the Colorado, Rio Grande, Indus, Ganges, Nile and Murray discharge very little water into the sea for months and even years at a time. Up to one-third of the world's major rivers and lakes are drying up, and the groundwater wells for three billion people are being affected. The overuse and diversion of water is largely to blame.

4.7.4 CASE STUDY: A dying lake in Iran

The largest lake in the Middle East and one of the largest salt lakes in the world is drying up. Since the 1970s, Lake Urmia in northern Iran has shrunk by nearly 90 per cent. In 1999, the lake's volume was 30 billion cubic metres; by 2018, this had reduced to 2 billion cubic metres, exposing extensive areas of salt flats (see **FIGURE 3 (a)** and **(b)**).

FIGURE 3 Lake Urmia (a) in 1998 and (b) in 2016

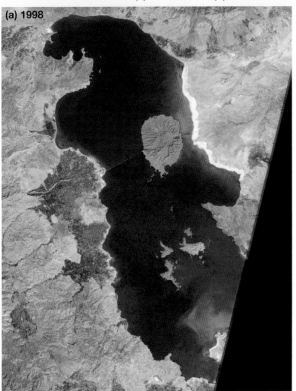

The lake was declared a Wetland of International Importance by the Ramsar Convention in 1971, and a UNESCO Biosphere Reserve in 1976. The lake and its surrounding wetlands serve as a seasonal habitat and feeding ground for migratory birds that feed on the lake's shrimp. This shrimp is the only thing, other than plankton, that can live in the salty water.

Lake Urmia is a **terminal lake**; the rivers that flow into the lake (some permanent and some ephemeral) bring naturally occurring salts. Because of the arid climate, high evaporation causes salt crystals to build up around the shoreline. **FIGURE 4** shows the rapid decline in the surface area of Lake Urmia from 2006 to 2013.

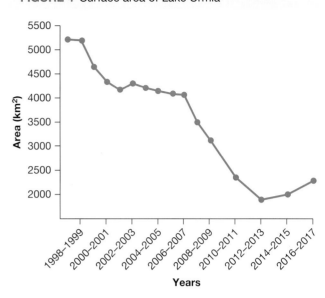

FIGURE 4 Surface area of Lake Urmia

Why is the lake drying up?

A combination of environmental, economic and social factors has been blamed for the large-scale changes in Lake Urmia. Prolonged drought and the illegal withdrawal of water by farmers who do not pay or who take more than their allocation are minor contributors to the problem.

Researchers have found that 60 per cent of the decline can be attributed to climate changes (increased frequency of drought and higher temperatures) and 40 per cent of the decline relates to water diversions and the increased demand for water in the region as the population has risen and the area under agriculture has tripled. The result is a form of 'socioeconomic drought' — a human-induced drought caused when the demand for water is greater than the available supply.

Impacts of this drought include:

- increased salinity of the shallow lake due to high evaporation (rates of between 600 mm and 1000 mm per year) and reduced freshwater flowing in via rivers (salt levels have increased from 160 g/L to 330 g/L)
- collapse of the lake's ecosystem and food chain (salt levels over 320 g/L are fatal to the shrimp that form the basis of its food chain)
- loss of habitat as surrounding wetlands dry up, which then reduces tourism to view wetland wildlife
- over 400 km² of exposed lakebed around its shores is nothing but salty deserts, unable to support native vegetation or food crops

FIGURE 5 Distribution of dams, existing and under construction, in the lake's catchment area. This level of diversions is unsustainable.

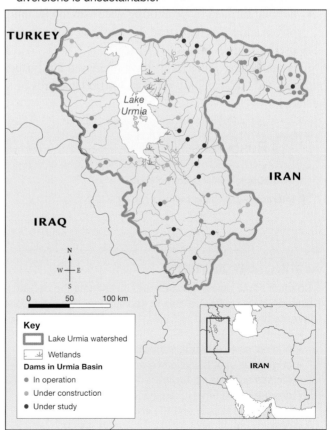

Source: United Nations Environment Programme. Vector Map Level 0 Digital Chart of the World.

- salt storms occuring, as wind blows salt and dust from the exposed, dry lakebed; the storms damage crops and are also a potential health hazard
- less water available to produce food crops.

The lake is divided in two by a causeway and bridge constructed to improve access across the lake (the bridge can be seen in **FIGURE 3 (b)**). However, there is concern that the nearly 1.5 km-long bridge does not allow sufficient mixing of water between the north and south sections of the lake. The bridge, completed in 2008, is already rusting as a result of the highly saline water.

The current situation

Lake Urmia and surrounds is an important region supporting a growing population of six million and its associated agricultural industries. Essentially, more water is required to flow into the lake to increase the water level, dilute the salt and maintain an ecological balance. This would require doubling the current water level.

In 2017, the government pledged an annual budget of US$460 million to help restore the lake and its surrounding wetlands. However only about US$5 million has actually been available. Other programs in place include:

- a water transfer scheme moving water from the Little Zab basin through tunnels and channels
- engineering works to help clear sediment clogging many of the rivers that feed into the lake
- releasing water from dams to flow into the lake
- constructing 13 treatment plants in the region to treat wastewater from urban areas and deliver it to the lake
- a development plan launched in 2017 to reduce consumption of potable water by 30 per cent by 2021 and to use desalinated water to meet 30 per cent of the water demands in South Iran
- trials of planting vegetation to reduce wind speed and salt storms
- helping farmers by promoting water-saving techniques and planting less water-thirsty crops, such as olives and saffron, instead of water-intensive sugar beet, as 85 per cent of the water in the region is used for agriculture.

In recent years, water levels have increased but further progress is limited by a lack of funding. In addition, the lake can only support 300 000 hectares of farmland while currently farming uses 680 000 hectares.

 **Resources**

4.7 INQUIRY ACTIVITY

Conduct some online research and investigate the decline of either Lake Chad in Africa, Owens Lake in the United States or the Aral Sea in Kazakhstan. Use the following as a research guide for elements to include in your investigative report:
- location
- original size and appearance of the lake/sea
- original uses of the lake/sea and surrounding area
- causes and rate of decline
- *changes* that have taken place
- impacts on people and the *environment*
- possible solutions.

Include annotated images, maps and data where possible.

Examining, analysing, interpreting
Classifying, organising, constructing

4.7 EXERCISES

Geographical skills key: GS1 Remembering and understanding **GS2** Describing and explaining **GS3** Comparing and contrasting **GS4** Classifying, organising, constructing **GS5** Examining, analysing, interpreting **GS6** Evaluating, predicting, proposing

4.7 Exercise 1: Check your understanding

1. **GS1** What are wetlands?
2. **GS1** In what ways are wetlands important?
3. **GS1** Outline three major threats to wetlands.
4. **GS1** Why is it sometimes necessary to divert water?
5. **GS1** Lake Urmia is a terminal lake. What does this mean?

4.7 Exercise 2: Apply your understanding

1. **GS2** Explain eutrophication, its causes, and the impact it has on wetland biomes.
2. **GS5** Discuss the advantages and disadvantages of diverting water from one drainage basin to another.
3. **GS3** Examine **FIGURES 3 (a)** and **(b)**. Compare the appearance of Lake Urmia in 1998 and 2016.
4. **GS5** Refer to **FIGURE 4**. Describe, with the use of data, the *changes* in the surface area of Lake Urmia.
5. **GS6** Do you think there is a future for Lake Urmia? How successful will the restoration project be? Explain your view.

Try these questions in learnON for instant, corrective feedback. Go to www.jacplus.com.au.

4.8 SkillBuilder: Reading topographic maps at an advanced level

What is reading a topographic map at an advanced level?

Topographic maps are more than just contour maps showing the height and shape of the land. They also include local relief and gradients and allow us to calculate the size of various areas. Reading this information requires more advanced skills.

Select your learnON format to access:

- an overview of the skill and its application in Geography (Tell me)
- a video and a step-by-step process to explain the skill (Show me)
- an activity and interactivity for you to practise the skill (Let me do it)
- questions to consolidate your understanding of the skill.

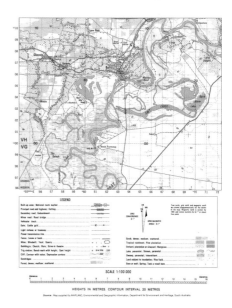

4.9 Putting water back — managing the Murray–Darling

4.9.1 Where has the water gone?

The rivers, lakes and wetlands of the Murray–Darling Basin (see **FIGURE 1**) make it Australia's most important inland water body. Decades of continually diverting water from its rivers and prolonged periods of drought have brought significant changes to the rivers, surrounding floodplains and surrounding wetlands. The amount of water taken out of the river system has increased five-fold over the past century. In addition, 90 per cent of floodplain wetlands in the Murray–Darling Basin have been lost because of human-induced changes to **river regimes**. The floodplains of the Murray River are now flooded once every 10–12 years compared to 3–4 years out of every five a century ago. Reduced flow has also meant that the mouth of the river blocks regularly, preventing the flushing out of pollutants and affecting the Lower Lakes wetlands.

FIGURE 1 Key features of the Murray–Darling Basin

Icon sites
1. Barmah–Millewa Forest
2. Gunbower–Koondrook–Perricoota Forest
3. Hattah Lakes
4. Chowilla Floodplains and Lindsay–Wallpolla Islands
5. Murray Mouth, Coorong and Lower Lakes
6. Murray River Channel

Murray–Darling Basin
Wetlands (Ramsar listed)

Source: © Commonwealth of Australia Geoscience Australia 2013. Murray Darling Basin Commission. Map by Spatial Vision.

4.9.2 Declining river health

One of the difficulties in managing the water resources of the Murray–Darling Basin has been the fact that there are four states and one territory that all use and manage the water in their own way. In the twentieth century, the management of the river switched focus from using the river for transport to expanding agriculture. This period saw a rapid rise in the amount of water withdrawn and a decline in the health of river ecosystems. Consequently, 20 out of 23 catchments within the Basin have 'poor' to 'very poor' ecosystem health. Contributing factors include:

- extensive clearing of native vegetation in the catchments
- introduction of exotic weeds and animals
- run-off of pollutants
- draining of wetlands.

In the twenty-first century we are now working towards a more sustainable approach to managing water, with a greater emphasis on balancing the competing needs of the community and river environments. In essence, it means improving the health of the rivers and wetlands and keeping more water in the system.

4.9.3 Attempts to restore the balance

There have been a number of government initiatives put in place over the years to reduce the amount of water being harvested from the river:

- *The Living Murray Program*. In an effort to try to improve the health of rivers, the Murray–Darling Basin Authority (MDBA) concentrates on maintaining the health of six **icon sites** (see **FIGURE 1**) by providing them with additional water from **environmental flows**. Water 'savings' have to be made elsewhere through improvements in water storage, distribution and irrigation methods. **FIGURES 2(a)**, **(b)** and **(c)** show the effects of an environmental flow on a stressed wetland.

FIGURE 2 Wanganella Swamp (Deniliquin, NSW) (a) before, (b) during and (c) after an environmental flow

- *The Basin Plan*, passed into law in 2012, allows the federal government, rather than four states and one territory to be responsible for the overall management of the Murray–Darling Basin. The Plan's aim is to 'increase the amount of water for the environment of the Murray–Darling Basin and ensure sufficient water for all users'. Thus, the Basin Plan aims to balance out social, economic and environmental demands on the water resources across the entire basin. Central to this aim is the need to provide sufficient water for the health of the river first, and then allocate water for other uses. The following were to be implemented to achieve this aim:
  - limits to the amount of water that could be withdrawn from the river system each year — 10 873 GL, compared with 13 623 GL extracted in 2009 (1 gigalitre equals 1000 Olympic-sized swimming pools)
  - a reduction (of between 27 and 45 per cent) in the amount of water allocated to different regions
  - a limit of 3334 GL of groundwater to be withdrawn each year
  - the establishment of an environmental watering plan to restore and protect the river and wetlands.
- Water is being saved through buying back farmer's water entitlements and through improved storage, distribution and irrigation methods. Long term, the goal is to allow sufficient movement of water to keep the mouth of the river open 90 per cent of the time.

4.9.4 How is the Basin Plan working?

The Basin Plan was formalised in 2012 but it took until 2017 for all state and federal governments to agree to the terms of how environmental water will be returned to the river system. An additional 450 GL (plus the original 2750 GL stated in the Plan) is to be returned, providing it does not have a negative socioeconomic impact on river communities. It may take up to a decade for people to see large-scale environmental changes from the improved flow. It is extremely difficult for water managers to balance the need for natural river flows while also protecting farmers' water needs and preventing large floods.

Menindee Lakes fish kill

The summer of 2019 saw more than one million fish killed by extremely poor water quality in the Menindee Lakes storage system (see **FIGURE 3**). Water levels in the lakes were extremely low and the system had stopped flowing. On average 4000 GL flows into the Barwon–Darling river system and then into the lakes where it can be released to flow downstream into South Australia. The lakes are very shallow and 20 to 30 per cent of the water evaporates each year. Management tries to avoid this by releasing water in a 'use it or lose it' mindset.

High temperatures and calm warm water provided ideal conditions for the growth of blue-green algae, which smothered the water

FIGURE 3 Fish killed at the Menindee Lakes

surface turning it quite green. A cold front moved across the region, dropping water temperatures, which served to kill off the algal bloom. The rapid growth of bacteria that feeds off dead algae used all the oxygen in the water, causing the fish to suffocate. Essentially insufficient water was available in the system to provide environmental flow.

Calculating how much water can be diverted from the river system is usually based on 'full rivers'; little planning is done to manage flow in times of drought. Current water licenses are also based on the height of a river. If one person leaves water in the river, the next person downstream can take more.

In 2018, in the wake of ongoing concerns about the management of the Basin, a Royal Commission was held to investigate the operations and effectiveness of the Murray–Darling Basin Plan. The Commission report, released in January 2019, contained 111 findings and 44 recommendations to overhaul the Plan, including adjusting the balance between irrigation and environmental flows, to give more water back to the environment. The nearly 750-page report highlighted the complexities and challenges of balancing environmental, social and economic needs in relation to a water source on which millions of people depend.

4.9 EXERCISES

Geographical skills key: GS1 Remembering and understanding **GS2** Describing and explaining **GS3** Comparing and contrasting **GS4** Classifying, organising, constructing **GS5** Examining, analysing, interpreting **GS6** Evaluating, predicting, proposing

4.9 Exercise 1: Check your understanding

1. **GS1** What are the factors that have contributed to the poor health of the Murray–Darling Basin?
2. **GS2** Describe how the management of water resources has ***changed*** over time.
3. **GS2** Refer to **FIGURE 1**.
   (a) What are icon sites and why are they important?
   (b) Where are they located?
4. **GS2** Why is there a need for ***environmental*** flows in the Murray–Darling Basin?
5. **GS2** Compare the three photographs in **FIGURE 2**. Describe the ***changes*** in the appearance of the wetland before, during and after an ***environmental*** flow.

4.9 Exercise 2: Apply your understanding

1. **GS6** Suggest why both the amount and timing of ***environmental*** flows is important for a healthy river.
2. **GS5** What would be one advantage and one disadvantage of the federal government taking control of the management of the Murray–Darling Basin over the individual states?
3. **GS6** What steps could an irrigation farmer take to adapt to a reduced allocation and more ***sustainable*** use of water?
4. **GS2** Explain how both natural events and human activities contributed to the Menindee Lakes fish kill.
5. **GS6** Suggest one way that water managers could try to prevent a fish kill occurring again.

Try these questions in learnON for instant, corrective feedback. Go to www.jacplus.com.au.

4.10 Thinking Big research project: Menindee Lakes murder! news report

online only

SCENARIO

As a reporter for a city newspaper, you have been sent to the small town of Menindee to investigate the recent massive fish kills in the Menindee Lakes. What caused this horrific event and what can be done to stop it happening again? It's your job to uncover the truth!

Select your learnON format to access:
- the full project scenario
- details of the project task
- resources to guide your project work
- and assessment rubric.

 Resources

ProjectsPLUS Thinking Big research project: Menindee Lakes murder! news report (pro-0213)

4.11 Review

4.11.1 Key knowledge summary

Use this dot-point summary to review the content covered in this topic.

4.11.2 Reflection

Reflect on your learning using the activities and resources provided.

 **Resources**

☑ **eWorkbook**   Reflection (doc-31767)

Crossword (doc-31768)

🧩 **Interactivity**   Inland water — dammed, diverted and drained crossword (int-7671)

KEY TERMS

aquifers layers of rock which can hold large quantities of water in the pore spaces

base flow water entering a stream from groundwater seepage, usually through the banks and bed of the stream

environmental flows the quantity, quality and timing of water flows required to sustain freshwater ecosystems

environmental impact assessment a tool used to identify the environmental, social and economic impacts, both positive and negative, of a project prior to decision-making and construction

ephemeral describes a stream or river that flows only occasionally, usually after heavy rain

eutrophication a process where water bodies receive excess nutrients that stimulate excessive plant growth

fertility rate the average number of children born per woman

flood mitigation managing the effects of floods rather than trying to prevent them altogether

green energy sustainable or alternative energy (e.g. wind, solar and tidal)

groundwater water held underground within water-bearing rocks or aquifers

icon sites six sites located in the Murray–Darling Basin that are earmarked for environmental flows. They were chosen for their environmental, cultural and international significance.

impervious a rock layer that does not allow water to move through it due to a lack of cracks and fissures

infrastructure the basic physical and organisational structures and facilities (e.g. buildings, roads, power supplies) needed for the operation of a society

micro hydro-dams produce hydro-electric power on a scale serving a small community (less than 10 MW). They usually require minimal construction and have very little environmental impact.

perennial describes a stream or river that flows permanently

rainwater harvesting the accumulation and storage of rainwater for reuse before it soaks into underground aquifers

recharge the process by which groundwater is replenished by the slow movement of water down through soil and rock layers

reservoir large natural or artificial lake used to store water, created behind a barrier or dam wall

river fragmentation the interruption of a river's natural flow by dams, withdrawals or transfers

river regime the pattern of seasonal variation in the volume of a river

subsidence the gradual sinking of landforms to a lower level as a result of earth movements, mining operations or over-withdrawal of water

terminal lake a lake where the water does not drain into a river or sea. Water can leave only through evaporation, which can increase salt levels in arid regions. Also known as an endorheic lake.

watertable upper level of groundwater; the level below which the earth is saturated with water

weir wall or dam built across a river channel to raise the level of water behind. This can then be used for gravity-fed irrigation.

wetland an area covered by water permanently, seasonally or ephemerally. They include fresh, salt and brackish waters such as rivers, lakes, rice paddies and areas of marine water, the depth of which at low tide does not exceed 6 metres.

---

5 Managing change in coastal environments

5.1 Overview

> Though it may not be obvious, coasts are constantly changing. How do natural and human processes contribute to this?

5.1.1 Introduction

The coast is home to 80 per cent of the world's population, and is a popular place to settle for reasons of climate, water resources, land for agriculture and industry, access to transportation systems, and recreation. Hence, it is essential to understand the changes that are occurring to coastal environments, and how they will affect human settlements. The changes are both natural and human-induced. They are sometimes short term (as a result of storms and tsunamis) and sometimes long term (climate change leading to rising sea levels). To cope with these changes, careful planning and management is needed to ensure a sustainable future for human activity at the coast.

Resources

☑ **eWorkbook**   Customisable worksheets for this topic

▤ **Video eLesson** Washed away (eles-1710)

LEARNING SEQUENCE

5.1 Overview

5.2 Understanding coastal landscapes

5.3 Challenges to coastal management

5.4 How do we manage coastal change?

5.5 **SkillBuilder:** Comparing aerial photographs to investigate spatial change over time

5.6 **SkillBuilder:** Comparing an aerial photograph and a topographic map

5.7 **Thinking Big research project:** Ecology action newsletter — Reef rescue

5.8 **Review**

To access a pre-test and starter questions and receive immediate, **corrective feedback** and **sample responses** to every question, select your learnON format at www.jacplus.com.au.

5.2 Understanding coastal landscapes

5.2.1 The importance of the coast

Coasts are a dynamic natural system. The forces of nature are constantly at work, either creating new land or wearing it away. Over 85 per cent of Australians live within 50 kilometres of the coast. As well as being a favoured place to live, the coast is the most popular destination for tourists and visitors.

All forms of human activities can have impacts on coastal landforms and the **ecosystems** of plant and animal life. Australia's coasts need to be managed to achieve goals of sustainable living for all who share this common environment. In addition, there is a need to balance the diverse viewpoints of human-induced development with conservation principles.

5.2.2 The coastal zone

The coastal zone may be broadly defined as the zone where the land meets the sea (see **FIGURE 1**). It includes an area of water known as coastal waters (waters within 3 nautical miles of the shore) and an area of land called the **hinterland**, which extends several kilometres inland of the coast. The Australian coast, which is approximately 37 000 kilometres in length, consists of many different environments such as plains, rivers and lakes, rainforests, wetlands, mangrove areas, estuaries, beaches, coral reefs, seagrass beds and all forms of sea life found on the adjoining continental shelf.

FIGURE 1 Typical Australian coastal scenario, Point Danger, Tweed Heads, New South Wales

In these varied coastal environments many of Australia's World Heritage sites are found, such as the Great Barrier Reef (see **FIGURE 2**), Lord Howe Island, Fraser Island and Shark Bay. The coast is also important for human settlement: urban complexes, ports and harbours. Many Aboriginal and Torres Strait

Islander people lived and continue to live in coastal communities. Historical evidence of middens, art sites, fish traps, stone and ochre quarries, and burial and religious sites show the long history of occupation of the Australian coast. It is important that we understand our coastal environments in order to ensure their sustainable management, to maintain marine biodiversity, and so that they may be preserved for the many generations to come.

FIGURE 2 The Great Barrier Reef was placed on the World Heritage List in 1981.

5.2.3 Coasts as natural systems

Coasts are natural systems consisting of landforms such as beaches, dunes and cliffs as well as biotic elements: aquatic and terrestrial plants and animals. Geographers refer to all these things as biophysical elements. Coasts are also said to be in a state of **dynamic equilibrium**, which means they are constantly changing. Examples of small-scale and short-term agents of change are the tides, ocean currents and wave action. Examples of large-scale agents of change (which happen over vast periods of geological time) include continental drift, uplift and sinking of land, movement of sea levels due to ice ages and warm periods, and creation of islands by volcanic activity.

5.2.4 Coastal landforms

There is a wide range of marine and terrestrial structures found at the coast. Many, such as the landforms of beaches, bays, dunes and cliffs, are familiar to us (see **FIGURE 3**). Others such as fiords are unique to polar regions, which were impacted by ice ages. Structures found under the sea are not so widely known and can include the continental shelf, canyons and trenches.

FIGURE 3 The range of coastal landforms

1. Dune blowouts occur when loose sand is blown from the dune because vegetation has been removed.
2. Waves are refracted (bent) towards a headland and release energy either side of it. Caves will be formed where weak rocks are eroded on each side of a headland.
3. Over time the caves will erode on either side of a headland and join to form an arch.
4. Erosion between low and high tides undercuts rocks and a rock platform develops. This undercut section eventually becomes weak and collapses, creating a cliff.
5. Longshore drift moves sand and other material along a beach. If this drift occurs mainly in one direction, sand may extend along the coastline forming a spit.
6. Estuaries are the parts of a river that are tidal and occur at the sea. They catch mud, sand and nutrients.
7. A lagoon is formed when a sandbar begins to develop, eventually closing an estuary.
8. Beaches are formed when material is brought to the shore by waves. The material can be sand, stones or pebbles.
9. Dunes are formed when sand on a beach is stabilised by vegetation.
10. Further erosion of the rock supporting the arch will cause it to collapse, leaving a stack.
11. If caves develop in places exposed to the sea and waves, water rushes in and can cause pressure to build at the back of the cave. If a section of rock in the roof of the cave is weak, part of the roof may collapse and a blowhole is formed.
12. A spit can sometimes join two land areas. This is called a tombolo.
13. Headlands are formed when coastal rocks are very hard and resist erosion from the waves. Softer rocks either side of the headland are eroded and transported elsewhere.

5.2.5 Ocean processes

Waves and the movement of coastal waters caused by tides, currents, rips, storm surges and tsunamis are ongoing processes that mould the coast, creating and destroying landforms and submarine forms. Tides, which are generated by the gravitational pull of the moon and the sun, cause changes in sea level; this can vary between low and high tide by as much as 10 metres, such as is found in Broome, Australia.

Waves are generated by winds out at sea and this creates what is called swell; the larger the fetch area or distance to the coast, the greater the potential height of the wave, particularly in storm conditions. Waves that are generated in storm conditions are called destructive waves, and these lead to increased erosion of coastal forms (see **FIGURE 4**). Waves that are generated in calm conditions are known as constructive waves, and these may build up sediments, giving rise to beaches and dunes (see **FIGURE 5**).

FIGURE 4 Destructive wave. The backwash is more powerful than the swash, and sand is stripped from beaches.

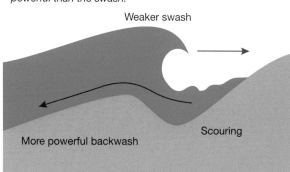

Destructive wave.
The backwash is more
powerful than the swash.

Weaker swash

More powerful backwash

Scouring

FIGURE 5 Constructive wave. The swash is more powerful than the backwash.

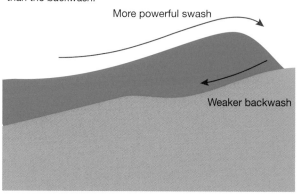

Constructive wave.
The swash is more powerful
than the backwash.

More powerful swash

Weaker backwash

As a wave approaches the coast it translates into a swash or forward movement, and to a backwash or return to the sea after its encounter with the land. Surfers look for the swash element of waves to carry them actively forwards (see **FIGURE 6**). If waves come into the coast at an angle, the swash will move up the beach at that angle, but the backwash returns under the action of gravity directly down the beach. This leads to what is called longshore drift (see **FIGURE 7**), which can move large quantities of sand along the coast and lead to spectacular forms such as spits, bars, barriers and tombolos.

FIGURE 6 How a wave breaks

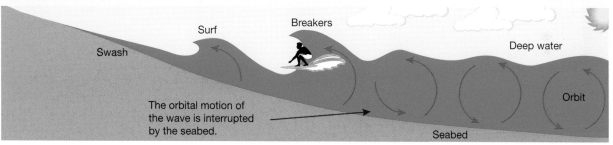

Swash

Surf

Breakers

Deep water

Orbit

The orbital motion of
the wave is interrupted
by the seabed.

Seabed

FIGURE 7 Longshore drift

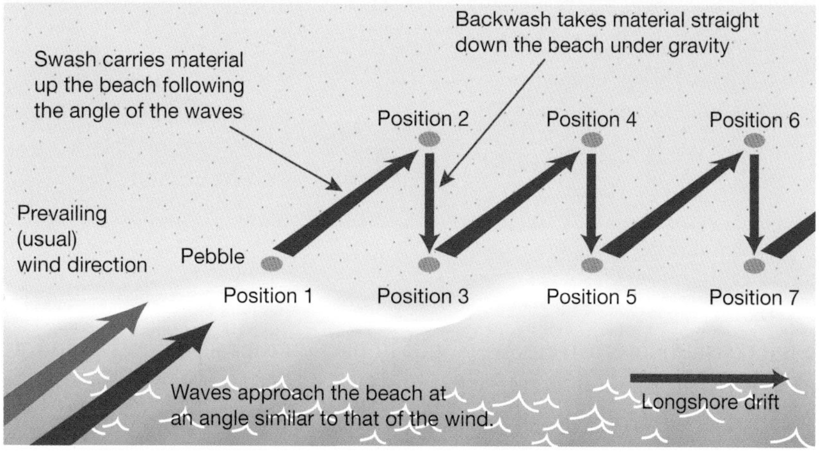

Backwash takes material straight
down the beach under gravity

Swash carries material
up the beach following
the angle of the waves

Position 2   Position 4   Position 6

Prevailing
(usual)
wind direction   Pebble

Position 1   Position 3   Position 5   Position 7

Waves approach the beach at
an angle similar to that of the wind.

Longshore drift

5.2.6 Erosional forms

As noted, erosion through the power of waves is significant in creating coastal landforms. For instance, during storm conditions it has been estimated that the hydraulic action or weight and pressure of water hitting the coast can amount to 10 tonnes per square metre. The mere contact of sea water with coastal rocks can lead to weathering, whereby rocks or loose sediments are either dissolved or abraded or worn away by the action of waves armed with pebbles and sediments. In some cases a section of the coast that is undermined by the sea may collapse; this is referred to as **mass wasting**. Erosional landforms include headlands, bays and bights, cliffs, platforms, caves, arches, blowholes and stacks.

Headlands and bays

Headlands and bays are created by a process of differential erosion. Some parts of the coast may be made of harder rocks or rocks that have fewer fractures, and these areas form headlands that tend to resist erosion. Bays, on the other hand, are composed of softer or more fractured rocks that are more easily eroded, leading to coastline retreat (see **FIGURE 8**).

FIGURE 8 Refraction results in a convergence of waves on a headland and divergence in a bay.

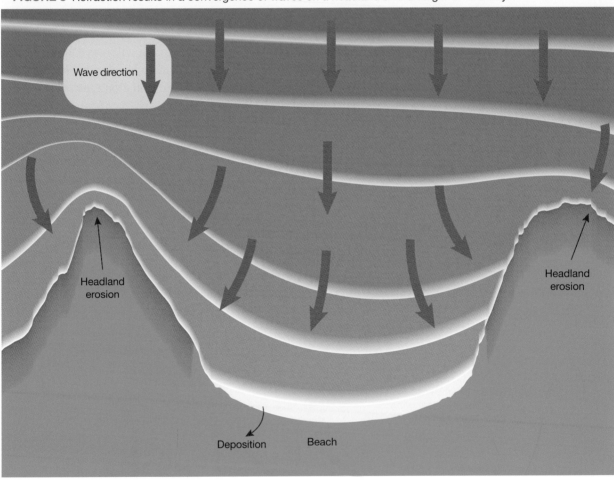

Cliffs and platforms

As waves approach the coast, their erosive powers will impact on rock structures. The force of the water acting on any raised rocky structure will create what is called a cliff. A notch, which is a zone where active wave energy is concentrated, will form at the base of the cliff, and this will lead to the cliff collapsing into the sea and the coastline retreating inland. With cliff retreat, as the active wave zone in the ocean waters becomes enlarged, a platform will develop (see **FIGURE 9**).

FIGURE 9 Formation of cliffs and rocky platforms

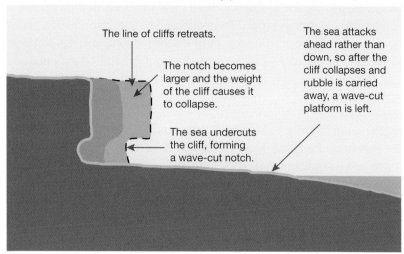

The line of cliffs retreats.

The notch becomes larger and the weight of the cliff causes it to collapse.

The sea undercuts the cliff, forming a wave-cut notch.

The sea attacks ahead rather than down, so after the cliff collapses and rubble is carried away, a wave-cut platform is left.

Caves, arches, blowholes and stacks

Differences in the hardness of rocks and its joints and fractures can be exploited by erosion to create features such as caves, arches (see **FIGURE 10**) and stacks. Caves are parts of the coast that have been more actively eroded from the surrounding rocky area, leaving a hollow in the cliff section. If part of the roof of the cave collapses, in-rushing waves may be channelled up this chimney structure, forcing water and air out at the horizontal land surface above (see **FIGURE 11**). Stacks are simply remnants of cliff areas that have resisted erosion and been left stranded out to sea. The Twelve Apostles and cliffs near Port Campbell in Victoria display many of these coastal forms.

FIGURE 10 Tasman's Arch, Tasmania

FIGURE 11 Blowhole at Quobba Point near Carnarvon, Western Australia

5.2.7 Depositional forms

Depositional coastal forms include beaches, sand dunes, bars and barriers, spits, sand islands, tombolos and lagoons. Processes that create these features are generally associated with constructive waves, movement and accumulation of sands due to longshore drift, and wind blowing sandy sediments onshore.

Beaches

Australia has many beaches, and we are all familiar with their structure. Sediments and sands from which beaches are formed come from materials eroded from cliffs and, more particularly, sediments brought down to the coast by rivers. Waves wash these sediments onto shallow sloping coastal platforms. As beaches are composed of soft materials, they are easily eroded in storms, but generally sands that are taken offshore will return when calmer conditions return and constructive waves can move the sediments back.

Dunes

At the back of the beach and in the zones above high tide, sandy sediments dry out. The wind can then pick these up and move them onshore and inland. Usually a fore dune will form (see **FIGURE 12**) and a depression will be found just inland behind it. If sufficient sand is available, a secondary dune may form, and in some cases dune fields may result with many lines of dunes. The development of dune vegetation that has adapted to survive in harsh conditions is important to stabilise dunes. This vegetation becomes larger and more varied as soil and freshwater conditions on the dune improve inland; this process is known as **coastal dune vegetation succession**. Dunes, due to their sandy structure, are particularly fragile and need to be managed carefully so that pedestrian traffic does not disturb stabilising plant life. Fraser Island in Queensland is composed of sandy sediments and covered with dunes that were built up during the last ice age.

FIGURE 12 Transect showing the beach and stable and well-vegetated dunes

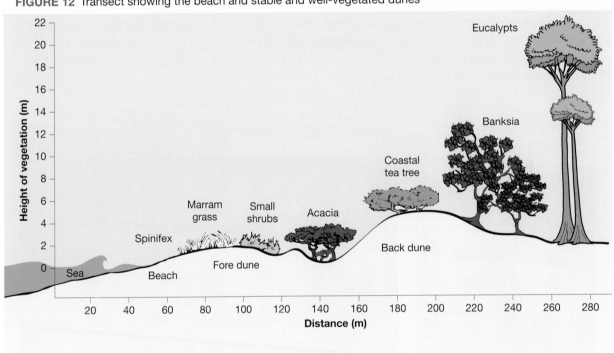

Other depositional coastal forms

Spits are sandy extensions of beaches formed by longshore drift currents (see **FIGURE 13**). There is a large spit at the mouth of the Noosa River in Queensland. Bars and barriers are sandy offshore structures that run parallel to the coast, and lagoons or wetlands may form behind them. The Coorong at the mouth of the Murray River in South Australia is an example of these features.

FIGURE 13 Coastal landforms created by transportation and deposition

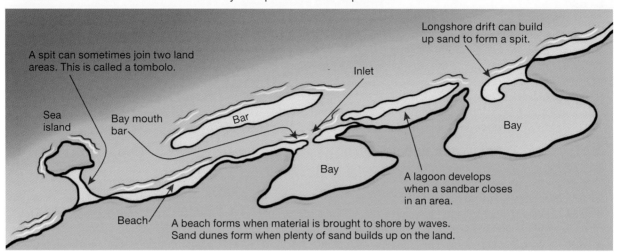

A spit can sometimes join two land areas. This is called a tombolo.

Longshore drift can build up sand to form a spit.

Inlet

Sea island

Bay mouth bar

Bar

Bay

Bay

A lagoon develops when a sandbar closes in an area.

Beach

A beach forms when material is brought to shore by waves. Sand dunes form when plenty of sand builds up on the land.

Explore more with myWorldAtlas

Deepen your understanding of this topic with related case studies and questions.
- Investigate additional topics > Oceans and coasts > **Coastal processes**

5.2 INQUIRY ACTIVITIES

1. Investigate the location of World Heritage sites in the coastal zone. On a world map, label each site and note the reason for its significance. **Classifying, organising, constructing**
2. Coastal landforms can be formed by processes of erosion and deposition. Using the internet, find examples of well-known erosional and depositional landforms and present them as a poster. **Classifying, organising, constructing**
3. Using a series of labelled sketches, explain how a rocky shore platform is formed. **Classifying, organising, constructing**
4. Fraser Island is located off the coast of Queensland. It has been recognised as the largest sand island in the world and contains many specialised landforms and unique coastal ecosystems. Once mined for sand and logged for timber, it has become a popular ecotourism destination. An issue for today is how to sustainably manage the impacts of ecotourism and protect the *environment* of this World Heritage listed island.
   (a) Research the past and present uses of Fraser Island.
   (b) When did it become a World Heritage site?
   (c) What are some of its unique features?
   (d) What does the future hold for Fraser Island? **Examining, analysing, interpreting**

5.2 EXERCISES

Geographical skills key: GS1 Remembering and understanding **GS2** Describing and explaining **GS3** Comparing and contrasting **GS4** Classifying, organising, constructing **GS5** Examining, analysing, interpreting **GS6** Evaluating, predicting, proposing

5.2 Exercise 1: Check your understanding

1. **GS2** Why are coasts important to people?
2. **GS2** Refer to **FIGURE 1**. What land use is found at Point Danger? How might this affect the dynamic nature of the coastal zone?
3. **GS1** What are the two main coastal processes that form coastal landforms?
4. **GS3** What are the differences between constructive and destructive waves? ▶

---

5. **GS2** Refer to the vegetation transect in **FIGURE 12**.
   (a) Why does the vegetation *change* as you move inland?
   (b) What adaptations does dune vegetation show in terms of being adapted to dry, windy coastal *environments*?
   (c) What would happen to the vegetation on the back dune if a blowout removed all the vegetation from the fore dune?
6. **GS2** Explain how longshore drift moves sand along a coastline. What is likely to happen on a beach if a council constructs a rock barrier at right angles to the beach?
7. **GS2** Explain the *changes* that high-energy waves can cause on coasts.

5.2 Exercise 2: Apply your understanding

1. **GS2** Explain why *environmental*, social and economic criteria must be applied to manage a coastal area such as that shown in **FIGURE 1**.
2. **GS6** Predict what might happen to the Great Barrier Reef if it wasn't listed on the World Heritage List.
3. **GS4** Draw and label a sketch of coastal depositional forms such as those shown in **FIGURE 3**. What role has transportation of materials by longshore drift had in the development of these features?
4. **GS6** Refer to the photograph of Tasman's Arch in **FIGURE 10**. What will happen to this coastal feature in the future? Explain how this *change* will occur.
5. **GS2** Discuss what happens over millions of years to eroded material that falls into the sea.
6. **GS2** Refer to the **FIGURE 8** wave refraction diagram. Why are headlands more vulnerable to erosion than bays?

Try these questions in learnON for instant, corrective feedback. Go to www.jacplus.com.au.

5.3 Challenges to coastal management

5.3.1 Human impacts on coasts

Human activities along the coast can interfere with natural coastal processes, resulting in significant changes to coastal environments. Our day-to-day lives across the globe contribute to global warming and sea-level rise, which in turn also affect our coastlines.

Human impacts on coastlines include the construction of ports, boat marinas and sea walls; changes in land use (for example, from a natural environment to agricultural or urban environments); and the disposal of waste from coastal and other settlements.

The *World Ocean Review* series is published by a not-for-profit company that seeks to raise awareness of issues relating to marine science and sustainable use of our oceans. The 2015 Review suggested that for oceans and coasts to be sustainably managed into the future, new environmental policies must be implemented. The issues identified in the review that need to be addressed include:

- **Marine pollution**
  - Toxic substances and heavy metals from industrial plants (liquid effluent and gaseous emissions)
  - Nutrients, in particular phosphate and nitrogen, from agricultural sources and untreated wastewater (eutrophication of coastal waters)
  - Ocean noise pollution from shipping and from growing offshore industry (exploitation of oil and natural gas reserves, construction of wind turbines, future mineral extraction)
- **Growing demand for resources**
  - Exploitation of oil and natural gas reserves in inshore areas and increasingly also in deep-sea areas, resulting in smaller or greater amounts of oil being released into the sea
  - Extraction of sand, gravel and rock for construction purposes
  - For the development of new pharmaceuticals: extraction of genetic resources from marine life such as bacteria, sponges and other life forms, the removal of which may result in damage to sea floor habitats
  - Future ocean mining (ore mining at the sea floor) which may damage deep-sea habitats
  - Aquaculture (release of nutrients, pharmaceuticals and pathogens)

- **Overfishing**
  - Industrial-scale fishing and overexploitation of fish stocks; illegal fishing
- **Habitat destruction**
  - Building projects such as port extensions or hotels
  - Clear-felling of mangrove forests
  - Destruction of coral reefs as a result of fishing or tourism
- **Bioinvasion**
  - Inward movement of non-indigenous species as a result of shipping transport or shellfish farming; changes in characteristic habitats
- **Climate change**
  - Ocean warming
  - Sea-level rise
  - Ocean acidification.

Source: World Ocean Review 2015

FIGURE 1 Destruction of mangrove habitats has enormous impacts on biodiversity.

The 2017 *World Ocean Review* explores the coastal habitat and the diverse expectations upon this habitat. It illustrates how the varied ecosystem services rendered by the coasts are being subjected to increasing pressure, and outlines measures that will be necessary in the future to respond effectively to the threats from both climate change and natural disasters.

Coastal environments have not always been managed sustainably. In the past, decision-makers generally had less understanding of the delicately balanced nature of many coastal ecosystems, and they had limited environmental worldviews about the use of coastal areas. Their aim was to develop coastal areas for short-term economic gains. This was based on the belief that nature's resources were limitless. Building apartment blocks and tourist resorts on sand dunes seemed like a good idea — little thought was given to the fragility of the coastline and the long-term suitability of such development in an environment subject to coastal storm erosion and, in more recent times, rising sea levels.

Over time, people have realised that sustainable coastal management requires an understanding of:
- the coastal environment and the effect of physical processes
- the effect of human activities within the coastal zone
- the different perspectives of coastal users
- how to achieve a balance between conservation and development
- how decisions are made about the ways in which coasts will be used
- how to evaluate the success of individuals, groups and the levels of government in managing coastal issues.

5.3.2 The threat of global warming

Perhaps the greatest threat to coasts today is rising sea levels. It is recognised that global warming is a result of the **enhanced greenhouse effect**, which is a human-induced phenomenon that is leading to the melting of polar ice caps and glaciers. Some of the changes to coastal environments that will result due to global warming include:
- increased intensity and frequency of storm surges and coastal flooding
- increased salinity of rivers and groundwaters resulting from salt intrusion
- increased coastal erosion
- inundation of low-lying coastal communities and critical infrastructure
- loss of important mangroves and other wetlands
- impacts on marine ecosystems such as coral reefs.

FIGURE 2 To bring attention to issues of global warming, in particular rising sea levels, a meeting was held on the sea floor by government representatives of the Republic of the Maldives.

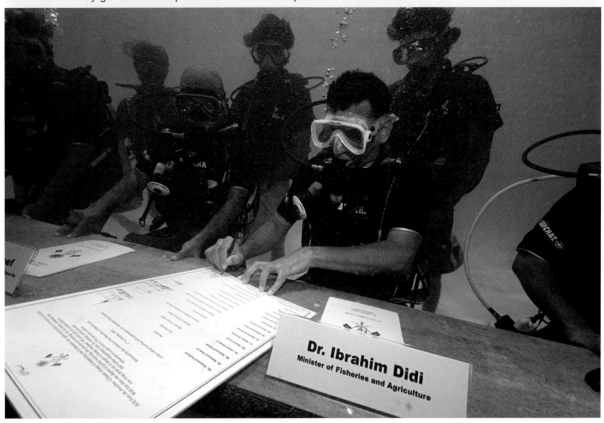

5.3.3 Our disappearing islands

As a result of climate change, it is currently anticipated that many low-lying islands will be flooded by the sea. Sea levels are currently estimated to be rising by about 2 to 3 millimetres each year. Melting glaciers and polar ice are adding to the water volume of the oceans; also, as the water warms, its volume increases. At current rates, many island groups in the Pacific and Indian Oceans will be almost completely inundated by 2050.

Coastal storms, tsunamis, flooding, inundation, erosion, deposition and saltwater intrusion into freshwater supplies present a combined threat to coastal regions. With stronger windstorms possible, many low-lying communities will be at risk from storm surges (see **FIGURE 3** and **TABLE 1**).

FIGURE 3 Low-lying islands in the Pacific under threat of disappearing due to climate change

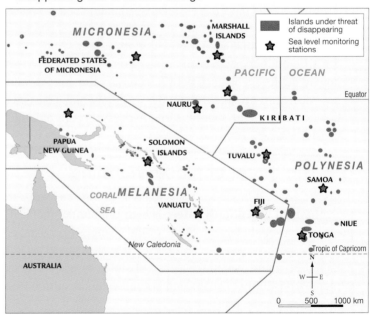

Source: MAPgraphics Pty Ltd, Brisbane.

TABLE 1 Selected Pacific Island nations, area and population

Island	Land area (km²)	National extent (km²)	Population	Gross income per person per year (US$)	Highest elevation (metres above sea level)
Kiribati	717	3 550 000	116 398	4280	80
Marshall Islands	181	2 131 000	53 127	5590	10
Tuvalu	26	900 000	11 192	5780	5
Australia	7 690 000	7 690 000	24 450 561	47 160	2229

People living on low-lying islands will be among the first wave of 'climate refugees'. Due to environmental change, mainly through rising sea levels, some people have already had to move, and many more could be without a home in our lifetime.

5.3.4 Rising sea levels in the Pacific

Many of the Pacific Islands are small and can in some cases be described as **atolls**. Their national boundaries, which include the waters and economic zones they control, extend over vast distances, but their land area is limited. Islands such as Tuvalu and the Marshall Islands in the south-west Pacific, which are only a few metres above sea level, are particularly vulnerable to rising sea levels and associated severe storm activity due to climate change.

The economies of these Pacific Islands are small-scale and earnings are not high, with a reliance on what limited natural resources occur on the islands and in the surrounding ocean waters. Although rainfall can be plentiful, not much water can be retained, due to sandy soils and low altitudes; streams are few and groundwater is scarce. Hence, any incursion by sea water can be devastating for agricultural produce (see **FIGURE 4**), the urban environment and tourism, which has more recently become a money earner for these islands.

FIGURE 4 Poulaka crops killed by salt water due to rising sea levels

Apart from the predicted rise in sea levels due to global warming, a secondary impact on the life of Pacific Islanders will be increases in the temperature of the sea, which will affect coral reefs and fish stocks that live in that environment. As for the Great Barrier Reef in Australia, bleaching and death of coral reefs can lead to the destruction of the whole aquatic ecosystem, and this will have devastating impacts on the Islanders' main diet, which is fish and other forms of seafood.

What can be done?

If food crops are destroyed by rising sea levels, **storm surges** and saltwater pollution, the Pacific Islanders do not have much scope for importing food due to their remoteness, high transport costs and low earnings of individuals. Combined with loss of seafood stocks, the Islanders will need to move to other islands to find a new home and livelihood. Clearly, this is an outcome that no-one wants to see. The task falls to the global community to instigate change to combat this impending humanitarian and ecological disaster.

The Pacific Islanders are strong advocates for the policies of the **Kyoto Protocol** and the subsequent **Paris Agreement** on climate change. Under the auspices of the United Nations Framework Convention on Climate Change (UNFCCC), these agreements have established a range of measures to reduce the impact of greenhouse gas emissions by introducing carbon trading schemes and energy-efficient forms of technology such as wind and solar power. One hundred and eighty-five countries have signed up to the Paris Agreement, which came into effect in November 2016. Signatories have agreed to ambitious targets for the reduction of greenhouse gas emissions, with an aim of keeping global temperature rise below 2 degrees Celsius, and attempting to limit it further still, to just 1.5 degrees Celsius. All parties are required to report regularly on their progress towards emissions targets. It is hoped that through these measures, the low-lying Pacific Islands can be spared a watery fate.

The leaders of the Pacific nations have spoken at the United Nations and many international climate change forums to raise awareness of their perilous situation and vulnerability to rising sea levels. They have also approached nations such as Australia and New Zealand to discuss the establishment of a future migration policy if global efforts to stem sea-level rise do not succeed.

DISCUSS

In small groups, consider the following statement: 'As Pacific neighbours, Australia and New Zealand should fulfil a duty and accept environmental refugees from Pacific islands in danger of flooding from sea-level rise'.

[Ethical Capability]

5.3.5 Rising sea levels in the Maldives

The Maldive Islands are located in the Indian Ocean, to the south-west of India (see **FIGURE 5**). There are about 1200 coral islands, grouped into 26 atolls. The average elevation across the islands is around 1.5 metres above sea level (the highest point in the island group is just 2.4 metres above sea level). Economically, the nation depends on tourism and the continuing appeal of its beautiful beaches.

FIGURE 5 Location of the Maldive Islands

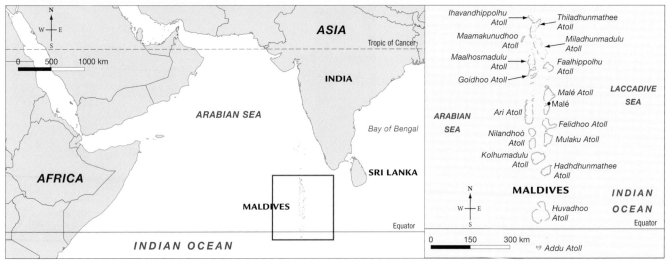

Source: Map by Spatial Vision.

The Boxing Day tsunami of 2004 showed how vulnerable the Maldives are, when the wave swept across many low-lying islands, causing widespread destruction of their fruit plantations. Eighty-two people died, while at least another 26 were reported missing, presumed dead. The number of deaths could undoubtedly have been much higher if not for the fact that most of the population lives in Malé, which is protected by a huge sea wall (see **FIGURE 6**).

FIGURE 6 Malé, the capital of the Maldives, occupies an entire island of its own. Why is there a need for a sea wall?

Sea wall

Only nine islands were reported to have escaped any flooding, while 57 islands faced serious damage to critical infrastructure, 14 islands had to be totally evacuated, and six islands were destroyed. A further 21 resort islands were forced to close because of serious damage. The total damage was estimated to be more than US$400 million, or some 62 per cent of the country's **gross domestic product** (GDP).

The impact of climate change

The long-term threat to the Maldives, however, as with the Pacific Island nations, is posed by global warming and the associated rise in sea levels. In 2013, the United Nations Intergovernmental Panel on Climate Change (IPCC) predicted that by 2100 sea levels would rise by up to one metre. A 2019 study by leading ice scientists suggested that the rise could actually be as high as two metres. Even using the more conservative estimate, many areas of the Maldives face significant threat. In the worst case, almost the entire nation of the Maldives would be effectively submerged.

What actions can save the islands?

The application of human–environment systems thinking in the form of various schemes is being examined by the Maldivian government, including moving populations from islands more at risk, building barriers against the rising sea, raising the level of some key islands and even building a completely new island. However, these approaches offer only short-term solutions. The long-term sustainable challenge is to deal with the basic problem: global warming itself. It is perhaps understandable that the Maldives was one of the first countries to sign the Kyoto Protocol and subsequent Paris Agreement, which the government **ratified** on the same day as signing.

Unless the international community agrees to an environmental worldview that incorporates changes to make large cuts in emissions, the problems facing the Islanders will worsen. Thousands will be forced to seek refuge in other countries. Without global action, eventually the Islanders will lose their country.

FIGURE 7 Sandbags protecting a home on the Maldives island Medu Fushi, damaged by the 2004 tsunami

5.3.6 The interconnection of coast and inland waters

Bangladesh's Sundarbans

The country of Bangladesh is a large **alluvial plain** crossed by three rivers: the Ganges, Brahmaputra and Meghna. Each river carries massive volumes of water from its source in the Himalayas, spreads out along the **deltaic plain**, and empties into the world's biggest delta, the Bay of Bengal. This makes Bangladesh's coastline one of the most flood-prone in the world.

Apart from flooding by rivers in the delta, sea-level rises caused by global warming will lead to the expansion of ocean waters and additional inflows from melting Himalayan snow. At current rates, the IPCC predicts rising sea levels will overtake 17 per cent of Bangladesh by 2050, displacing at least 20 million people.

The Sundarbans region, a World Heritage site, is just one area of Bangladesh at risk from increased flooding. The Sundarbans are the largest intact mangrove forests in the world. Mangroves protect against coastal erosion and land loss. They play an important role in flood minimisation because they trap sediment in their extensive root systems. Mangroves also defend against storm surges caused by tropical cyclones or king tides, both common in the Sundarbans.

The Sundarbans also provide a breeding ground for birds and fish, as well as being home to the endangered Royal Bengal tiger. By sheltering juvenile fish, the mangrove forest provides a source of protein for millions of people in South Asia. Recently, the Sundarbans have also attracted a growing human population as Bangladeshis flee overcrowding in the capital city, Dhaka, or flooding and poverty in rural areas.

Increasing human occupation poses a severe threat to the Sundarbans. Most Bangladeshis rely on wood as a source of energy, and mangroves are being cleared to make charcoal for cooking. Aquaculture industries also have a negative impact. Mangroves are cleared to accommodate huge ponds for fish breeding, which quickly become polluted by antibiotics, waste products and toxic algae. This damage to the Sundarbans destroys Bangladesh's natural defence against flooding.

FIGURE 8 Flooding in Bangladesh — some causes

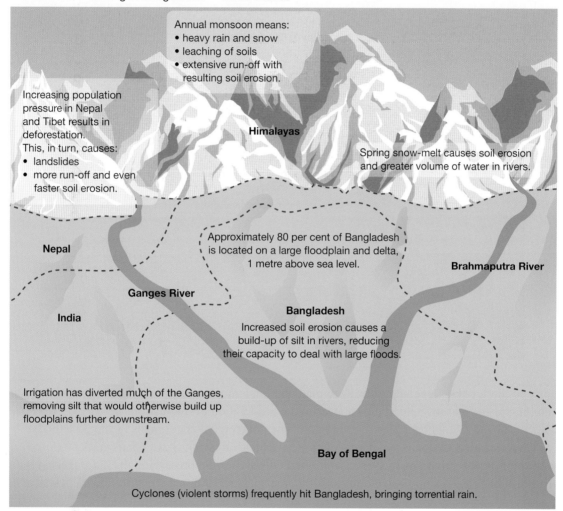

Annual monsoon means:
• heavy rain and snow
• leaching of soils
• extensive run-off with resulting soil erosion.

Increasing population pressure in Nepal and Tibet results in deforestation. This, in turn, causes:
• landslides
• more run-off and even faster soil erosion.

Himalayas

Spring snow-melt causes soil erosion and greater volume of water in rivers.

Nepal

Approximately 80 per cent of Bangladesh is located on a large floodplain and delta, 1 metre above sea level.

Brahmaputra River

Ganges River

India

Bangladesh

Increased soil erosion causes a build-up of silt in rivers, reducing their capacity to deal with large floods.

Irrigation has diverted much of the Ganges, removing silt that would otherwise build up floodplains further downstream.

Bay of Bengal

Cyclones (violent storms) frequently hit Bangladesh, bringing torrential rain.

The impact of flooding

The increase in temperature, which has led to an increased melting of glaciers and snow inland in the Himalayas, will exacerbate the existing problems of flooding in Bangladesh. Climate change also causes shifts in weather patterns. If the monsoon season (from June to October) coincided with an unseasonal snow-melt, flooding would occur on a scale never before seen, especially with the event of tropical cyclones. Land would be lost and people displaced. Many islands fringing the Bay of Bengal are already under water, producing 'climate refugees' — people who have to seek refuge, fleeing their uninhabitable lands.

The 1991 Bangladesh cyclone was among the deadliest tropical cyclones on record (see **FIGURE 9**). The cyclone struck the Chittagong district of south-eastern Bangladesh with winds of around 250 km/h. The storm forced a six-metre storm surge inland over a wide area, killing at least 138 000 people and leaving as many as 10 million homeless. In 1998, more than 75 per cent of the total area of the country was flooded and 30 million people were made homeless. Over a thousand people were killed and the flooding caused contamination of crops and animals. Unclean water resulted in cholera and typhoid outbreaks. In 2004, the country flooded again. This was similar to the 1998 flood, with two-thirds of the country under water. Again, in early October 2005, dozens of villages were inundated when rain caused the rivers of north-western Bangladesh to burst their banks. Further floods occurred in 2015 and 2017.

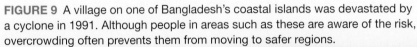

FIGURE 9 A village on one of Bangladesh's coastal islands was devastated by a cyclone in 1991. Although people in areas such as these are aware of the risk, overcrowding often prevents them from moving to safer regions.

Because of these risks, Bangladesh needs to plan and implement management strategies based on an understanding of the reasons behind the changes and consideration of interactions between environmental, economic and social factors operating in the region. The government encourages farming methods that avoid deforestation, and new standards relating to vehicle emissions have been set. A proposed economic solution is ecotourism, as it attracts foreign currency while preserving the natural ecosystems and promoting sustainable development and responsible management of its vital environmental resources.

DISCUSS

To what extent do you think economic goals and objectives are important with respect to **environmental** goals such as societal **change** and reducing greenhouse gas emissions? What policy direction would you push if you were in a position of influence in government?   **[Ethical Capability]**

 Resources

🔹 **Interactivity**   Bangladesh awash (int-3297)
🔹 **Google Earth**   The Sundarbans

5.3 INQUIRY ACTIVITY

In groups, discuss the following statements concerning the impacts of climate *change* in the Pacific and Indian Oceans using *environmental*, economic and social criteria. Alternatively, you may wish to use a SWOT analysis to help in evaluating each statement.

- Australia and New Zealand should be prepared to resettle the 100 000 people of the islands of Kiribati if sea level rises create a 'climate refugee' problem.
- Increasing the height of the sea wall around the island of Malé will solve the threats of rising sea levels in the Maldives and ensure tourism into the future. **Evaluating, predicting, proposing**

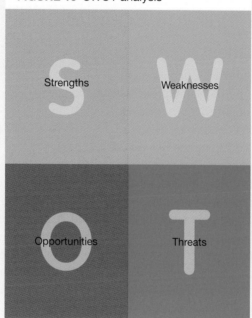

FIGURE 10 SWOT analysis

Strengths

Weaknesses

Opportunities

Threats

5.3 EXERCISES

Geographical skills key: GS1 Remembering and understanding **GS2** Describing and explaining **GS3** Comparing and contrasting **GS4** Classifying, organising, constructing **GS5** Examining, analysing, interpreting **GS6** Evaluating, predicting, proposing

5.3 Exercise 1: Check your understanding

1. **GS1** What are some impacts that people have on coastal areas?
2. **GS2** Select one of the impacts of rising sea levels on coasts identified in this subtopic and explain why this would be a problem for a selected coastal settlement in Australia.
3. **GS2** Many islands in the Pacific and Indian Oceans are under threat from rising sea levels. Why is this so?
4. **GS1** What can the governments and Pacific Island peoples do, both in the short term and long term, to solve the problems they will face due to climate *change*?
5. **GS1** How is climate *change* threatening water supplies and affecting food resources?
6. **GS1** How do mangroves minimise the impact of floods and coastal erosion?
7. **GS1** What are two reasons for mangroves being cleared in the Sundarbans?
8. **GS1** What are 'climate refugees'?

5.3 Exercise 2: Apply your understanding

1. **GS5** Refer to **FIGURE 8**. Explain how the geography of Bangladesh makes it so vulnerable to the threat posed by climate *change*.
2. **GS2** How can ecotourism play a role in preserving Bangladesh's ecosystems?
3. **GS6** Suggest other parts of the world you can think of that might also be threatened if sea levels were to rise by about one metre over the next 100 years. Justify your views.
4. **GS1** List the factors that are displacing Bangladeshis and forcing them to move to the Sundarbans.

5. **GS5** Refer to **FIGURE 8**.

5. **GS5** Refer to **FIGURE 8**.
    (a) Describe how cyclones can contribute towards flooding in Bangladesh.
    (b) List some short-term and long-term actions that neighbouring nations Tibet, India and Nepal could implement to lessen the impact of flooding in **places** like Bangladesh.
    (c) Divide a table into three columns with the headings 'Food production', 'Transport' and 'Settlement', and list the consequences of flooding for each category.

Try these questions in learnON for instant, corrective feedback. Go to www.jacplus.com.au.

5.4 How do we manage coastal change?

5.4.1 Changing coastlines

Coastal areas are not static or fixed places, and as such they are subject to two main agents of change. These can be defined as natural environmental processes and human-induced processes. In terms of natural environmental processes, where deposition processes dominate, coasts have been growing. However, where erosion processes dominate, coastal land is lost to the sea. Wherever people have imposed their structures, in the form of housing, harbour works and the like on coasts, there is a need to manage or at least moderate the processes of coastal change in a sustainable manner.

In 2016, 41 per cent of the world's population, or some three billion people, lived within 100 kilometres of the coast. Most of the world's megacities are located on the coast. According to predictions made by the *World Ocean Review* in 2010 and more recently in 2015 and 2017, at least one billion people who live in low-lying coastal areas could experience inundation and/or erosion of their lands into the future. This change to coasts is seen as stemming essentially from climate change which, as a largely human-induced event, is leading to rising seas, associated flooding in major delta regions, and more frequent severe storm events, such as hurricanes (see **FIGURE 1**). A consequence will mean an increase in what are known as 'climate refugees' — people who will have to relocate due to coastal changes.

FIGURE 1 Zones across the world subject to threats associated with climate change

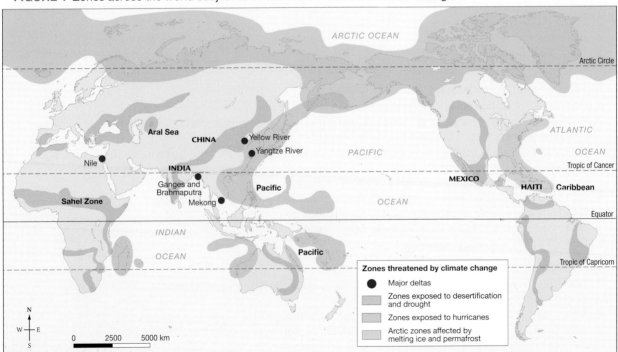

Source: UNEP (United Nations Environment Programme).

5.4.2 Protecting the coast

The protection of the coast through management programs is a costly business that aims to overcome problems associated with land loss, waterlogging and incursions of **groundwater salinity**.

The Netherlands, a country with two-thirds of its land below sea level, has proven that protecting the coastline is possible through a large investment of capital. The most common form of coastal protection in the Netherlands are **dykes** to hold back the sea; however, a recent addition is **floating settlements** that can rise and fall as sea levels change (see **FIGURE 2**).

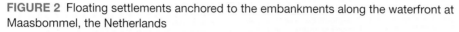

FIGURE 2 Floating settlements anchored to the embankments along the waterfront at Maasbommel, the Netherlands

5.4.3 Coastal management in Australia

If coastlines are to be protected, a wide range of strategies must be employed to combat changes to the coastline and, in particular, flooding of low-lying areas and increased erosion of beaches and bluffs. The techniques shown in **TABLE 1** are used in Australia.

TABLE 1 Possible management solutions to reduce the impacts of sea level rise and erosion

Solution	Description	Diagram	Advantages	Disadvantages
Beach nourishment	The artificial placement of sand on a beach. This is then spread along the beach by natural processes.	Established vegetation – shrubs and sand grasses; Initial nourishment designed for 10 years; Fencing; Sea level; Existing profile	Sand is used that best matches the natural beach material. Low environmental impact at the beach	The sand must come from another beach and may have an environmental impact in that location. Must be carried out on a continuous basis and therefore requires continuous funds

(continued)

TABLE 1 Possible management solutions to reduce the impacts of sea level rise and erosion *(continued)*

Solution	Description	Diagram	Advantages	Disadvantages
Groyne	An artificial structure designed to trap sand being moved by longshore drift, therefore protecting the beach. Groynes can be built using timber, concrete, steel pilings and rock.	Groyne	Traps sand and maintains the beach	Groynes do not stop sand movement that occurs directly offshore. Visual eyesore
Sea wall	A structure placed parallel to the shoreline to separate the land area from the water	Coastal vegetation / Sea wall	Prevents further erosion of the dune area and protects buildings	The base of the sea wall will be undermined over time. Visual eyesore Will need a sand nourishment program as well High initial cost Ongoing maintenance and cost
Offshore breakwater	A structure parallel to the shore and placed in a water depth of about 10 metres	Sheltered area protected from erosion / Wave breaks on breakwater, reducing much of its energy	Waves break in the deeper water, reducing their energy at the shore.	Destroys surfing amenity of the coast Requires large boulders in large quantities Cost would be extremely high
Purchase property	Buy the buildings and remove structures that are threatened by erosion.	House threatened by erosion / Sea level	Allows easier management of the dune area Allows natural beach processes to continue Increases public access to the beach	Loss of revenue to the local council Possible social problems with residents who must move Exposes the back dune area, which will need protection Cost would be extremely high Does not solve sand loss

5.4.4 CASE STUDY: Shifting sands on the Gold Coast

Have you ever wondered where all the sand on a beach comes from and where it goes? Wind, wave and current action is responsible for moving sand on and off a beach and along a coastline. Human-induced environmental changes along the coast can interrupt these natural processes, often creating long-term problems. One such example is located on the Gold Coast in Queensland where a human-centred — that is, a technical 'we can fix it' — viewpoint was taken.

The sand that ends up on the beach actually comes from inland sources, the weathering and erosion of rock and soils. Sand is then washed into rivers and transported downstream to eventually arrive at the coast. At the mouth of the Tweed River, at the southern end of the Gold Coast, sand would often block the river mouth, making it difficult for boats to pass through. As a result, a pair of **training walls** was constructed (see **FIGURE 3**) 400 metres out to sea, to keep the mouth open.

FIGURE 3 The Tweed River training walls

These, however, effectively interrupted the longshore drift current (see subtopic 5.2), which moves sand along the coast.

Further north from the river mouth, natural wave action, especially during storms, continued to strip sand from the beaches. However, without new sand arriving in the longshore drift current, the beaches eventually eroded. Local residents and tourists had lost their beach (see **FIGURE 4**).

The sand destined for the beach was effectively trapped at the southern end of the training wall, where it built out the Letitia Spit by 250 metres (see **FIGURE 5**).

How was the problem solved?

In response, various attempts have been made to restore the Gold Coast beaches. Rock walls and groynes were built to trap sand further up the coast. However, they made the problem worse by interfering further with the longshore drift.

Eventually a sand bypass system was installed (see **FIGURE 5**). It pumped sand from the build-up on the southern end of the training walls and piped it north to the eroded beaches. Each year, 500 000 cubic metres of sand have to be moved. This has been successful in maintaining the beaches, but it is an expensive and ongoing management technique.

Coastal protection works such as groynes, sea walls and training walls are usually built to protect human-built structures such as buildings and roads against erosion. However, they usually reduce the ability of coastal processes to adjust naturally, often exacerbating the problem and actually accelerating erosion!

FIGURE 4 The effect of destructive storm waves on Duranbah Beach

FIGURE 5 Sand bypass system

Longshore drift

West Snapper Rocks outlet

East Snapper Rocks primary outlet

Rainbow Bay

Kirra outlet

Greenmount Beach

Duranbah Beach

Duranbah outlet

Training wall

Tweed River

Training wall

Coolangatta

Water intake

Sand collection jetty

Control building

Tweed Heads

Letitia Spit

Source: Spatial Vision.

5.4.5 CASE STUDY: Sustainable development on the Sapphire Coast

The main pressures on many coastal systems relate to the development of towns and tourist facilities. Careful management can enable growth of urban areas while at the same time protecting the natural coastal features.

The 'Sapphire Coast' in south-east New South Wales is a popular tourist destination because of its array of beautiful beaches, stunning scenery and mild, sunny weather. Merimbula is a coastal resort town in this Sapphire Coast region. Similar to any other popular coastal location, it experiences natural changes as well as the pressures relating to development. **FIGURE 6** clearly shows the settlement areas that have been established close to the ocean and lake in this region.

FIGURE 6 Aerial view over Merimbula Lake

The natural landform features along this coastline include a series of headlands separated by bay head beaches. Merimbula Lake has formed from a slow and gradual build-up of a sand barrier, leaving only a narrow channel for salt water to enter and fresh water to exit. The shallow and sheltered waters of the lake provide an ideal environment for oyster farming and recreation. The **FIGURE 7** topographical map shows the narrow entrance to Merimbula Lake. It also shows the various settlement areas and the geographical elements of the region that have influenced this development.

You can learn more about this area by completing the **Merimbula** worksheet and the **Predict changes around Merimbula** interactivity in the Resources tab.

FIGURE 7 Topographic map extract of Merimbula

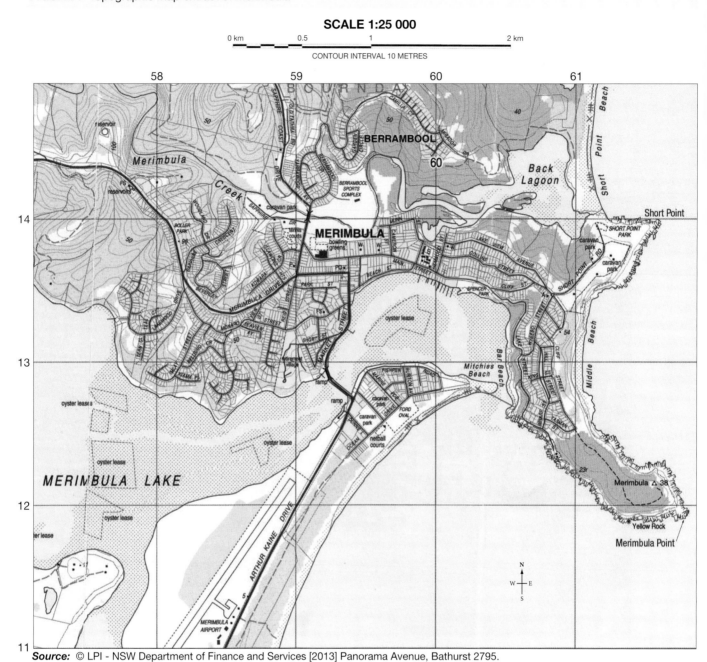

SCALE 1:25 000

0 km   0.5   1   2 km

CONTOUR INTERVAL 10 METRES

Source: © LPI - NSW Department of Finance and Services [2013] Panorama Avenue, Bathurst 2795.

Key

Built-up area		Closed forest: 80–100% crown cover. Open forest: 50–8-% crown cover	
Major road: paved, unpaved		Woodland: 20–50% crown cover	
Secondary road: paved, unpaved		Perennial lake	
Minor road: paved, unpaved		Intertidal flat. Sand	
Vehicular track		Perennial stream	
Walking track		Large dam or weir	
Bridge		Jetty or wharf	
Werong △ 1215 Survey landmark (with height)		Rocky shoreline	
• 846 Spot height		Building, small. Building, large.	
1000 Contours		Ambulance station. Police station.	
176r Cliff, with relative height		Fire station. Post office.	
Levee or dyke		Local government	
		County	

5.4 INQUIRY ACTIVITIES

1. In the USA state of North Carolina, it is now illegal to build coastal structures to protect houses close to the shoreline. Structures such as rock walls and groynes may offer protection but disrupt the natural movement of sand along the coast.
   (a) Is this approach fair to those whose houses are threatened by storms and sea-level rising? Why?
   (b) What would be the arguments for and against the idea of using ratepayer or taxpayer funds to build coastal structures to protect the houses threatened by storms and sea level rising?
   (c) Is it equitable for all those people who use the coast? Write a page outlining and explaining your views.
   [Ethical Capability]
2. Complete the **Merimbula** worksheet in the Resources tab to learn more about the *environmental* management of this coastal location. **Examining, analysing, interpreting**

5.4 EXERCISES

Geographical skills key: GS1 Remembering and understanding **GS2** Describing and explaining **GS3** Comparing and contrasting **GS4** Classifying, organising, constructing **GS5** Examining, analysing, interpreting **GS6** Evaluating, predicting, proposing

5.4 Exercise 1: Check your understanding

1. **GS2** Refer to **FIGURE 1**. Write a brief description of the particular climate *change*–related threats faced by various nations and regions of the world.
2. **GS1** Why does the Netherlands spend money on coastal protection?
3. **GS1** List the various ways that sand can be moved in a coastal region.
4. **GS1** What is the purpose of training walls at a river mouth?
5. **GS1** Study **FIGURE 3**. What effect have these training walls had on preventing a sand build-up in the Tweed River mouth?
6. **GS2** Examine **FIGURE 5**. Describe the direction of the longshore drift in this region.
7. **GS5** Refer to **FIGURE 5**. What is the approximate distance that sand has to be pumped from the sand collection jetty to the Kirra outlet?
8. **GS6** One coastal geographer has stated that to manage the beaches along the Gold Coast, 'ecologically, non-intervention would be a better and more *sustainable* option'. This is an earth-centred rather than human-centred attitude.
   (a) Do you agree or disagree with this viewpoint?
   (b) How could you manage this stretch of the coast without using engineering methods such as the sand bypass system?
9. **GS6** Refer to the section 5.4.4 case study. How effective has the management of sand along this section of the Gold Coast been in terms of *environmental* and economic criteria? ▶

5.4 Exercise 2: Apply your understanding

1. **GS2** Explain how a coastal defence system such as a dyke works.
2. **GS6** What impact would sea level rise and erosion have on future food security?
3. **GS6** What might be the impact of sea level rise and coastal erosion on the tourist industries of the Gold Coast area of Australia? What strategies of coastal protection mentioned in this topic could help solve the problems, and how might they work?
4. **GS3** Evaluate the strengths and weaknesses of two of the management strategies shown in **TABLE 1**.
    (a) Which strategy would have the least *environmental* impact?
    (b) Which strategy would have the greatest economic impact or be the most costly to maintain?
    (c) Which strategies could improve social amenities such as tourism and recreation in coastal areas? Give reasons for your answer.
5. **GS6** Refer to **FIGURES 3** and **5**. What *changes* would you expect to see along this section of coast if the training walls on the Tweed River were removed?

Try these questions in learnON for instant, corrective feedback. Go to www.jacplus.com.au.

5.5 SkillBuilder: Comparing aerial photographs to investigate spatial change over time

What is an aerial photo?

Aerial photos are images taken above the Earth from an aircraft or satellite. Two images taken at different times, from the same angle, and placed side by side, show change that has occurred over time. Comparing aerial photographs is useful because each photograph captures details about a specific place at a particular time.

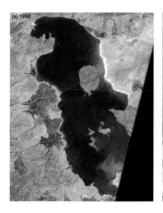

Select your learnON format to access:

- an overview of the skill and its application in Geography (Tell me)
- a video and a step-by-step process to explain the skill (Show me)
- an activity and interactivity for you to practise the skill (Let me do it)
- questions to consolidate your understanding of the skill.

on Resources

Video eLesson   Comparing aerial photographs to investigate spatial change over time (eles-1750)

Interactivity   Comparing aerial photographs to investigate spatial change over time (int-3368)

5.6 SkillBuilder: Comparing an aerial photograph and a topographic map

online only

What comparisons can be made between aerial photographs and topographic maps?

Comparing an aerial photograph with a topographic map enables us to see what is happening in one place. Photographs and maps may be from the same date but they may also be from different dates, and will thus show different information.

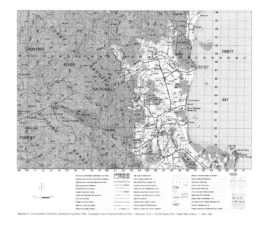

Select your learnON format to access:

- an overview of the skill and its application in Geography (Tell me)
- a video and a step-by-step process to explain the skill (Show me)
- an activity and interactivity for you to practise the skill (Let me do it)
- questions to consolidate your understanding of the skill.

 Resources

 Video eLesson   Comparing an aerial photograph and a topographic map (eles-1751)

Interactivity   Comparing an aerial photograph and a topographic map (int-3369)

5.7 Thinking Big research project: Ecology action newsletter — Reef rescue

online only

SCENARIO

You are a member of your school's ecology action group. Each term the group publishes a newsletter highlighting various environmental issues. This month your focus is on Australia's iconic Great Barrier Reef — how can we protect the reef from environmental threats and ensure its health for now and all time?

Select your learnON format to access:

- the full project scenario
- details of the project task
- resources to guide your project work
- and assessment rubric.

 Resources

ProjectsPLUS Thinking Big research project: Ecology action newsletter — Reef rescue (pro-0214)

5.8 Review

5.8.1 Key knowledge summary
Use this dot point summary to review the content covered in this topic.

5.8.2 Reflection
Reflect on your learning using the activities and resources provided.

 **Resources**

eWorkbook	Reflection (doc-31769)
	Crossword (doc-31770)
Interactivity	Managing change in coastal environments crossword (int-7672)

KEY TERMS

alluvial plain an area where rich sediments are deposited by flooding

atoll a coral island that encircles a lagoon

coastal dune vegetation succession the process of change in the plant types of a vegetation community over time — moving from pioneering plants in the high-tide zone to fully developed inland area vegetation

deltaic plain flat area where a river(s) empties into a basin

dyke an embankment constructed to prevent flooding by the sea or a river

dynamic equilibrium when the input of a coastal system such as winds and waves moving sediments onshore is equal to the output that moves sediments offshore, the system is said to be in a steady state. It is therefore not unstable and it has a dynamic equilibrium.

ecosystems systems formed by the interactions between the living organisms (plants, animals, humans) and the physical elements of an environment

enhanced greenhouse effect increasing concentrations of greenhouse gases in the Earth's atmosphere, contributing to global warming and climate change

floating settlements anchored buildings that float on water and are able to move up and down with the tides

gross domestic product (GDP) the value of all the goods and services produced within a country in a given period, usually discussed in terms of GDP per capita (total GDP divided by the population of the country)

groundwater salinity presence of salty water that has replaced fresh water in the subsurface layers of soil

hinterland the land behind a coast or shoreline extending a few kilometres inland

Kyoto Protocol an internationally agreed set of rules developed by the United Nations aimed at reducing climate change through the stabilisation of greenhouse gas emissions into the atmosphere

mass wasting the movement of rock and other debris downslope in bulk, due to a destabilising force such as undermining compounded by the pull of gravity

Paris Agreement United Nations Framework Convention on Climate Change (UNFCCC) agreement outlining steps to reduce greenhouse gas emissions and tackle global warming

ratify to formally consent to and agree to be bound by a treaty, contract or agreement

storm surge a temporary increase in sea level from storm activity

training walls a pair of rock walls built at a river's mouth to force the water into a deeper and more stable channel. The walls improve navigation and reduce sand blockages.

6 Marine environments — are we trashing our oceans?

6.1 Overview

Exactly how much plastic ends up in oceans and waterways, and why should we care if it does?

6.1.1 Introduction

Imagine you are on a beach. You are looking out to sea at the endless, constantly moving mass of water that stretches to the horizon. Why does it move, how does it move, what lies beneath?

Life on Earth would not be possible without our oceans. Humans are interconnected with the oceans, which provide or regulate our water, oxygen, weather, food, minerals and resources. Oceans also provide a surface for transport and trade and a habitat for 80 per cent of all life on Earth. Our oceans are under threat; as we use them to extract resources and dump waste, we destroy them. This image shows just a tiny

fraction of the many thousands of tonnes of marine debris floating at sea. The health of our oceans is at risk. In this topic, we look at this problem in more detail.

on **Resources**

- [ ] **eWorkbook** Customisable worksheets for this topic
- **Video eLesson** Thrown overboard (eles-1711)

LEARNING SEQUENCE

6.1 Overview
6.2 Motion in the ocean
6.3 Travelling trash — marine pollution and debris
6.4 Cleaning up our mess
6.5 **SkillBuilder:** Using geographic information systems (GIS) `online only`
6.6 Oil and water — a toxic mix
6.7 **SkillBuilder:** Describing change over time `online only`
6.8 **Thinking Big research project:** 'Plastic not-so-fantastic' media campaign `online only`
6.9 **Review** `online only`

To access a pre-test and starter questions and receive immediate, **corrective feedback** and **sample responses** to every question, select your learnON format at www.jacplus.com.au.

6.2 Motion in the ocean

6.2.1 What are ocean currents?

Why doesn't water at the equator get hotter and hotter and water at the poles get colder and colder? The answer is ocean currents. Currents are movements of water from one region to another, often over long distances and time periods. Currents effectively interconnect the world's oceans and seas. They are critically important for 'stirring' the waters and transporting heat, oxygen, carbon dioxide, salts, nutrients, sediments and marine creatures.

In January 1992, a ship sailing from Hong Kong to the United States lost a shipping crate containing 28 000 plastic bath toys at sea during a storm. The toys drifted off in the currents, the first ones eventually reaching the Alaskan coast in November of that year. Nearly 30 years later, many are still floating! Tracking these toys has enabled scientists to improve their understanding of ocean currents.

Knowledge of currents is vital for navigation, shipping, search and rescue, and the dispersal of pollutants. The direction that currents take is influenced by a number of factors, including the Earth's rotation, the shape of the sea floor, water temperature, salinity levels and the wind.

6.2.2 Different types of ocean currents

Surface currents

The action of winds blowing over the surface of the water sets up the movement of water in the top 400 metres of the ocean, creating surface currents. These currents flow in a regular pattern, but they can vary in depth, width and speed. Caused by the rotation of the Earth, the **Coriolis force** deflects currents into large circular patterns called gyres, which flow clockwise in the northern hemisphere and anticlockwise in the southern hemisphere (see **FIGURE 1**). Surface currents make up about 10 per cent of water movements in the ocean; deep water currents powered by **thermohaline** circulation make up the other 90 per cent.

Deep water currents

The Global Ocean Conveyor Belt (also known as the Great Ocean Conveyor Belt) is the largest of the thermohaline-driven ocean currents (see **FIGURE 1**). Warm water, which holds less salt and is less dense than cold water, travels from the equator near the surface into higher latitudes. There it loses some of its heat to the atmosphere. The current mixes with colder Arctic waters and this cold, salty water becomes more dense and sinks, flowing as a deep ocean current. This creates a continual looping current that moves at a rate of 100 millimetres per second and may take up to 1000 years to complete one loop. The quantity of water moved in the Global Ocean Conveyor Belt is more than 16 times the water volume of all the world's rivers.

Upwellings and downwellings

The movement of cold-water currents from the deep sea to the surface is called an upwelling. This is shown in **FIGURE 2(a)**. Regions where these occur are very productive fishing grounds as the upwellings bring nutrients from the seabed, which provide food for plankton, which are often the start of marine food chains. Over 50 per cent of the world's fish are caught in these areas.

Downwellings, shown in **FIGURE 2(b)**, occur when currents sink, taking oxygen and carbon dioxide from the atmosphere with them. These currents essentially 'stir up' the water and help distribute heat, gases and nutrients.

FIGURE 1 The Global Ocean Conveyor Belt and the five main ocean gyres

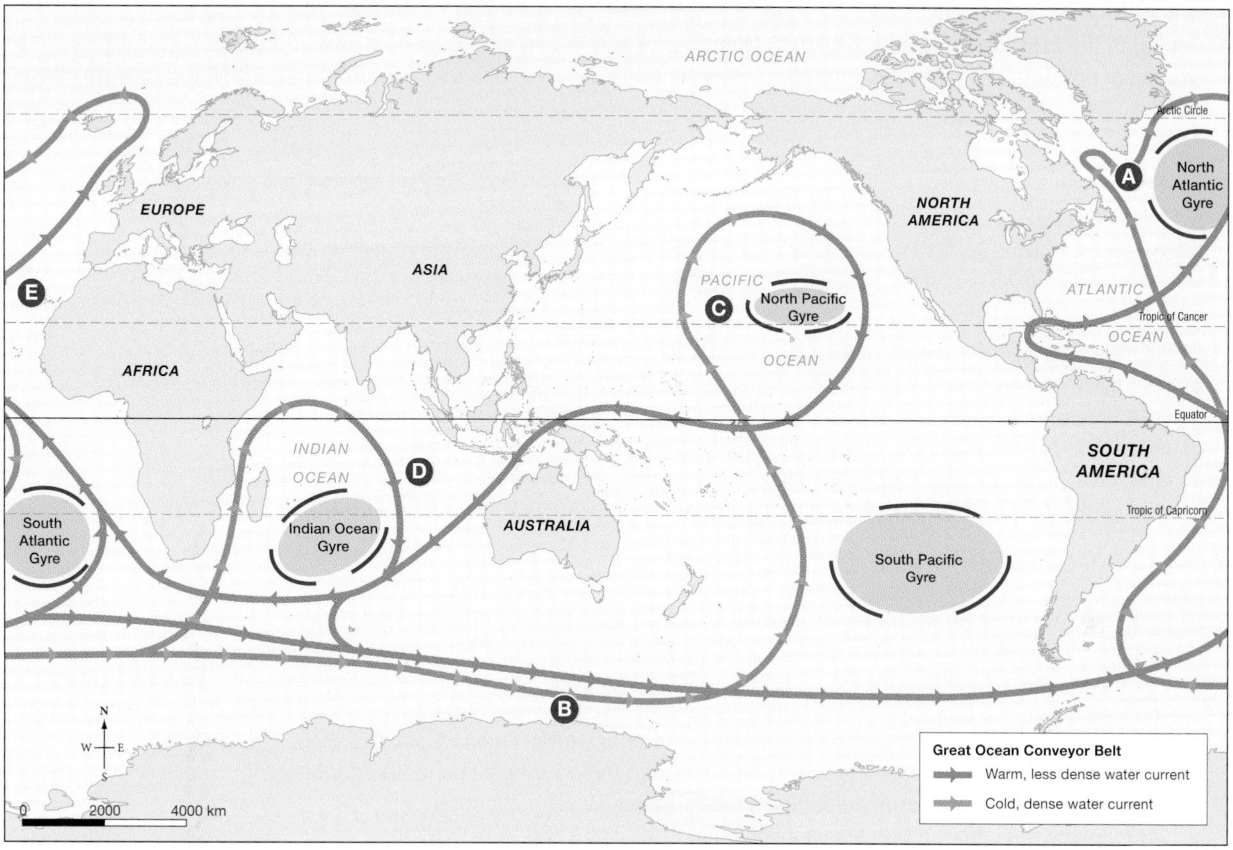

ARCTIC OCEAN

EUROPE

ASIA

NORTH
AMERICA

Arctic Circle

A North
Atlantic
Gyre

ATLANTIC

Tropic of Cancer

OCEAN

E

PACIFIC

C North Pacific
Gyre

OCEAN

AFRICA

Equator

SOUTH
AMERICA

INDIAN

OCEAN

D

Indian Ocean
Gyre

AUSTRALIA

Tropic of Capricorn

South
Atlantic
Gyre

South Pacific
Gyre

B

Great Ocean Conveyor Belt

Warm, less dense water current

Cold, dense water current

N
W——E
S

0   2000   4000 km

Source: Map by Spatial Vision.

FIGURE 2 (a) Upwelling and (b) downwelling

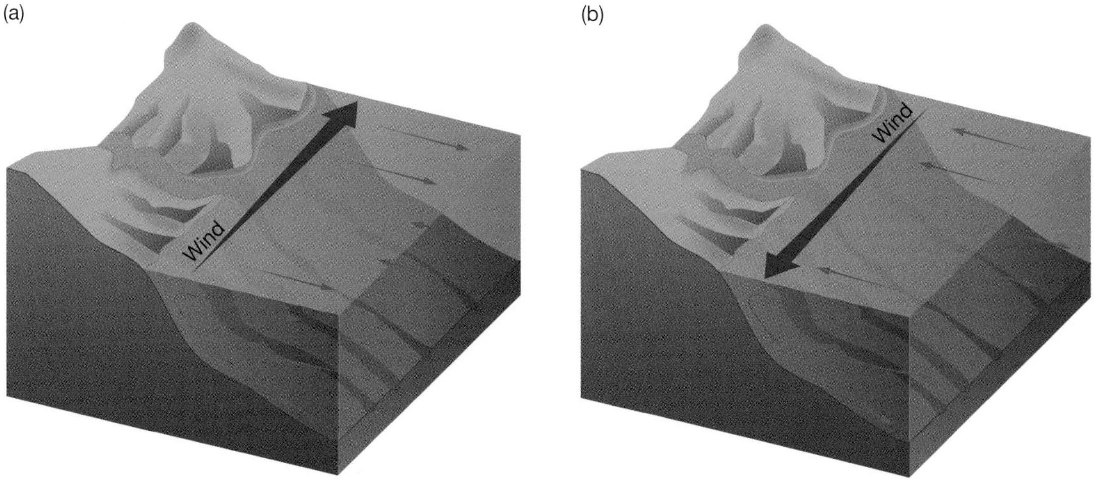

(a)

Wind

(b)

Wind

 **Resources**

Interactivity Motion in the ocean (int-3298)

Research the *interconnection* between El Niño events and the Humboldt current (cold upwelling) on the west coast of South America. Summarise your findings. **Examining, analysing, interpreting**

6.2 EXERCISES

Geographical skills key: GS1 Remembering and understanding **GS2** Describing and explaining **GS3** Comparing and contrasting **GS4** Classifying, organising, constructing **GS5** Examining, analysing, interpreting **GS6** Evaluating, predicting, proposing

6.2 Exercise 1: Check your understanding

1. **GS1** Why do ocean currents form? What is the driving force behind surface and thermohaline currents?
2. **GS2** Why are upwellings and downwellings important for marine *environments*?
3. **GS1** Refer to **FIGURE 1**. Describe the location of the five main ocean gyres.
4. **GS1** What factors influence the direction that ocean currents take?
5. **GS1** Looking at **FIGURE 1**, how does the Global Ocean Conveyor Belt *interconnect* the world's oceans?

6.2 Exercise 2: Apply your understanding

1. **GS1** Why do you think ocean currents are described as 'conveyor belts'?
2. **GS6** Suggest what *changes* might happen to the Global Ocean Conveyor Belt if there was a significant melting of the polar ice caps.
3. **GS5** Why doesn't water at the equator keep getting hotter and water at the poles keep getting colder? Use your knowledge of currents to write an explanation for a younger student.
4. **GS1** Refer to **FIGURE 1**. Describe the route taken by the Global Ocean Conveyor Belt. At each of the locations marked A–E, name the ocean, the direction the current is taking, the continent it is passing, and its thermohaline features (warm, cold, higher salt content, lower salt content).
5. **GS3** Outline the different processes involved in upwellings and downwellings.

Try these questions in learnON for instant, corrective feedback. Go to www.jacplus.com.au.

6.3 Travelling trash — marine pollution and debris

6.3.1 What is marine pollution?

Accidentally or deliberately, the oceans receive millions of tonnes of man-made pollutants each year, which are collected in currents and swirled around the oceans.

Marine pollution is any harmful substance or product that enters the ocean. Most are human pollutants including fertilisers, chemicals, sewage, plastics and other solids, including more than 500 shipping containers per year.

Close to 80 per cent of marine pollutants start off on land and are either washed or deposited into rivers, from where they make their way to the coast. Even industrial air pollution can be returned to the Earth's surface via rainfall (see **FIGURE 1**).

6.3.2 What is marine debris?

What happens to that empty drink can or plastic bag that misses the bin? There's a good chance it might wash down the gutter, into the drain and out to sea, never to be seen again. The world's largest rubbish dump is not on land, it is in the ocean.

Marine debris, the most prolific form of marine pollution, is litter and other solid material that washes or is dumped into the oceans, much of which is plastic (see **FIGURE 2**). The special features of plastic that make it such a useful product — it is light, cheap to produce and disposable — also make it a major problem for the ocean.

Plastics have revolutionised almost every aspect of society; over 320 million tonnes (nearly the weight of the entire human population) are produced each year. On a global scale, less than one-fifth of all plastic is recycled. In Australia in 2018, only 12 per cent was recycled; the rest was either shipped overseas or ended up in landfill. Much is unaccounted for, lost in the environment and eventually washed out to sea, often ending up in the gut or wrapped around the neck of marine creatures, or even buried in Arctic ice.

FIGURE 1 The sources of marine pollution

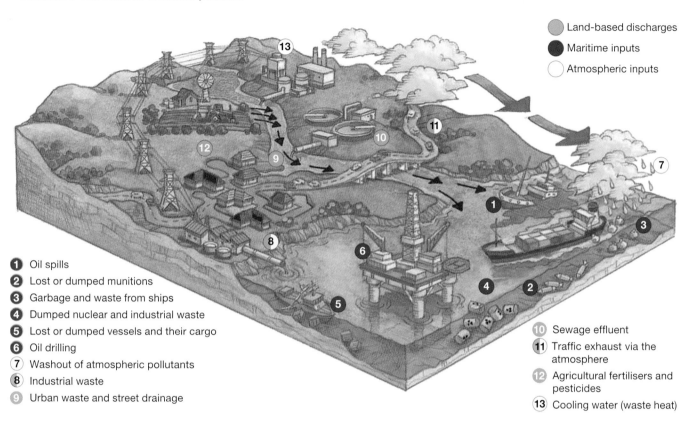

● Land-based discharges
● Maritime inputs
○ Atmospheric inputs

① Oil spills
② Lost or dumped munitions
③ Garbage and waste from ships
④ Dumped nuclear and industrial waste
⑤ Lost or dumped vessels and their cargo
⑥ Oil drilling
⑦ Washout of atmospheric pollutants
⑧ Industrial waste
⑨ Urban waste and street drainage

⑩ Sewage effluent
⑪ Traffic exhaust via the atmosphere
⑫ Agricultural fertilisers and pesticides
⑬ Cooling water (waste heat)

FIGURE 2 Top ten marine debris items

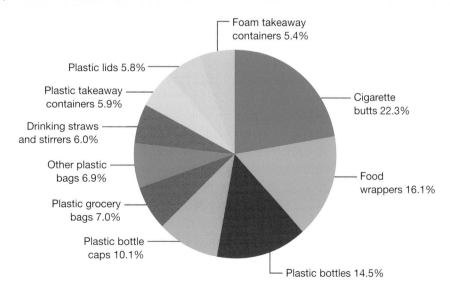

Foam takeaway containers 5.4%
Plastic lids 5.8%
Plastic takeaway containers 5.9%
Drinking straws and stirrers 6.0%
Other plastic bags 6.9%
Plastic grocery bags 7.0%
Plastic bottle caps 10.1%
Plastic bottles 14.5%
Food wrappers 16.1%
Cigarette butts 22.3%

Researchers estimate that 8 million tonnes of plastic end up in the oceans each year — imagine a garbage truck dumping a load of plastic in the ocean every minute of every day of every year. Some 60 per cent of plastic is less dense than sea water. When washed or dumped into the ocean, the buoyant material is easily moved by wind and surface currents, degraded into smaller and smaller pieces, or finally losing buoyancy and sinking. Surface currents and wind can also move debris back on to the coast, where it can become buried in sand or swept out to sea again. A survey of Australia's coastline found that plastics made up 74 per cent of marine litter.

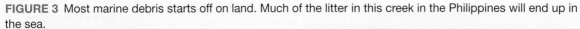
FIGURE 3 Most marine debris starts off on land. Much of the litter in this creek in the Philippines will end up in the sea.

Unlike most other litter, plastics generally are not biodegradable. The technological features of plastic mean that when it is exposed to constant wind, waves, salt and sunlight, it breaks down into tiny fragments known as microplastics (20–50 microns in diameter, thinner than a human hair), which can float or sink to the seabed. Samples taken from selected sites in the Mediterranean Sea and the Atlantic and Indian Oceans have shown microplastics as deep as 3000 metres and in concentrations 1000 times higher than those found floating on the surface. Microplastics make up 85 per cent of plastic wastes along shorelines.

As well as microplastics we also have microbeads and microfibres. Microbeads are tiny solid plastic fragments, less than 5 mm in diameter, intentionally added to exfoliate and cleanse in rinse-off personal care products such as toothpaste, body wash and cleansers. The beads, as many as 100 000 per shower, flow straight down the bathroom drain and into the sewer. Wastewater treatment plants cannot filter such fine particles and so they ultimately end up out at sea. They are easily ingested by even the smallest sea creatures and are passed up through the food chain. Microbeads are not biodegradable and are impossible to remove from the marine environment.

Synthetic textiles pose a similar problem. When clothes made of microfibres such as lycra, acrylic, polarfleece or nylon are washed, between 600 000 and 17.7 million microfibres per wash are shed and end up in wastewater.

6.3.3 What are ghost nets?

Ghost nets are lost or discarded fishing nets and fishing gear that can drift in the oceans for many years, trapping marine life and sea birds. It is estimated that one tonne of ghost fishing gear is accidently lost or deliberately dumped into the ocean every minute. Nets may be lost because of bad weather, sea conditions, vandalism, or as a result of irresponsible or illegal fishing operations. Discarded fishing gear alone makes up approximately 46 per cent of the plastics found in the Great Pacific Garbage Patch (see section 6.3.4) and is responsible for a large death rate among marine creatures.

6.3.4 Where do we find the most marine debris?

Marine debris can be found in all oceans and along coastlines, from heavily populated regions such as the Caribbean Sea to the isolated Pitcairn Islands territory in the South Pacific. Here on the tiny uninhabited Henderson Island, more than 3000 kilometres from any major population centre, plastic debris has accumulated at a rate of 671 items per square kilometre of beach.

In 2018, the five countries that produced the most marine plastic were located in Asia (see **FIGURE 4**). Research has also found that 90 per cent of all marine plastic pollution originates from just ten river systems, eight of which are located in Asia (see **FIGURE 5**). These rivers all support high populations in their catchment areas and do not necessarily have the infrastructure to collect, recycle and dispose of plastic waste before it enters the sea.

FIGURE 4 Quantity of plastic released into the ocean for selected countries

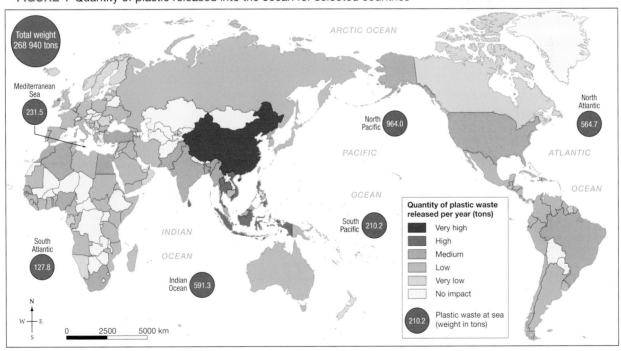

Source: Ministry of Environment & World Economic Forum; Statista.

FIGURE 5 River systems producing the most plastic marine litter

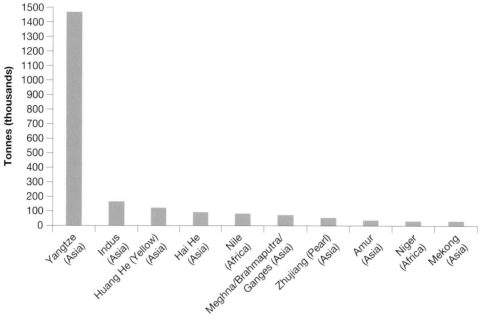

The Great Pacific Ocean Garbage Patch

A swirling mass of plastic waste, microplastics and other rubbish has been growing in the middle of the North Pacific Ocean, thousands of kilometres from the nearest coastline. Why is it there and how did it get there? Discarded waste from the east coast of Asia and west coast of the United States gets swept up in the North Pacific gyre. Within the marine environment, the slow-moving currents and winds push material into the calmer centre of the gyre, where it accumulates. It can take a year for material to reach the centre of the gyre from Japan and five years from the United States. The accumulation of debris has earned this region the name the 'Great Pacific Ocean Garbage Patch' (see **FIGURE 6**). It is the largest of the five offshore plastic accumulation zones in the world's oceans, all corresponding with the major gyres. Very little garbage is visible on the surface; rather, it is a thick soupy mass of minute pieces of plastic, with an average depth of 10 metres. The extent of the patch in 2018 was around 1.6 million square kilometres. This is an area almost three times the size of France. Suspended within the patch is an estimated 1.8 trillion plastic pieces — around 250 pieces for every person on Earth!

FIGURE 6 Location of the Great Pacific Ocean Garbage Patch

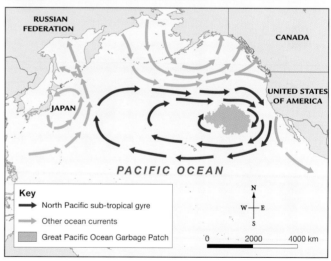

Source: The Ocean Cleanup.

6.3.5 Where are ghost nets a problem around Australia?

Marine debris occurs around all coastlines, including Australia. It is a major problem in northern Australia, particularly in the Gulf of Carpentaria. Here densities of nets can reach up to three tonnes per kilometre per year, among the highest densities in the world. The coastlines in this region are pristine environments and support six of the world's seven marine turtles. Turtles make up 96 per cent of marine creatures captured in the nets. Over 90 per cent of the debris that collects is derived from the fishing industry, most of it originating from South-East Asia, with the remaining 4 per cent coming from Australia. Most of the nets come from the Arafura Sea, an important fishing ground, especially for the Indonesian fishing industry. More than 62 per cent of the nets are trawling nets — the Arafura Sea being the only region of Indonesian waters where trawling is not banned. Industrial-scale fishing has doubled in this region since 2009. Under the influence of the south-east trade winds and north-west monsoon winds (see **FIGURE 7**), a circular gyre pattern develops, which allows the build-up of ghost nets to develop, similar to the Great Pacific Ocean Garbage Patch.

FIGURE 7 Distribution of ghost net hot spots around northern Australia

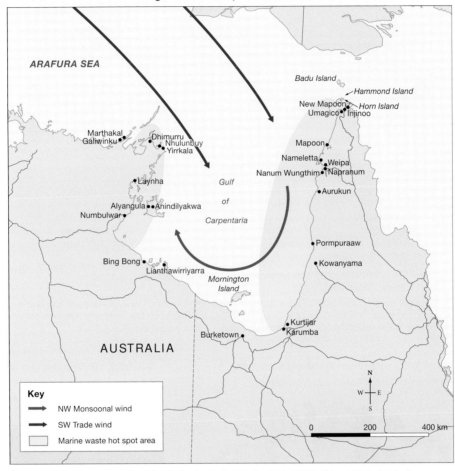

Source: © Commonwealth of Australia Geoscience Australia 2013. Ghost Nets Australia, www.ghostnets.com.au/index.html.

Further to Australia's north, the Danajon Banks in the Philippines is an extensive reef system habitat that supports a rich marine life. In recent years, overfishing has depleted fish stocks driving fishers to set more fishing nets to increase their chances of a haul. Discarded fishing nets accumulate on beaches, coral reefs and mangrove swamps or drift in large rafts, trapping marine life.

6.3.6 Impacts of marine debris

Environmental impacts

FIGURE 8 gives estimates of the length of time some marine debris takes to decompose. Most plastics undergo photodegradation, which happens more slowly in water than on land because of the cooler temperatures and reduced exposure to the sun. As the plastics break down into smaller particles, they 'thicken' the water and can release toxins. If the particles are less than 5 millimetres in diameter, sea creatures can consume them, in turn, these creatures are eaten by bigger creatures and so on up the food chain. Marine animals such as mussels, which filter seawater, take up the microplastics, which can release toxins into their tissues. Small floating pieces of debris are often mistaken for food and are scooped up by seabirds and fed to their chicks (see **FIGURE 9**).

Plastic pollution is believed to affect at least 800 marine species, with estimates of up to 100 000 marine mammal deaths, including whales, dolphins, porpoises, seals and sea lions, and one million seabird deaths each year. Deaths are mostly caused by eating plastic (see **FIGURE 10**), starvation, suffocation, infection or drowning due to becoming tangled.

FIGURE 8 Time periods for the decomposition of marine litter

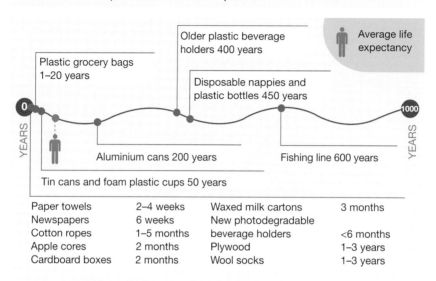

Paper towels	2–4 weeks	Waxed milk cartons	3 months
Newspapers	6 weeks	New photodegradable	
Cotton ropes	1–5 months	beverage holders	<6 months
Apple cores	2 months	Plywood	1–3 years
Cardboard boxes	2 months	Wool socks	1–3 years

Note: Estimated individual item timelines depend on product composition and environmental conditions.

Source: South Carolina Sea Grant Consortium, South Carolina Department of Health and Environmental Control (DFHC) — Ocean and Coastal Resource Management, Centers for Ocean Sciences Education Excellence (COSEE) — Southeast and NOAA 2008.

FIGURE 9 Foreign objects found in the stomach of a seabird. How many different items can you identify?

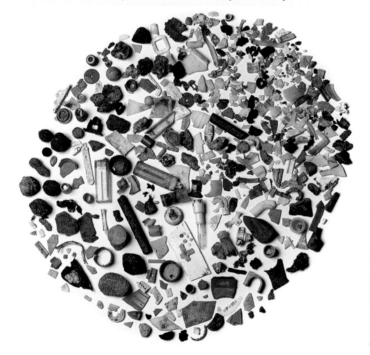

FIGURE 10 Mistaken for jellyfish floating in the ocean, discarded plastic bags are eaten by hundreds of marine species.

Twelve out of 25 of the most important species of fish for human consumption have been discovered to have consumed microplastics. These tend to lodge in the gut rather than being found in the muscle tissue that humans consume.

Small marine creatures, such as barnacles, which normally spend their lives attached to rock, coral or coconut shells, can 'hitch a ride' on marine debris. The arrival of pest species in new locations can seriously affect ecosystems as they compete with native species for food or habitat.

Impacts on people

While the fishing industry contributes to marine debris, the industry itself is also affected by the litter. A survey in northern Scotland found that 92 per cent of fishers had continual problems with marine debris in their nets, snagging nets on rubbish, and that they avoided some fishing grounds because of their high litter concentrations.

Because of the action of currents, garbage discarded in one country can end up on the beaches of another country thousands of kilometres away. Thus the impacts of marine litter on people are mostly found in coastal regions. Impacts include the rising cost of clearing debris from beaches, loss of tourism revenue, and debris interfering with boating and aquaculture.

DISCUSS

The most effective ways of reducing marine pollution have to start on land. What are our obligations and duties as global citizens to reduce waste and pollution?

[Ethical Capability]

 Resources

 Interactivity Garbage patch (int-3299)

6.3 EXERCISES

Geographical skills key: GS1 Remembering and understanding **GS2** Describing and explaining **GS3** Comparing and contrasting **GS4** Classifying, organising, constructing **GS5** Examining, analysing, interpreting **GS6** Evaluating, predicting, proposing

6.3 Exercise 1: Check your understanding

1. **GS1** What are the two biggest contributors to marine pollution across the world's ocean *space*?
2. **GS5** Refer to **FIGURE 1**.
   (a) Give an example of a pollutant from each of the following sources of marine pollution:
       (i) atmospheric-based, (ii) land-based, (iii) marine-based.
   (b) Which of the three sources makes up the largest component of marine pollution?
3. **GS1** Refer to **FIGURE 5**. Which river carries the greatest amount of plastic waste, and where is it located?
4. **GS5** Refer to **FIGURE 2**. How would these items compare to a survey of marine litter conducted 50 years ago? What do you think has *changed* the most?
5. **GS2** Refer to **FIGURE 4**. Describe the distribution of those countries that produce medium to high (pink–red) quantities of plastic marine waste.
6. **GS1** Why are ghost nets a problem in northern Australia?
7. **GS1** Why are fishing nets an *environmental* problem?
8. **GS2** Refer to **FIGURE 7**. On which side of the Gulf would you expect ghost nets to build up:
   (a) during the north-west monsoon season
   (b) during the south-east trade wind season?

▶

6.4 Cleaning up our mess

6.4.1 Think global ...

The way we consume and discard our resources has created one of the biggest environmental challenges in the world. Our throw-away society has literally thrown all our waste into the oceans! Doug Woodring, a co-founder of ocean clean-up charity Project Kaisei, says, 'The water in our oceans is like blood for our planet. If we continue to fill it with toxic materials such as plastic, it will be to the detriment of life on Earth'.

Marine debris might start as a local issue but it can contribute to a global problem as debris may travel great distances from its original source, crossing both geographic and political boundaries. We will only significantly reduce marine debris if we control its land-based sources. Communities and governments need to develop effective waste reduction schemes if we want to manage our oceans sustainably. It has been suggested that if no action is taken, pieces of plastic in the ocean will outnumber fish by 2050! Every year, people around the world use 500 billion disposable plastic bags — around 65 for every person on Earth. On average, a person uses a plastic bag for just 12 minutes, but it may then take many, many years for it to decompose (if ever). The scale of the waste issue is huge.

FIGURE 1 What message is this advertisement sending?

WHAT GOES IN THE OCEAN GOES IN YOU.

Can't we just scoop it up?

Scooping up marine debris is not as easy as it sounds. Firstly, debris such as that found in the Great Pacific Ocean Garbage Patch is constantly moving in response to shifts in winds and currents. Secondly, much of the garbage is in the form of minute particles that become microplastics, suspended beneath the ocean's surface. One organisation, *Ocean Cleanup* have developed a device that can be towed by ships into the Patch. The device floats in a large U-shape with an impermeable screen draped beneath large floating arms. The debris can be caught up in the screen, while marine life can swim beneath. Winds and waves propel the device forward at a faster rate than the current-driven debris, allowing for it to be collected. The debris can then be brought back to land for processing, recycling or resale. The project is in its early days, with one ocean trial having been completed. Lessons learned from the trial will be incorporated in the system's design, with hopes for improved performance to reach their aim of reducing the amount of larger plastic items in the garbage patch by 50 per cent within five years.

What can be done at the national and international scale?

In 2017, 173 countries signed a United Nations resolution to eliminate plastic pollution in the sea, although this has yet to become a legally binding treaty. As seen in **FIGURE 2**, at least 32 countries have already placed bans on single-use plastic bags. When Ireland first introduced a plastic bag levy back in 2002, bag usage dropped by 90 per cent. Kenya has gone to extremes, with fines of US$40 000 for anyone producing, selling, or even just carrying a plastic bag. This has been successful in terms of creating a cleaner environment, but manufacturers and retailers are struggling to adapt and find affordable and environmentally suitable alternatives.

FIGURE 2 Plastic bag bans around the world

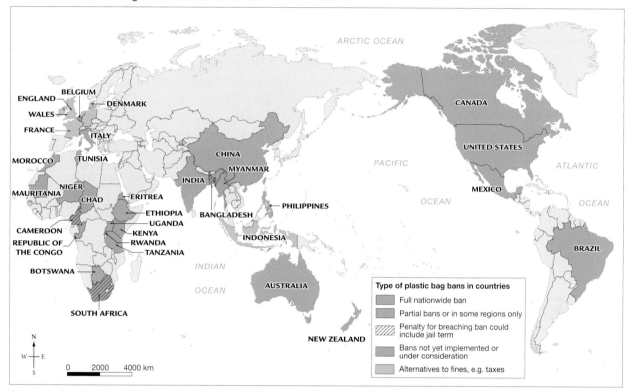

Source: UNEP; Greenpeace, national governments.

Several countries, such as Britain, Canada, the United States and the Netherlands have now banned the use of microbeads in many products, particularly body scrubs and toothpaste. Other products such as sunscreens and lipsticks have not been included in the ban.

In 2017, the UN Environment launched the Clean Seas campaign, with the aim of eliminating the major sources of marine litter by 2022. The focus is single-use plastics and microplastics. The campaign urges governments to introduce policies to reduce plastic waste, encourages companies to reduce plastic packaging and asks citizens to change their consumption and waste practices.

International agreements such as he International Convention for the Prevention of Pollution from Ships (known as MARPOL) forbid ships to dispose of plastic into the sea, and they are not allowed to dispose of food waste within 12 nautical miles of land. Although such regulations are an important step in tackling the issue of marine pollution, they are extremely difficult to police and have no impact on the amount of waste entering the ocean from land-based sources. It is worth noting that in 2019, the world's largest cruise line company, Carnival, was fined A$28 million for knowingly discharging plastic and food waste in the Bahamas and grey water in Alaska's Glacier Bay National Park. The company may be further penalised by being banned from entering ports in the United States.

What can communities and organisations do?

Numerous organisations are dedicated to improving our marine environment. These include:

- *Clean Up the World*. Established in 1993, this group works with communities to foster environmental stewardship, through activities such as urban recycling projects, waste recovery, tree planting and habitat restoration projects, education programs, rubbish removal and clean up, and care of marine environment events.
- *Friends of the Earth (FOTE).* As part of their efforts to promote a better and healthier environment by 2030, FOTE promote a 'Plastic Free Friday', where followers join efforts to avoid the use of plastic products for one whole day each week.
- *the Flipflopi expedition.* The founders of this organisation set out to construct a sailing boat ('The FlipFlopi') from ten tonnes of collected marine debris (including many thousands of rubber thongs!). They launched the boat in September 2018, with a 500-kilometre trip along the East African coast from Kenya to Tanzania, aiming to show the potential of reused plastic and to raise awareness of plastic marine pollution in a unique way.
- *Greenpeace.* A global organisation that promotes environmental issues, one of Greenpeace's largest campaigns against plastics is a global petition to encourage government action against single-use plastic.
- *International Coastal Cleanup.* Since the late 1980s, volunteers from around the world have participated in *Cleanup* events – cleaning up and collecting data on debris found along coastlines, lake shores and river banks, to help prevent litter from entering the oceans and to build awareness of the problem of marine pollution. Their 2018 report highlighted the results of global cleanup events held in 2017: across more than 100 countries, 9 285 600 kilograms of waste (a total of 20 824 689 individual items) were collected. The data collected over the years has contributed to new littering laws (see **FIGURE 4** for an example).

FIGURE 3 Thousands of volunteers all around the world participate in Clean Up days in an effort to reduce marine pollution.

FIGURE 4 Marine pollution restrictions in the United States

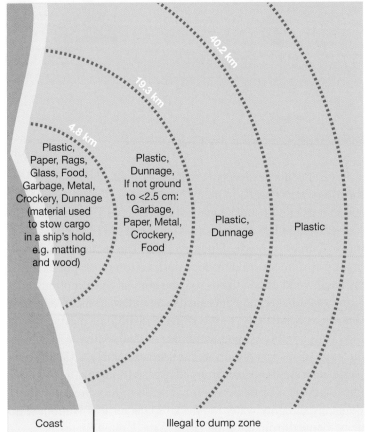

4.8 km

19.3 km

40.2 km

Plastic, Paper, Rags, Glass, Food, Garbage, Metal, Crockery, Dunnage (material used to stow cargo in a ship's hold, e.g. matting and wood)

Plastic, Dunnage, If not ground to <2.5 cm: Garbage, Paper, Metal, Crockery, Food

Plastic, Dunnage

Plastic

Coast

Illegal to dump zone

What can manufacturers do?

In recent years, manufacturers have become much more environmentally aware, with new biodegradable packaging materials and improved recycling methods developed. Several of the large beauty and personal toiletry companies have voluntarily removed plastic microbeads from their products, replacing them with natural, biodegradable 'beads' made from products such as jojoba, coffee, salt and oats.

Examples of individual progress include:

- McDonald's and Swiss food manufacturer Nestlé have pledged to make all their plastic packaging 100 per cent recyclable or reusable by 2025.
- Furniture maker IKEA has pledged to phase out single-use plastic products from its stores and restaurants by 2020.
- Car manufacturer Volvo has pledged that, from 2025, one-quarter of the plastic used in new cars will come from recycled plastic sources.
- Coca-Cola has created a 'World Without Waste' campaign to address its own contribution to much of the increase in global litter. Coca-Cola has stated that, by 2030, it will recycle a used bottle or can for every new one sold. It has also pledged to increase the amount of recycled content in plastic bottles from 13 to 50 per cent by 2030.

FIGURE 5 By 2025 all McDonald's packaging will be recyclable or reusable.

6.4.2 ... act local!

What can fishers do?

Numerous schemes are being trialed around the world to combat the problem of fishing debris. These include:

- a 'Fishing for Litter' scheme set up in Scotland where fishers and port authorities have collaborated to collect all litter caught in nets. Instead of throwing this litter overboard, the debris is collected and brought back to port for managing. Between 2005 and 2015, Scottish fishers collected more than 900 tonnes of marine litter.
- recreational fishers in the United States can recycle fishing lines back to the manufacturer via collection points. Since the scheme started in 1990, it has prevented more than 15 million kilometres of fishing line potentially entangling wildlife.
- on beaches around Port Phillip Bay in Victoria, labelled containers are available for fishers to deposit used fishing lines and other fishing paraphernalia.

What can *you* do?

The Surfrider Foundation in Australia and the United States is responsible for the 'Rise Above Plastics' campaign. The aim of the campaign is to get people to think about how they can make a difference and prevent marine debris. They suggest ten ways to reduce your personal plastic footprint. There are also many innovative ways to recycle plastic products that can be found on YouTube, such as converting plastic bags to rope or handbags.

Ten ways to reduce your personal plastic footprint

1. Choose to reuse when it comes to shopping bags and bottled water. Use cloth bags and metal or glass reusable bottles if possible.
2. Refuse single-serving packaging, excess packaging, straws and other 'disposable' plastics. Carry reusable utensils in your bag, backpack or car.
3. Reduce everyday plastics such as sandwich bags and juice cartons by replacing them with a reusable lunch bag or box that includes a thermos.
4. Bring a reusable cup with you to the café, restaurant or juice bar. This is a great way to reduce the use of lids, plastic cups and/or plastic-lined cups.
5. Go digital! No need for plastic CDs, DVDs and jewel cases when you can buy your music and videos online.
6. Seek alternatives to the plastic items you use.
7. Recycle. If you must use plastic, try to choose #1 (PETE) or #2 (HDPE), which are the most commonly recycled plastics. Avoid plastic bags and polystyrene foam as both typically have very low recycling rates.
8. Volunteer at a beach clean-up. Surfrider Foundation Chapters hold clean-ups monthly or more frequently.
9. Support plastic bag bans, polystyrene foam bans and bottle recycling bills.
10. Spread the word. Talk to your family and friends about why it is important to 'rise above plastics'!

6.4.3 Cleaning up the ghost nets

In the Philippines, 'Networks', an initiative supported by the London Zoological Society, is working with local fishers to harvest discarded nets. Former fishers dive for nets, and then on shore they are compressed into tight cubes and shipped to Slovenia, in Europe, where the nets are turned into nylon yarn and then shipped to the United States to be woven into carpets. Instead of being dumped, the nets have value and can be sold. Jobs are created and an environmental threat is averted.

In Australia, the ghost nets issue is being tackled by GhostNets Australia, an alliance of over 22 Indigenous communities in remote coastal places of Western Australia, Queensland and the Northern Territory, funded by the federal government. Since it was established in 2004, over 13 000 ghost nets have been captured by locally trained rangers (see **FIGURE 6**).

Often, helicopters are used to spot the ghost nets washed ashore, which are then checked for trapped wildlife. Live turtles are tagged and data recorded before they are returned to the sea. Nets are dragged up above the high tide line to be identified, collected and disposed of later. The project works on a '6R' principle:

FIGURE 6 Captured trawler nets being collected by rangers

1. **R**emove ghost nets from waters and coastline of the Gulf of Carpentaria.
2. **R**ecord the number, size, type and location of nets.
3. **R**escue animals trapped in nets.
4. **R**eport the activities that the community has done to increase awareness.
5. **R**educe the number of nets in the Gulf by working together.
6. **R**esearch factors that influence the distribution, movement and impact of ghost nets.

This program is part of a Caring for our Country initiative in the region, which promotes stewardship of Indigenous customary lands and seas.

What can be done with the debris?

Traditionally, fishing nets were made of more eco-friendly materials, such as flax or hemp, but now they are usually made of nylon, which makes them stronger, cheaper and more buoyant. However, this means they take a very long time to break down. Nets can also range in size from 30 cm to 6 km long! There are three options for disposing of the waste: burning, placing in landfill, or recycling. Each, however, has disadvantages, and all methods require the difficult task of collecting the waste.

Disadvantages of burning fishing nets include:

- burning plastic is illegal in most countries
- after burning, the residue is a huge, heavy, immovable mass of melted plastic, which is a visual eyesore
- health risks associated with burning plastic.

Disadvantages of disposing of fishing nets in landfill include:

- expense of transporting the waste over large distances to a landfill site
- often waste is burned in tips, and these tips are close to settlements.

Disadvantages of recycling or reusing fishing nets include:

- remoteness of and distances to recycling plants (South Australia and Taiwan have plants big enough to cope with fishing nets)
- expense of transporting the waste over large distances
- the need for large machinery to chop plastic into manageable pieces
- the need to find a local use for the recycled waste material.

GhostNets Australia's solution

While only a partial solution to the large quantity of nets accumulating, GhostNets Australia promotes the reuse of nets by providing local artists with netting material. The artists use traditional weaving techniques to create artworks (see **FIGURE 7**). This type of cottage industry brings economic and social benefits as well as raising awareness of the problem of marine debris.

FIGURE 7 Woven basket made of recycled fishing nets

6.4 INQUIRY ACTIVITIES

1. If you have access to a beach, walk along the high tide line and see if you can collect and identify different forms of marine litter. Collate and record your findings. What were the most common forms of litter you identified? Where have they come from? **Examining, analysing, interpreting**

2. How can you reduce your school's plastic footprint? Brainstorm ideas as a class and then develop one idea in detail. How can you promote your idea? You may like to create a slogan and poster, address a school group or assembly or write a proposal to the school administration. **Evaluating, predicting, proposing**

3. Undertake a plastic bottle survey at home. Check your kitchen, laundry and bathroom and count the number of plastic bottles, jars and other containers you find (only count containers). Collate your results, in graph form, with other students in your class and then write a summary of your findings. **Examining, analysing, interpreting**

4. Research information on the different types of fishing nets used: gill, purse, seine and trawl nets.
   (a) Construct a table to list the advantages and disadvantages of each from a fishing and an *environmental* perspective.
   (b) Which net design might prove to be the most damaging to the *environment* if lost or discarded? **Examining, analysing, interpreting**

6.4 EXERCISES

Geographical skills key: GS1 Remembering and understanding **GS2** Describing and explaining **GS3** Comparing and contrasting **GS4** Classifying, organising, constructing **GS5** Examining, analysing, interpreting **GS6** Evaluating, predicting, proposing

6.4 Exercise 1: Check your understanding

1. **GS1** What items is it illegal to dump within a 4.8 km zone of the coast in the United States?
2. **GS1** Identify five ways that you can reduce your plastic footprint.
3. **GS1** What are two important initiatives that have been undertaken at the national or international level to reduce marine pollution?
4. **GS2** Why is an understanding of local wind patterns useful to GhostNets rangers?
5. **GS2** Why is transporting ghost nets to South Australia for recycling not a viable option?
6. **GS2** Evaluate the *environmental*, economic and social aspects of the GhostNets program.

6.4 Exercise 2: Apply your understanding

1. **GS6** Is the saying, 'Think global, act local' applicable to marine pollution? Justify your answer.
2. **GS6** How successful would an international agreement where all countries decide to reduce land-based marine pollution be? What would be the advantages and disadvantages?
3. **GS4** Construct a table, similar to the one below, to evaluate each of the proposals to help reduce ocean debris.

Response	Economic criteria (e.g. cost)	Social criteria (e.g. time and effort required)	Environmental criteria (e.g. effectiveness)
Individual actions			
Manufacturers			
International community			

4. **GS5** What conclusions can you draw from the table you created in question 3?
5. **GS6** Suggest ways in which you could encourage others to reduce their plastic footprint.

Try these questions in learnON for instant, corrective feedback. Go to www.jacplus.com.au.

6.5 SkillBuilder: Using geographic information systems (GIS)

What is GIS?

GIS is a computer-based system of layers of geographic data. Just as an overlay map allows you to interchange layers of information, GIS allows you to turn layers on and off to make comparisons between pieces of data.

Select your learnON format to access:

- an overview of the skill and its application in Geography (Tell me)
- a video and a step-by-step process to explain the skill (Show me)
- an activity and interactivity for you to practise the skill (Let me do it)
- questions to consolidate your understanding of the skill.

on Resources

Video eLesson Using geographic information systems (GIS) (eles-1752)

Interactivity Using geographic information systems (GIS) (int-3370)

6.6 Oil and water — a toxic mix

6.6.1 Sources of marine pollution

You have probably seen images of birds covered in sticky oil, usually as a result of the most dramatic type of marine pollution: oil spills and shipping accidents. The impact of oil on ocean and coastal ecosystems is often localised over a relatively small area, but may last for many years.

Almost all of the Earth's supply of oil and natural gas is found in deep underground reservoirs. Reservoirs can be under a landmass, under the seabed and under continental shelves. Extracting oil from the seabed accounts for nearly 30 per cent of the world's production. Offshore drilling takes place on huge floating platforms, in waters up to 2 kilometres deep and as far as 300 kilometres from the coast. More than 50 per cent of countries around the world drill for offshore oil and gas.

The most obvious and visible kinds of marine oil spills usually involve tanker accidents, or leaks from offshore oil rigs. However, oils enter the ocean from a variety of sources, with both natural and land-based sources accounting for a much larger proportion than disasters (see **FIGURE 1**). There has been a decrease in the number of tanker accidents in recent years, mostly due to improved ship design and greater safety methods. However, with more ships and supertankers being built, the potential risk of an accident is still high.

FIGURE 1 Main sources of marine oil pollution

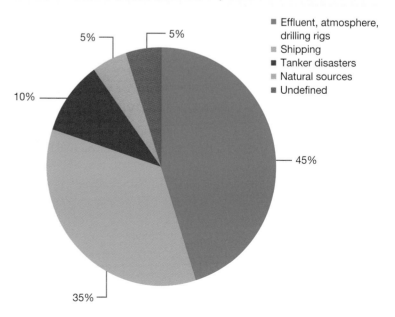

- Effluent, atmosphere, drilling rigs
- Shipping
- Tanker disasters
- Natural sources
- Undefined

5%
5%
10%
45%
35%

6.6.2 What happens to oil in the ocean?

Each oil spill is different and there are various physical, chemical and biological factors that will influence the behaviour of spilt oil. The type of oil, temperature of the water, wave and current action, and the nutrient content of the water are all critical influences. The stages in the breakdown of oil can be seen in **FIGURE 2**.

FIGURE 2 What happens to oil in the ocean?

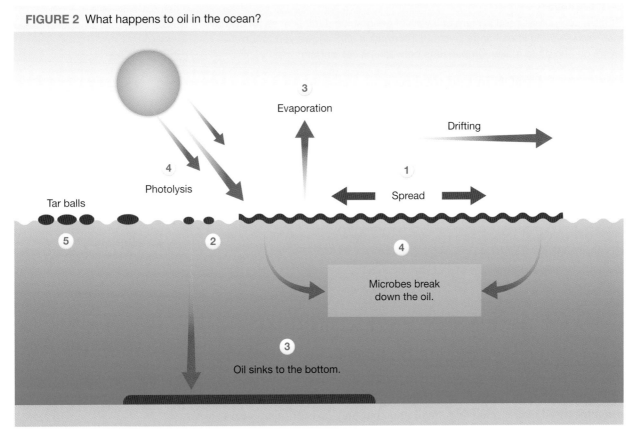

1. When oil is released into the ocean it immediately forms large slicks that float on the surface. It can take only 10 minutes for one ton of oil to disperse over a radius of 50 m and be 10 mm thick.
2. After a few hours, weathering by wind and waves breaks down the slick into narrow bands, or windrows, that float parallel to the wind. The oil may be less than 1 mm thick but can now cover 12 km^2. After the slick thins down it breaks up into fragments and fine droplets that can be transported over larger distances.
3. Some of the oil evaporates or sinks.
4. Some of the oil can be chemically broken down by sunlight or bacteria (photolysis).
5. Finally the oil solidifies into tar balls (clumps), which are more resistant to bacterial decomposition.

6.6.3 Impact of oil on the environment

Oil spills can result in both short- and long-term environmental change, with some damage lasting for decades. A spill in open waters is usually less destructive than a spill near coastal waters, where most fish and bird breeding takes place. Oil pollution is less visible in the open ocean, especially once it disappears from the surface, but it is still capable of being moved via ocean currents.

Coastlines

The geography of the coastline can influence the degree of impacts from an oil spill. Impacts are less on exposed coasts due to strong wave action. A long, sheltered, sandy coastline is vulnerable as the oil can soak into the sand, which is extremely difficult to clean. Mangroves, salt marshes and extensive sandbanks are also sensitive as the oil soaks into the fine sediments and can be taken up by plants. This affects wildlife that live in this habitat, and the loss of vegetation increases the risk of coastal erosion, as shown in **FIGURE 3(a)** and **(b)**. Coral reefs are possibly the most vulnerable to oil spills, and they are extremely slow to recover.

FIGURE 3 (a) Oil damage to wetland habitat (b) The same area of wetlands one year after the oil spill

Wildlife

Any oil on the surface of the sea will kill birds that swim and dive for their food there. Feathers covered in oil rob birds of waterproofing and insulation. Ingesting the oil can poison them. Oil spills also damage coastal nesting and breeding grounds. Oil can block the blow holes of marine mammals such as whales, dolphins and seals, making breathing difficult. If oil coats their fur, they become vulnerable to hypothermia. Animals' food supply is also poisoned by floating oil. Fish, especially shellfish, suffer immediate effects of an oil accident. Reduced reproduction, birth defects and other abnormalities develop in the next generation of wildlife exposed to oil spills, creating a longer-term impact.

6.6.4 Cleaning up oil spills

Prevention

The most important way to deal with oil spills is to prevent them from happening in the first place. The UN MARPOL treaty was established in 1983 to deal with the growing problem of marine pollution. Individual countries have also established new rules and regulations. For example, since 2015 all tankers operating in United States waters must be double hulled, so that if the outer hull is damaged, the inner hull can still hold the fuel. The oil industry must also now have detailed response plans for cleaning up any spills.

Remediation

When a spill does occur, the clean-up will generally involve a combination of approaches. Methods will depend on factors such as location, weather and the type of spill. It is extremely difficult to contain and clean up any oil spill, and many early methods often caused more environmental damage than the oil itself. For instance, consider the impact of a high-pressure hose on a fragile ecosystem, or the spraying of toxic chemicals to absorb the oil. A range of remediation methods are used to clean up oil spills (see **FIGURE 4**).

FIGURE 4 Different methods of cleaning up oil spills

1. Boats with **booms** attached skim oil off the water's surface.
2. Oil collected by booms is then burned off the surface of the water.
3. **Bioremediation** techniques use microorganisms and fertilisers to break the oil down into less harmful compounds.
4. Boats and planes spray chemical dispersants, similar to detergents, on the oil to break it down into droplets.
5. Manual/mechanical methods: People with rakes and spades as well as heavy equipment physically remove oil from along beaches.
6. Natural processes: Often the impact of cleaning up is greater than the oil damage itself in fragile environments. Over time, naturally occurring microorganisms, sunlight and wave action will slowly break the oil down.
7. Absorbent material such as hay, wood shavings and even human hair (collected from hairdressers and stuffed into nylon casings) can be used to help mop up oil (see **FIGURE 5**).

6.6.5 CASE STUDY: The Gulf of Mexico oil spill

The Gulf of Mexico is rich in natural resources and, in particular, oil. More than 4000 active offshore oil well platforms are distributed along the northern region, attached to thousands of pipelines delivering oil and gas to the mainland. Ninety per cent of America's offshore drilling takes place here. In 2010, it was the site of the world's second biggest oil spill.

On 20 April 2010, an explosion on the Deepwater Horizon drilling platform, located 74 kilometres off the Gulf coast, caused the rig to burn and sink, killing 11 workers (see **FIGURE 6**).

As a result, oil began leaking into the Gulf from ruptures in the drilling pipe, more than 2000 metres below the surface, creating the largest spill in American history. Over the course of 87 days, an estimated 4.9 billion barrels of oil were released into the sea (see **FIGURE 7**). Over 35 per cent, or 2650 kilometres, of the Gulf coast was fouled.

FIGURE 5 Recycled human hair turned into oil-absorbent logs for soaking up oil spills

FIGURE 6 The Deepwater Horizon oil rig on fire

FIGURE 7 Mapping the extent of the Gulf oil spill (a) 4 May, (b) 28 May and (c) 21 July

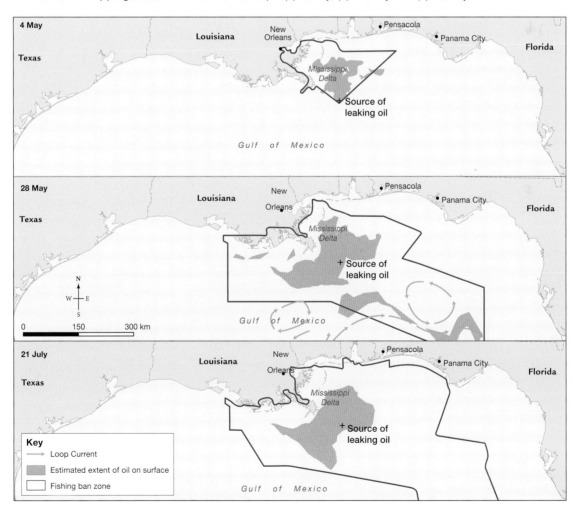

Source: Conservation Biology Institute, Publication Date: 7/5/2010 Map by Spatial Vision.

The role of ocean currents

There was a risk of the oil being caught up in the Gulf Loop Current, which then had the potential to feed oil into the Gulf Stream Current, a powerful ocean conveyor belt that carries warm water north along the east coast of America, moving at a rate of approximately 80–160 km/hour. Had this happened, the extensive wetlands of the Florida Keys and important tourist beaches would have been badly affected. The Loop Current is constantly shifting, and fortunately, during the period of the spill, it actually shifted out of the danger area.

What was the damage?

Scientists were able to track and map the oil spill on a daily basis. The distribution of oil changed regularly, largely due to weather and tide conditions and the efforts of clean-up teams. The fact that the oil was leaking in deep water certainly reduced some of the impact, as did the relatively calm weather during this time. For the first month, the oil stayed mostly at sea, but by June, the oil had reached the Louisiana wetlands to the north-west and the Florida coastline to the north-east. Early records of dead or injured turtles and dolphins can be seen in **FIGURE 8** and **TABLE 1**.

FIGURE 8 The effect of the Gulf oil spill on the marine ecosystem, 17 May to 15 August 2010

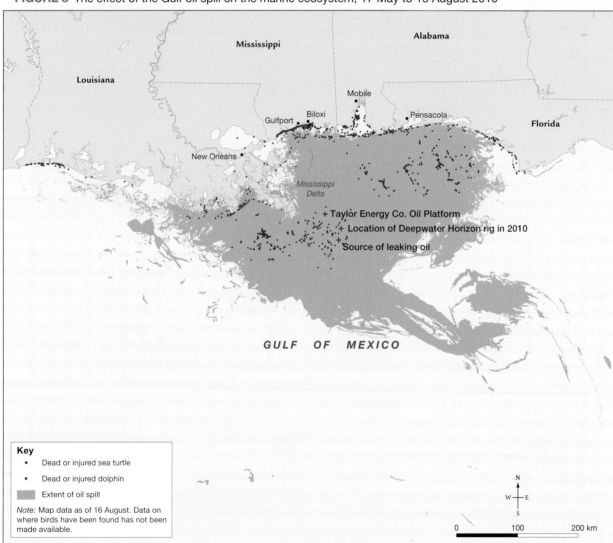

Source: Conservation Biology Institute, Publication Date: 7/5/2010. Map by Spatial Vision.

TABLE 1 The effect of the Gulf oil spill on sea turtles, dolphins and birds as of 2 November 2010

	Found alive	Found dead	Total found
Sea turtles			
Visibly oiled	456	18	474
Not visibly oiled	79	319	398
Dolphins			
Visibly oiled	2	4	6
Not visibly oiled	7	92	99
Birds			
Visibly oiled	2079	2263	4342
Not visibly oiled	0	3827	3827

Source: © National Oceanic and Atmospheric Administration, www.noaa.gov.

After the accident, the number of dolphin deaths rose from an average of 63 deaths per year to a high of 335 in 2011. Deaths then averaged 200 per year until 2015 and have continued to decline. The long-term impacts on dolphin populations is still unknown.

Animal death counts would have been higher if not for the efforts of rescue teams who collected 25 000 turtle eggs and relocated them to Florida's Atlantic coast. In the same time period, inland flooding of farmland provided alternative wetlands for migratory birds that normally would have inhabited the coastal wetlands. Most of the 1600 kilometres of coastline have now been cleaned up.

Impacts on people

To minimise the risk to humans of eating contaminated seafood, more than 150 000 km^2 of federal waters were closed to fishing. The Gulf fishing industry was estimated to have lost $247 million from the fishing ground closures. The fishing industry is open again and is carefully monitored. Research in 2018 noted that oil contamination in fish has been declining over time but the diversity of fish species was lowest in those areas of the gulf with the greatest number of oil rigs.

The initial loss to the tourism industry in 2013 was US$22.7 million, but the industry quickly picked up and in 2017 the coastal areas of Alabama were earning US$4.4 billion in tourist dollars. Since the spill, BP has paid out more than $20 billion in damages claims to state governments, individuals and businesses, in addition to providing funding for restoration projects.

How was the Gulf cleaned up?

FIGURE 9 shows the results of the clean-up following the Gulf oil spill, 103 days after the accident. Favourable weather conditions at the time enabled authorities to put some defensive measures in place, including more than 4000 kilometres of booms, to protect coastal land.

In all, an estimated 6.4 million litres of dispersants were used on the spill. Scientists believe that nearly 50 per cent of the oil spilled and nearly 100 per cent of the methane gas released has stayed deep in the ocean. As much as 3200 square kilometres of ocean floor is thought to be polluted.

A disaster unfolding

Another potential disaster is unfolding in the Gulf of Mexico. In 2004, Hurricane Ivan swept through the region and created an underwater avalanche that buckled and sank an oil rig owned by Taylor Energy, located 19 kilometres off the coast of Louisiana (seen in **FIGURE 8.**) A mixture of steel and leaking oil was buried in 45 metres of mud on the sea bed. Since then it has been leaking up to 700 barrels of oil per day, a rate that will surpass the volume lost in the Deepwater Horizon spill, also making it the longest oil spill in US history. To date there has not been any evaluation of the environmental consequences. More than US$200 million has been spent on the clean-up but only one-third of the wells have been capped or plugged. The task is near impossible with many of the leaking wells buried in thick mud. Drilling runs the risk of striking a pipe or well, creating a catastrophe. The only other alternative is to build some sort of structure to contain the oil on the surface.

FIGURE 9 The clean-up of the Gulf of Mexico oil spill

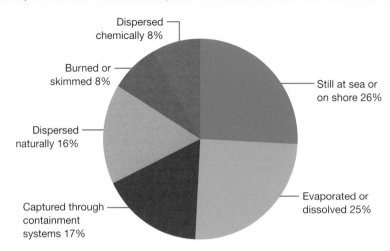

An ongoing threat

Because of the high number and distribution of oil wells in the Gulf of Mexico (see **FIGURE 10**), the threat of future accidents remains. In fact, there are more than 4000 oil platforms and nearly 80 000 kilometres of active and inactive pipelines carrying oil across the sea bed to the shore. The frequency of storms and hurricanes in the Gulf region is high; in the four years from 2004 to 2008, 150 oil rigs were battered or destroyed by extreme weather events. Today, the energy companies employ more sophisticated weather forecasting techniques and implement safety programs, closures and evacuation of personnel well in advance of approaching hurricanes.

FIGURE 10 Distribution of oil wells in the northern Gulf of Mexico

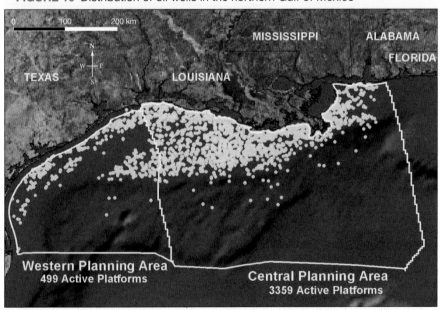

Source: National Oceanic and Atmospheric Administration, Office of Ocean Exploration and Research, U.S. Department of Commerce. Adapted by Spatial Vision.

In Florida, it has been estimated that the annual value of tourism and fishing along the state's eastern Gulf coast is three times higher, and considerably more sustainable, than the value of any oil or gas that might be found there.

However, oil continues to be a very important fuel and raw material for many products ranging from asphalt to chemicals, synthetic materials and plastics. The Gulf of Mexico supplies the United States with 17 per cent of its total US crude oil and 5 per cent of its gas. Consequently, oil resources will continue to be explored and developed. Very careful planning and management will be needed to prevent the environmental changes that an oil rig disaster can bring.

DISCUSS

'Oil is an essential part of our modern life. It is worth the risks to the environment to ensure ongoing supply of this important resource.' Discuss your views on this statement in small groups.

[Critical and Creative Thinking Capability]

Resources

 eWorkbook   Oil spill (doc-31789)

Interactivity   Oil slick (int-3300)

 Google Earth   Gulf of Mexico

6.6 INQUIRY ACTIVITIES

1. Complete the **Oil spill** worksheet to investigate this topic further.   **Examining, analysing, interpreting**
2. Research the potential impact of oil spills on the Great Barrier Reef. How does the Great Barrier Reef Marine Park Authority manage the park *sustainably* to prevent oil spills?   **Examining, analysing, interpreting**
3. One of the biggest *environmental* disasters from a marine oil spill was the spill by the *Exxon Valdez* ship off the Alaskan coast in 1989. Research the disaster and then create a newspaper front page account of the accident. Include a location map, cause of the accident, examples of the impacts and methods used to clean it up. Annotated photographs could be used for illustration. What are the similarities and differences between the *Exxon Valdez* and Gulf oil spills?   **Examining, analysing, interpreting**

6.6 EXERCISES

Geographical skills key: GS1 Remembering and understanding **GS2** Describing and explaining **GS3** Comparing and contrasting **GS4** Classifying, organising, constructing **GS5** Examining, analysing, interpreting **GS6** Evaluating, predicting, proposing

6.6 Exercise 1: Check your understanding

1. **GS1** What percentage of the world's oil comes from the seabed?
2. **GS2** Examine **FIGURE 2**. Why is it important that oil spills are treated as quickly as possible?
3. **GS2** Examine **FIGURES 3(a)** and **(b)** and describe the *changes* that you can see in the two *environments*.
4. **GS6** Suggest one *environmental*, one economic and one technological factor that can contribute to marine oil pollution.
5. **GS1** List the ways in which oil creates *environmental change* in the ocean.
6. **GS3** Compare some of the advantages and disadvantages of drilling for oil in the ocean compared to drilling for oil on land.
7. **GS1** List three things that helped reduce the *environmental changes* of the Gulf of Mexico oil spill.
8. **GS1** What weather conditions could have worsened the Gulf disaster?
9. **GS2** Why is there is no single solution to cleaning up oil spills?

▶

6.6 Exercise 2: Apply your understanding

1. **GS4** Examine **FIGURE 2**. Select the conditions from those listed below that would be most likely to encourage the rapid breakdown of an oil spill. Justify each of your choices.
   - Cold ocean water/warm ocean water
   - Calm seas/choppy seas
   - Ready supply of bacteria/limited supply of bacteria
   - High level of oxygen in the water/low level of oxygen in the water
   - High number of bacteria-eating organisms/low number of bacteria-eating organisms
2. **GS3** Construct a table to suggest the possible advantages and disadvantages of the seven methods of remediation shown in **FIGURE 4**. Consider the influence of the following factors: weather conditions, timing, location of treatment area (at sea or on coast), size of area to be treated, *environmental* impacts, practicality, economic viability and social justice.
3. **GS2** Study the maps in **FIGURE 7**. Why was there a need for a fishing ban in the region?
4. **GS2** Study **FIGURE 8**. What was the furthest distance from the oil source that dead or injured marine creatures were found?
5. **GS5** Study **FIGURE 9**, which shows data about the status of the Gulf oil spill clean-up, 103 days after the incident.
   (a) What percentage of the oil spill had been treated at this time?
   (b) What percentage of the oil had dispersed naturally or evaporated?
   (c) Why do you think only a small percentage was chemically dispersed?
6. **GS2** Imagine that you spill a whole bottle of cooking oil on your kitchen floor. Describe three different remediation methods that you could use to clean up the spill. Select from booms, skimmers, bioremediation, manual/mechanical, dispersants and absorbers. Would one method be more effective than another? Give reasons.

Try these questions in learnON for instant, corrective feedback. Go to www.jacplus.com.au.

6.7 SkillBuilder: Describing change over time

What is a description of change over time?
A description of change over time is a verbal or written description of how far a feature moves, or how much it alters, over an extended period.

Select your learnON format to access:
- an overview of the skill and its application in Geography (Tell me)
- a video and a step-by-step process to explain the skill (Show me)
- an activity and interactivity for you to practise the skill (Let me do it)
- questions to consolidate your understanding of the skill.

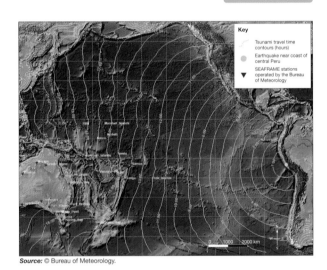

Source: © Bureau of Meteorology.

Resources

Video eLesson Describing change over time (eles-1753)

Interactivity Describing change over time (int-3371)

6.8 Thinking Big research project: 'Plastic not-so-fantastic' media campaign

SCENARIO

As a film producer, you have been asked to produce a 2–3-minute commercial to raise public awareness on the issue of plastic waste in our oceans. The commercial will be shown on prime-time television and across social media platforms.

Select your learnON format to access:

- the full project scenario
- details of the project task
- resources to guide your project work
- and assessment rubric.

 Resources

 ProjectsPLUS Thinking Big research project: 'Plastic not-so-fantastic' media campaign (pro-0215)

6.9 Review

6.9.1 Key knowledge summary

Use this dot-point summary to review the content covered in this topic.

6.9.2 Reflection

Reflect on your learning using the activities and resources provided.

 Resources

 eWorkbook Reflection (doc-31771)

Crossword (doc-31772)

 Interactivity Marine environments crossword (int-7673)

KEY TERMS

bioremediation the use of biological agents, such as bacteria, to remove or neutralise pollutants

booms floating devices to trap and contain oil

Coriolis force force that results from the Earth's rotation. Moving bodies, such as wind and ocean currents, are deflected to the left in the southern hemisphere and to the right in the northern hemisphere.

thermohaline relating to the combined influence of temperature and salinity

7 Sustaining urban environments

7.1 Overview

How far can our urban environments spread before they become unsustainable?

7.1.1 Introduction

Urban environments provide homes, places of work and all the conveniences of modern-day life for their citizens. They are often a magnet for people living in small rural townships, as goods and services abound and social and economic opportunities for a better life are seen as more possible in the big cities.

The complexity of urban environments can be seen in a modern city such as Shanghai, with all its multi-layered buildings, bridges, roadways, electricity, water supplies and services. The need to deal with the huge amounts of waste generated by the population of a city of this size is a concern for its urban planners and managers. To ensure viability into the future, sustainable solutions to the wide range of problems that exist in big cities must be found.

ON Resources

☑ **eWorkbook**	Customisable worksheets for this topic
🎞 **Video eLesson**	Sprawling cities (eles-1712)

LEARNING SEQUENCE

7.1 Overview
7.2 Cities' impact on the environment
7.3 The development of urban environments
7.4 Case studies in urban growth: Melbourne and Mumbai
7.5 **SkillBuilder:** Constructing a land use map
7.6 **SkillBuilder:** Building a map with geographic information systems (GIS) `online only`
7.7 Factors in urban decline
7.8 Future challenges for sustainable urban environments
7.9 **Thinking Big research project:** Slum improvement proposal `online only`
7.10 **Review** `online only`

To access a pre-test and starter questions and receive immediate, **corrective feedback** and **sample responses** to every question, select your learnON format at www.jacplus.com.au.

7.2 Cities' impact on the environment

7.2.1 Interconnections between urban and biophysical environments

The earliest forms of **urban environments** consisted of shelters to protect people from the elements and provide security from the attacks of predators. From these simplest forms, the highly complex modern urban environment has developed.

All forms of urban environments are interconnected with the **biophysical environment**. The 'bio' elements are all forms of plant and animal life including people and all their activity and industry. The 'physical' elements are the atmosphere, hydrosphere and lithosphere or Earth's surface.

The biophysical elements impose limits on the development and sustainability of all forms of urban environment. Conversely, the urban environment imposes significant human-induced change on the biophysical world. The understanding of this interconnection is particularly important in a world of increasing human numbers and pressure for resources on the biophysical environment (see **FIGURE 1**).

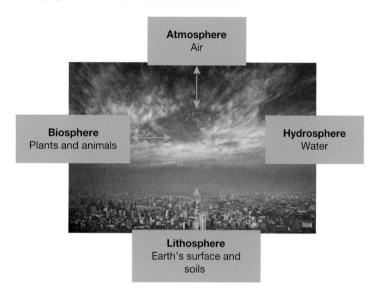

FIGURE 1 Interaction between the urban environment and the biophysical environment

Atmosphere
Air

Biosphere
Plants and animals

Hydrosphere
Water

Lithosphere
Earth's surface and soils

7.2.2 Effects on the atmosphere

Where sources of potentially dangerous gaseous emissions are high, such as from buildings, transport systems and industry, atmospheric pollution can be a problem. Examples such as hazy conditions, photochemical smogs, light and noise pollution, and acid rain are significant problems that need to be addressed. Thus, the development of clean-air policies controlling emissions of gases into the atmosphere is important. Examples of such measures include the introduction of lead-free petrol, banning the burning of household waste, and emission control systems on factory furnaces.

Cities and industries have huge demands for energy, and the by-product of this is heat. The 'heat island effect', whereby urban environment structures such as buildings and roads absorb heat from the sun, raises the temperature of the city environment compared to rural surrounds (see **FIGURE 2**).

The production of greenhouse gases such as carbon dioxide and methane by urban environments is recognised as probably the greatest contemporary climate issue. Global warming leading to climate change is largely the result of emissions of these gases into the atmosphere, particularly in large urban centres.

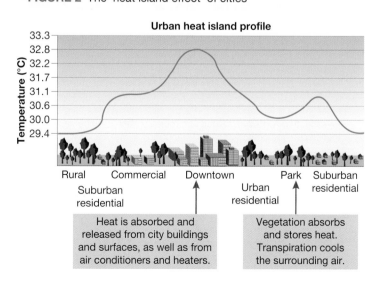

FIGURE 2 The 'heat island effect' of cities

Urban heat island profile

Temperature (°C): 33.3, 32.8, 32.2, 31.7, 31.1, 30.6, 30.0, 29.4

Rural | Commercial | Downtown | Park | Suburban residential
Suburban residential | | Urban residential

Heat is absorbed and released from city buildings and surfaces, as well as from air conditioners and heaters.

Vegetation absorbs and stores heat. Transpiration cools the surrounding air.

7.2.3 Effects on the hydrosphere

As the urban environment is closely dependent on the hydrosphere, it is not surprising that **water security** and **water rights** are important management objectives for a sustainable future. One of the most important aims for urban planners is to ensure the supply of clean water and to manage the waste water from cities.

In general, all urban centres are trying to find increasing supplies of water for domestic and industrial consumption from rivers, groundwater and, more recently, desalinisation sources.

Infrastructure in the form of dams, pipelines, and artesian waters at the local level and major water management schemes such as the Snowy Mountains Scheme in Australia are ways that water is gathered. Water pollution caused by urban environments is also important as polluted waters are a risk to all life forms in any environment. Considerations for biomes and ecosystems of rivers, wetlands and swamps in terms of protecting habitats and maintaining biodiversity is also a major management aim.

FIGURE 3 Like those of many rivers in the world, the banks and channel of the Seine River have been heavily modified by people.

7.2.4 Effects on the lithosphere

The 'tar and cement' structures that are our cities often cover vast areas of land (the built-up areas of greater Melbourne, for instance, cover more than 2000 square kilometres, and its overall planning boundary encompasses more than 7500 square kilometres). The associated problems include the disposal of the enormous amount of waste that cities produce, and the impacts on agriculture, plants and animal life in adjacent habitats and ecosystems.

Urban environment surfaces, such as footpaths, roads and carparks, generate two to six times more run-off than a natural surface. Rain that falls on roads and carparks can be contaminated with petroleum residues and other pollutants, which can then find their way into waterways.

FIGURE 4 Dubai International Airport's Terminal 3 is the largest airport terminal in the world; its carpark offers space for 100 000 cars. Consider the run-off generated by this surface.

Explore more with my**World**Atlas

Deepen your understanding of this topic with related case studies and questions.
- Investigating Australian Curriculum topics > Year 7: Place and liveability > **Polluted cities**
- Investigate additional topics > Urbanisation > **Mexico City**

 Resources

Interactivity Urban impacts on the environment (int-3301)

7.2 Exercise 1: Check your understanding

1. **GS1** What is the 'bio' part of the biophysical *environment*?
2. **GS2** Give reasons why urban *environments* can have such a major impact on the Earth's atmosphere.
3. **GS2** Why do most of the large urban centres of the world have high-rise buildings?
4. **GS2** What is meant by the *heat island effect*?
5. **GS2** Why are water supplies a problem for large cities?

7.2 Exercise 2: Apply your understanding

1. **GS6** How might rising sea levels, predicted to be a result of global warming, affect the *place* and *space* of a city such as New York?
2. **GS6** How will the supply of fresh water affect the development of cities in the future?
3. **GS5** Consider how the development of extensive public transport systems impacts on the environment. How can such systems lead to a more *sustainable* urban *environment*?
4. **GS3** Contrast the *sustainability* of car use with public transport in terms of positive and negative effects on the *environment*.
5. **GS6** Suggest ways that the biodiversity of plant species might be increased in an urban area.

Try these questions in learnON for instant, corrective feedback. Go to www.jacplus.com.au.

7.3 The development of urban environments

7.3.1 Expansion of cities

The earliest cities emerged five to six thousand years ago in Mesopotamia (present-day Iraq), Egypt, India and China. These cities became centres for merchants, craftspeople, traders and government officials, and were dependent on agriculture and domesticated animals from surrounding rural areas.

Before 1800, over 90 per cent of the world's population lived in rural agriculture-based societies. With the **Industrial Revolution**, people began to move from rural areas to find employment in the factories of the rapidly expanding industrialised cities. In 1850, only two cities in the world — London and Paris — had a population above one million. By 1900 there were 12, by 1950 there were 83, and by 1990 there were 286. In 2018, more than 500 cities had populations of a million or more people; over half the global population now lives in urban areas.

FIGURE 1 The region between the Euphrates and Tigris Rivers in Mesopotamia is often called the 'cradle of civilisation'; it is here that the first urban centres developed.

Megacities

The term '**megacity**' commonly refers to urban settlements of 10 million inhabitants or more. Currently, about 530 million people live in 33 megacities across the world. By 2030, the world is projected to have 43 megacities that will be home to more than 750 million people.

Of the 20 largest megacities, only six are located in highly developed industrialised countries: Tokyo–Yokohama, Seoul–Incheon, New York, Osaka–Kobe–Kyoto, Moscow and Los Angeles. Three-quarters of the world's megacities are in developing countries; they include gigantic **conurbations** such as Jakarta, Manila and Karachi (see **FIGURES 3** and **4**).

FIGURE 2 Distribution of the world's population, 2018

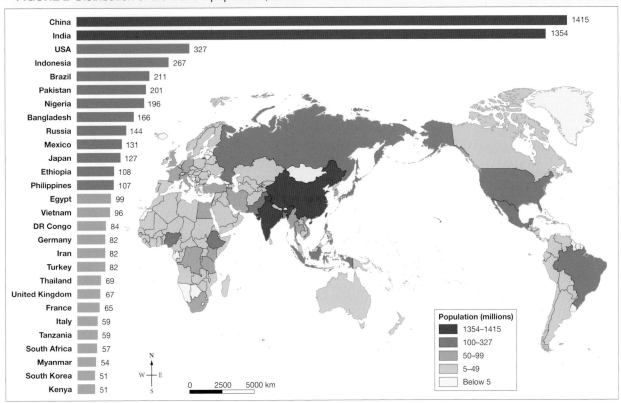

Source: United Nations, Department of Economic and Social Affairs, Population Division (2017). World Population Prospects: The 2017 Revision, Medium Variant.

FIGURE 3 Percentage of urban population and urban agglomerations, 2018

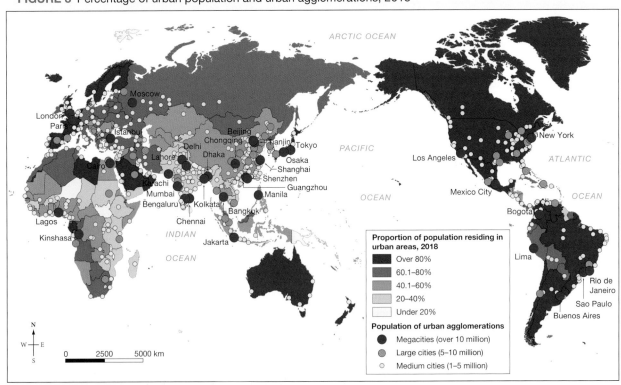

Source: United Nations Department of Economic and Social Affairs; Worldatlas.com.

Asia has by far the greatest proportion of the world's large urban area population. Regions such as Oceania and Africa are less urbanised (see **FIGURES 3, 4** and **5**). For example, in Papua New Guinea (Oceania) and Burundi (East Africa), only 10 per cent of the population is urbanised, whereas in Singapore this figure is 100 per cent.

FIGURE 4 Urban areas with more than 10 million population, 2019

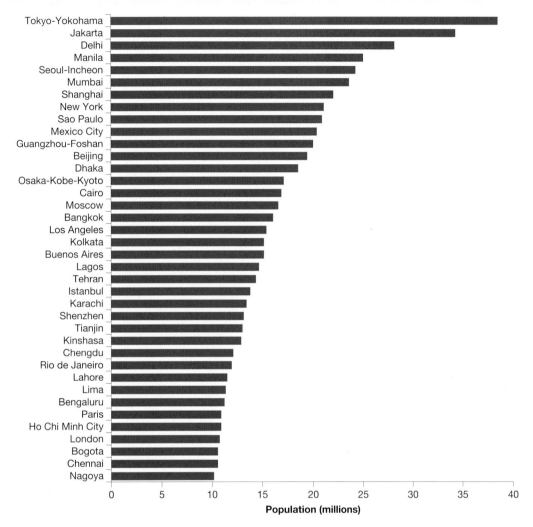

FIGURE 5 Proportion of global built-up urban area population, by region, 2019

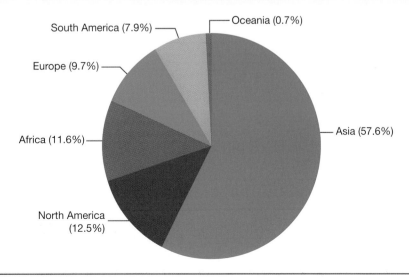

7.3.2 Impacts of the growth of urban environments

The United Nations predicts that by the year 2050, 70 per cent of the population in developed nations and 40 per cent in **developing nations** will live in large urban complexes. Rapid growth in city populations has led to problems such as urban sprawl, traffic congestion and air and water pollution, with significant impacts on the natural environment. Social problems such as unemployment; inadequate housing, **infrastructure**, water, sewerage and electricity supplies; pollution; and the spread of slums and crime are further problems. In addition, with prospects of climate change through global warming, many of the world's coastal cities are under threat from rising sea levels. The application of **human–environment systems thinking** will be the key to evaluating and solving these economic and social issues.

The United Nations estimates that a staggering 90 per cent of the worlds' population growth is taking place in the cities of the developing nations. For many people in these countries, pressures such as extreme poverty, famine and civil unrest often 'push' them away from rural areas towards cities, to which they are 'pulled' by the promise of jobs, shelter and protection.

FIGURE 6 Contrasts in urban development in the capital of the Philippines, Manila

Explore more with my**World**Atlas

Deepen your understanding of this topic with related case studies and questions.
• Investigate additional topics > Urbanisation > **World urbanisation**
• Investigate additional topics > Urbanisation > **Urbanisation in Australia**

7.3 EXERCISES

Geographical skills key: GS1 Remembering and understanding **GS2** Describing and explaining **GS3** Comparing and contrasting **GS4** Classifying, organising, constructing **GS5** Examining, analysing, interpreting **GS6** Evaluating, predicting, proposing

7.3 Exercise 1: Check your understanding

1. **GS1** When and where did the first cities develop?
2. **GS1** Why did cities experience rapid growth and development after the Industrial Revolution?
3. **GS2** What factors are driving the process of urbanisation in the world?
4. **GS2** In which regions of the world is urbanisation occurring most quickly? Why?
5. **GS2** What are some of the major economic and social issues facing rapidly developing cities in the world?

7.4 Case studies in urban growth: Melbourne and Mumbai

7.4.1 CASE STUDY: Ever-sprawling Melbourne

What does a city do when it runs out of room? Across the world cities are expanding at a rapid rate, bringing unprecedented change to the built and natural environments. To accommodate growing populations there is a need for more housing and the infrastructure to support so many people. How can this be done?

Today, the Melbourne metropolitan area sprawls over a huge area of more than 7000 square kilometres and has a population of over 4.9 million. It is the fastest growing Australian capital city, increasing at a rate of 130 000 people per year. By the year 2050, this population will increase to over 8 million, with a need for more than 500 000 additional households. To house this number, urban planners essentially have two choices. One option is the **urban infilling** of land in the inner and middle suburbs. This can be done by dividing older larger blocks into smaller new blocks, by the **urban renewal** of old industrial sites, or by building up and increasing population density with medium- or high-rise apartments, as seen in **FIGURE 1**.

The second option is to expand space by extending the city outwards into a zone known as the **rural–urban fringe**. **FIGURE 2** shows the predicted population growth for Melbourne. Note the location of those suburbs expected to have the greatest population increases.

FIGURE 1 These apartments are an example of high-rise housing in a large-scale urban renewal project at Docklands in Melbourne.

FIGURE 2 Future population growth for Melbourne

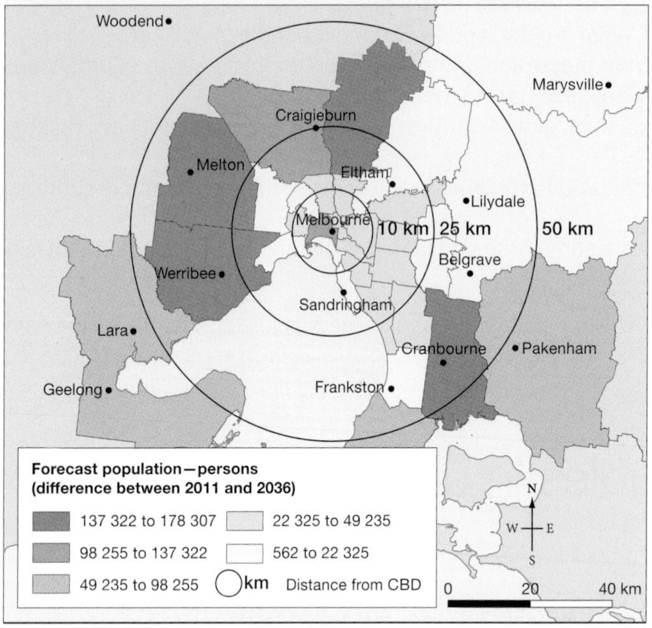
Woodend
Marysville
Craigieburn
Melton
Eltham
Lilydale
Melbourne
10 km
25 km
50 km
Belgrave
Werribee
Sandringham
Lara
Cranbourne
Pakenham
Geelong
Frankston
Forecast population—persons
(difference between 2011 and 2036)
137 322 to 178 307
22 325 to 49 235
98 255 to 137 322
562 to 22 325
49 235 to 98 255
km Distance from CBD
N
W
E
S
0
20
40 km

Source: © The State of Victoria, Department of Environment and Primary Industries 2013 © Commonwealth of Australia Geoscience Australia 2013.

Increasing density in established suburbs

To increase the density of housing in older suburbs, one concept is to establish activity centres. These consist of higher density housing in specific locations, where people shop, work, meet, relax and live in the local environment. These centres are focused on existing infrastructure, transport networks, popular shopping centres, employment opportunities and community facilities. New housing tends to be medium-rise apartments (three to five storeys) built along main transport routes.

Impacts of changes on the rural–urban fringe

The rural–urban fringe is typically the urban zone that is undergoing the most rapid change. Former farmland, often market gardens and orchards, are sold off and new housing and industrial estates are built. These are usually low-density, planned estates sometimes built around a theme or geographical feature such as a built lake or wetland. Urban expansion into the rural–urban fringe brings environmental, economic and social impacts.

Cost of infrastructure

A major problem of **urban sprawl** is the cost to provide infrastructure (for example, roads and other transport systems and services such as water, gas and electricity) to new areas on the rural–urban fringe. In recent years, new suburbs planned for Melbourne have included Diggers Rest, near Sunbury, Lockerbie, near Greenvale, Manor Lakes, in Wyndham, Merrifield West, in the city of Hume, and Rockbank North, near Melton. In 2019, the Victorian government rezoned yet more land to enable 50 000 new homes to be built in new suburbs across Melbourne's north, north-west and south-east. These new suburbs require the development of infrastructure such as shopping complexes, medical centres, open spaces, schools and recreation facilities.

Loss of fertile farm lands

Arguably the largest issue associated with urban sprawl is the loss of fertile farmlands. The Casey Council, 48 kilometres south-east of Melbourne's city centre, initially resisted moves for subdivision of farmlands at Clyde, arguing that the sandy loam soils that produce most of Melbourne's fruit and vegetables should be set aside for growing food, not houses (see **FIGURE 3**).

FIGURE 3 Growing over our food

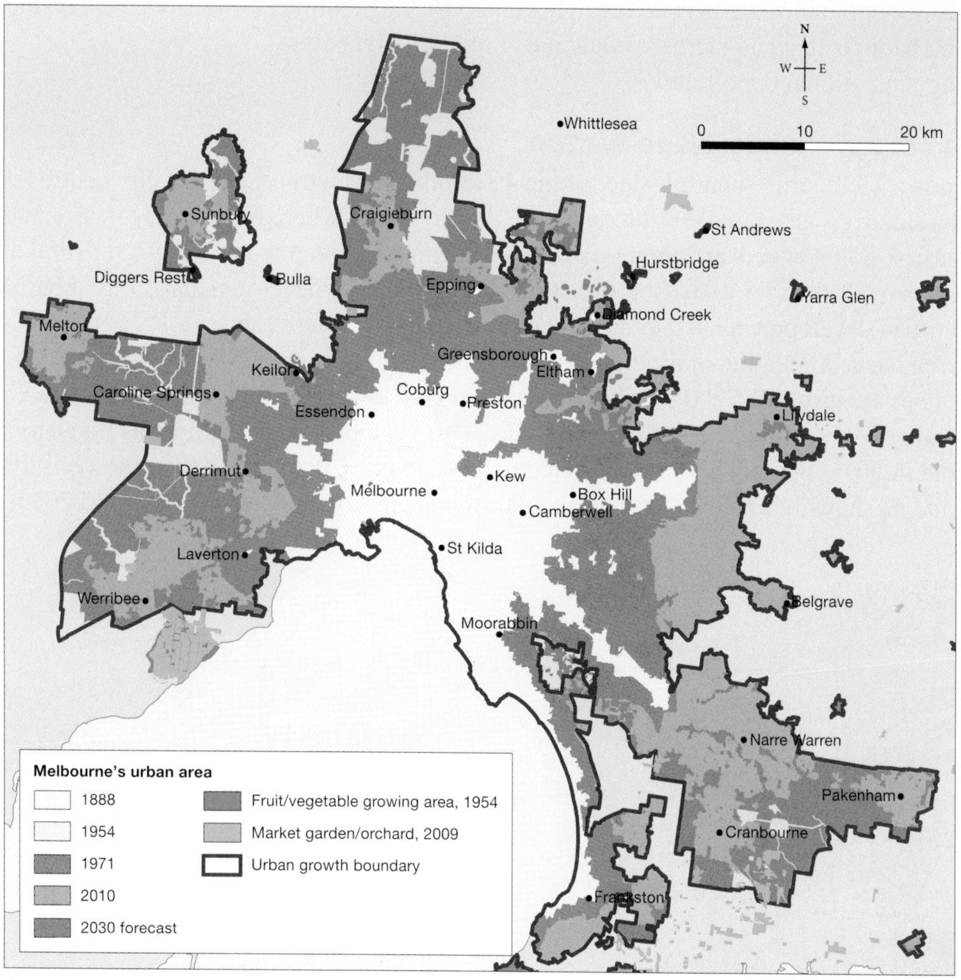

Source: © The State of Victoria, Department of Environment and Primary Industries 2013.

As one newspaper article put it: 'We've already built over the best soils in this state — the soils around Melbourne. Why would you keep building over it and subdividing it when in the next 50 years we're facing an era of incredible uncertainty and major changes to climate, to fuel supplies and to energy markets?'

Some local farmers had different viewpoints on this matter because rezoning their properties into the urban boundary immediately boosted the value of their land. These farmers preferred to relocate further out and reap the financial gains from the sale of the land for housing.

Inevitably, the pressure of population and the expansion of nearby suburban development has meant that many of the farms are now gone.

Loss of green spaces

The expansion of urban areas can significantly alter the natural environment. Clearing of vegetation can reduce habitat and biodiversity. Natural drainage and topography can be altered, with streams redirected or even converted to pipes. Today there is a growing awareness of the need to preserve environments for the important functions they provide for wildlife and people. As such, planners now try to incorporate and retain as much of the natural environment as possible when developing housing estates.

'Green zones' are open landscapes set aside to conserve and protect significant natural features as well as resources such as farms, bushland and parks. They ensure habitats for native flora and fauna are preserved.

The construction or expansion of wetlands in the rural–urban fringe can have many benefits, including:

- acting as flood retention basins and receiving and purifying stormwater run-off from residential areas
- providing habitats that can increase plant and animal biodiversity
- providing recreational opportunities.

Rural–urban fringe change: Narre Warren

Narre Warren is a Melbourne suburb located some 40 kilometres south-east of the city centre (shown in **FIGURE 3**). The suburb is an example of urban-fringe development, where blocks vary in size and are more affordable for first-home buyers and young families. This area has seen what was a semi-rural town become part of a major growth corridor to the south-east of Melbourne. The development of this suburb, along with other nearby urban developments, led to the loss of farmlands over the years.

Narre Warren has a population of around 26 600, with an average age of 35 years. According to 2016 census data, there are just over 7200 families and 9500 private dwellings in the suburb. Narre Warren sits within the local government area of the City of Casey. This growing municipality was home to around 314 000 people in 2016, but this figure is expected to grow to more than 549 000 by the year 2041. **FIGURES 4(a)** and **(b)** show the changes to the Narre Warren environment between 1966 and 2013.

FIGURE 4(a) Topographic map extract of Narre Warren, 1966

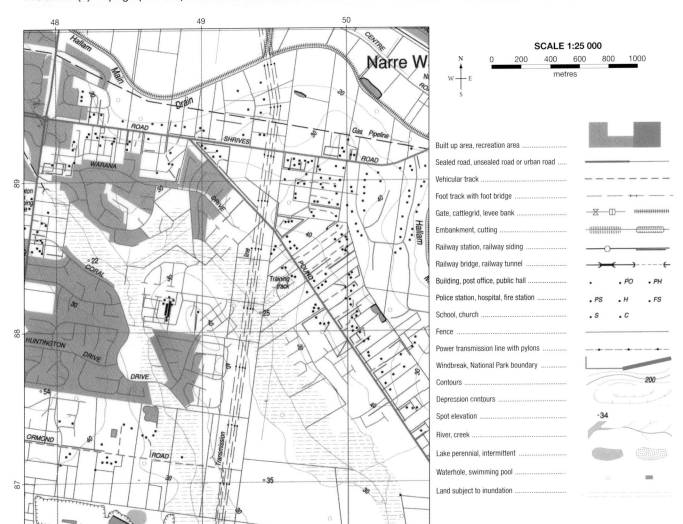

Source: © Vicmap Topographic Mapping Program / Department of Environment and Primary Industries.

FIGURE 4(b) Topographic map extract of Narre Warren, 2013

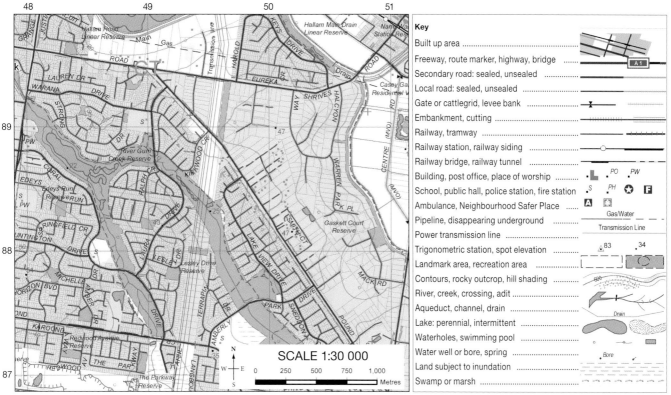

Source: © Vicmap Topographic Mapping Program / Department of Environment and Primary Industries.

7.4.2 CASE STUDY: The growth of Mumbai

Located in the western Indian state of Maharashtra, Mumbai is the most populous city in India and the ninth most populous city in the world, with a population of over 20 million in 2019. In 2009, Mumbai was named an **alpha world city**. Although the richest city in India, with the highest **gross domestic product** (GDP) of any city in south, west or central Asia, it also has much sub-standard housing and many of its residents live in squalor.

The large numbers of people and rapid population growth have contributed to serious social, economic and environmental problems for Mumbai. Mumbai's business opportunities, and its potential to offer a higher standard of living, attract migrants from all over India seeking employment and a better way of life. In turn, this has made the city a melting pot of many communities and cultures. In 2019, the population density was estimated to be around 20 000 persons per square kilometre, and the living space just 8 square metres per person.

Despite government attempts to discourage the influx of people, the city's population grew by more than 4 per cent between 2001 and 2011, and by a staggering 61.8 per cent between 2011 and 2019 (see **TABLE 1**). The number of migrants to Mumbai from outside Maharashtra during the ten-year period from 2001 to 2011 was over one million, which amounted to 54.8 per cent of the net addition to the population of Mumbai.

Many newcomers end up in abject poverty, often living in **slums** or sleeping in the

TABLE 1 Population growth in Mumbai

Census	Population	% change
1971	5 970 575	—
1981	8 243 405	38.1
1991	9 925 891	20.4
2001	11 914 398	20.0
2011	12 478 447	4.7
2019*	20 185 064	61.8

Source: *2019 data: Mumbai Metropolitan Region Development Authority. All other data is based on Government of India Census, conducted every 10 years.

streets. By 2017, an estimated 62 per cent of the city's inhabitants lived in slum conditions. Some areas of Mumbai city have population densities of around 46 000 per square kilometre — among the highest in the world.

Challenges

Mumbai suffers from the same major urbanisation problems that are seen in many fast-growing cities in developing countries: widespread poverty and unemployment, urban sprawl, traffic congestion, inadequate sanitation, poor public health, poor civic and educational standards, and pollution. These pose serious threats to the quality of life in the city for a large section of the population. Automobile exhausts and industrial emissions, for example, contribute to serious air pollution, which is reflected in a high incidence of chronic respiratory problems. With available land at a premium, Mumbai residents often reside in cramped, relatively expensive housing, usually far from workplaces and therefore requiring long commutes on crowded public transport or clogged roadways (see **FIGURE 5**). Although many live in close proximity to bus or train stations, suburban residents spend a significant amount of time travelling southwards to the main commercial district.

FIGURE 5 Mumbai train rush hour

The Dharavi slum

Dharavi, Asia's second-largest slum, is located in central Mumbai. Stretching across 220 hectares of land, it is home to more than one million people (see **FIGURE 6**). In Dharavi, it is estimated that there is only one toilet for every 200 people. This results in floods of human excrement during the monsoon season. Much of the water becomes contaminated because of this, and death rates tend to be significantly higher in Mumbai's slums than in upper- and middle-class areas.

FIGURE 6 Dharavi

Dharavi's recycling entrepreneurs

Hidden amid Dharavi's labyrinth of ramshackle huts and squalid open sewers are an estimated 20 000 single-room factories, employing around a quarter of a million people and turning over a staggering US$1 billion each year through recycling and other trades such as the production of pottery, textiles and leather goods.

In developed countries, communities recycle because there is the understanding that it contributes to sustaining the planet's resources. However, for some of the poorest people in the developing world, recycling often isn't a choice, but rather a necessity of life.

In India, it is estimated that anywhere between 1.5 million and 4 million people make their living by recycling waste. At least 300 000 of these live and work in Mumbai. These people are known as 'ragpickers' and are made up of India's poorest and most marginalised groups (see **FIGURE 7**). The ragpickers wade through piles of unwanted goods to salvage easily recyclable materials such as glass, metal and plastic, which are then sold to scrap dealers who process the waste and sell it on either to be recycled or to be used directly by the industry.

Due to the lack of formal systems of waste collection, it falls to Mumbai's ragpickers to provide this basic service for fellow citizens. Without them, solid waste and domestic garbage would not even be collected, let alone sorted or recycled. Despite many of the social and ethical controversies surrounding the recycling industry in India, Dharavi is seen as the 'ecological heart of Mumbai', recycling up to 85 per cent of all waste material produced by the city, an excellent example of human–environment systems thinking in action.

FIGURE 7 Ragpickers

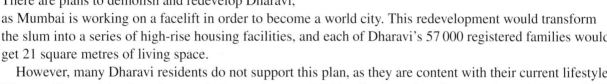

An uncertain future

There are plans to demolish and redevelop Dharavi, as Mumbai is working on a facelift in order to become a world city. This redevelopment would transform the slum into a series of high-rise housing facilities, and each of Dharavi's 57 000 registered families would get 21 square metres of living space.

However, many Dharavi residents do not support this plan, as they are content with their current lifestyle. Most residents of the slum do not mind squatting near Mahim Creek, and prefer not to have their own flush toilets. Most are working and making a living, and many have lived their entire lives in Dharavi and do not want to trade their culture for a redeveloped life.

DISCUSS

Most Australians would probably perceive slums negatively. Comment on whether the role of ragpickers and the recycling that takes place in Dharavi support a more positive perception of slums.

[Critical and Creative Thinking Capability]

 Explore more with my**World**Atlas

Deepen your understanding of this topic with related case studies and questions.
• Investigating Australian Curriculum topics > Year 8: Changing nations > **Urbanisation in Australia**

---

on Resources

 Interactivity Changes on the rural–urban fringe (int-3302)

7.4 INQUIRY ACTIVITIES

1. Refer to the 'Constructing a land use map' SkillBuilder in subtopic 7.5.
    (a) Create a map to show the main land uses in Narre Warren in 1966. Include other key features such as main roads and railway lines.
    (b) Using tracing paper, make an overlay map of built-up areas from the 2013 map, and attach your overlay to your base map. Complete your map with full BOLTSS.
    (c) Study your completed map and overlay.
        i. What was the main land use in this area in 1966?
        ii. What is the main land use for the area in 2013?
        iii. Study both maps and describe three other *changes* in land use in Narre Warren from 1966 to 2013.

Classifying, organising, constructing

2. Working with a partner or in small groups, undertake a fieldwork investigation of your local area in terms of:
   - the types of dwellings
   - transport facilities and issues
   - shopping and other community services available
   - amount of open spaces and parkland, and associated recreation facilities
   - how 'liveable' it is. (Consult your local council or conduct a survey of local residents).

   Document your findings in a report, including maps, photographs and data (e.g. tables, pie charts), and listing any references. **Examining, analysing, interpreting**

3. Search for and watch some of the YouTube videos about Dharavi. Create a one-page infographic detailing life in the slum. Consider questions such as:
   - How many people live there and what are living conditions like?
   - What work is done here?
   - What are the risks to health and what could be done to improve the situation?

   Classifying, organising, constructing

4. Find out more about the natural and human influences on the development of cities. Examples for research could include Canberra, Australia (a planned city); Cape Town, South Africa (a port city); Rotenburg, Germany (a walled city); Geneva, Switzerland (where a river meets a lake); Johannesburg, South Africa (near a mining site); Chicago, United States (where north–south and east–west railway routes cross); Jerusalem, Israel (an ancient religious city); Bath, England (located at the site of a natural supply of mineral waters).

   Examining, analysing, interpreting

7.4 EXERCISES

Geographical skills key: GS1 Remembering and understanding **GS2** Describing and explaining **GS3** Comparing and contrasting **GS4** Classifying, organising, constructing **GS5** Examining, analysing, interpreting **GS6** Evaluating, predicting, proposing

7.4 Exercise 1: Check your understanding

1. **GS1** What are the two main ways that additional housing can be established in an expanding city?
2. **GS2** Study **FIGURE 2**. Describe the location of the suburbs of Melbourne that are expected to show the greatest increase. What is the average distance of these suburbs from Melbourne's CBD?
3. **GS5** Refer to **FIGURE 3**.
   (a) What has happened to the areas of market gardening (fruit and vegetable farming) between 1954 and 2009? Use distances and directions in your answer.
   (b) What would be the benefits of market gardening being located close to urban areas?
   (c) Predict the future location of this land use in Melbourne in 2030.
4. **GS2** Suggest how urban planners can reduce some of the *environmental*, social and economic impacts of expansion into the rural–urban fringe.
5. **GS5** Study **FIGURE 4(a)**. What evidence is there to suggest that this area is part of the rural–urban fringe?
6. **GS1** Outline the various challenges that Mumbai faces.

7.4 Exercise 2: Apply your understanding

1. **GS3** Construct a table to compare the advantages and disadvantages of living in an inner-city high-rise apartment and a housing estate on the rural–urban fringe.
2. **GS5** Refer to **FIGURE 4(b)**.
   (a) The area at GR485880 is subject to flooding (inundation). Use evidence from the maps to suggest two reasons why it is flood prone.
   (b) How have planners used this flood-prone land when designing this housing estate?
3. **GS5** Study **FIGURE 4(b)**. List and give grid references for any new forms of infrastructure established. Consider schools, shopping centres, parks and transport.
4. **GS6** Suggest one human and one *environmental* factor that make Narre Warren suitable for a housing estate.
5. **GS6** What would be the economic, social and *environmental* benefits of ragpickers?
6. **GS3** List the advantages and disadvantages of replacing slums with high-rise low-income housing.

Try these questions in learnON for instant, corrective feedback. Go to www.jacplus.com.au.

7.5 SkillBuilder: Constructing a land use map

What is a land use map?

A land use map may be drawn from a topographic map, an aerial photograph or a plan, or during fieldwork. It shows simplified information about the uses made of an area of land.

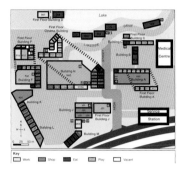

Select your learnON format to access:

- an overview of the skill and its application in Geography (Tell me)
- a video and a step-by-step process to explain the skill (Show me)
- an activity and interactivity for you to practise the skill (Let me do it)
- questions to consolidate your understanding of the skill.

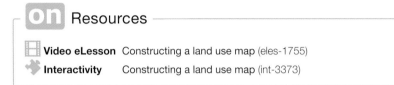

Resources

Video eLesson Constructing a land use map (eles-1755)

Interactivity Constructing a land use map (int-3373)

7.6 SkillBuilder: Building a map with geographic information systems (GIS)

What is GIS?

A geographic information system (GIS) is a computer-based system that consists of layers of geographic data. Just as an overlay map allows you to interchange layers of information, GIS allows you to turn layers on and off to make comparisons between data.

Select your learnON format to access:

- an overview of the skill and its application in Geography (Tell me)
- a video and a step-by-step process to explain the skill (Show me)
- an activity and interactivity for you to practise the skill (Let me do it)
- questions to consolidate your understanding of the skill.

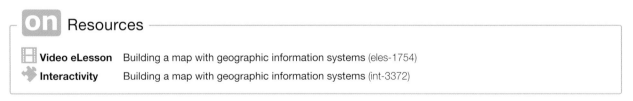

Resources

Video eLesson Building a map with geographic information systems (eles-1754)

Interactivity Building a map with geographic information systems (int-3372)

7.7 Factors in urban decline

7.7.1 Environmental factors

Over time, all forms of urban environments will deteriorate with age and require renovation or renewal. Extreme atmospheric events such as cyclones, hurricanes and tornadoes, which exhibit strong winds and flooding rains, can have devastating short-term impacts on urban environments. Longer term events such as **desertification** and climate change can also have negative impacts.

Movements of the earth such as those due to earthquakes and volcanic eruptions can also destroy urban environments. One well-documented example of such events is the eruption of Mt Vesuvius in Italy in 79 CE, which completely buried the cities of Pompeii and Herculaneum under volcanic ash (see **FIGURE 1**).

Tsunamis are also a significant hazard that can lead to the destruction of settlements. On 22 December 2018, for example, the coastline regions of Banten and Lampung in Indonesia were devastated by a 3-metre tsunami triggered by an underwater landslide following the volcanic eruption of Anak Krakatau.

Hundreds of lives were lost and many villages were destroyed. Residents were forced to relocate until restoration works could be completed. Similarly, as a result of a massive earthquake and resultant tsunami in Japan in 2011, towns such as Otsuchi underwent significant change. Thousands of people simply left the region — the lack of employment opportunities and the risk associated with living in a disaster-prone region combining to drive people to move elsewhere.

FIGURE 1 Ruins of the city of Pompeii with Mt Vesuvius in the background

7.7.2 Human factors

Human factors, which include changes in the social, economic and political elements of a region, can also be a cause of the decline of cities and their urban environment. The destructive effects of war on the social fabric and economy of a nation, which have significant impacts on urban environments, are one example.

Angkor, the capital of the Khmer Empire in Cambodia, and thought to be the largest city in the world at the time, was abandoned in the fifteenth century due to a combination of wars and a series of droughts. The destruction of its economy, which was based on management of water and rice production, meant the city was no longer viable. The elaborate Khmer temples constructed in the twelfth century (see **FIGURE 2**) have now become popular tourist attractions; more than two million people visit these sites each year.

FIGURE 2 Main temple complex, Angkor Wat, Cambodia

In modern times, there are many examples of towns and cities with extensive urban environments that have declined. Some reasons for change include depletion of mineral supplies and mining operations, changes in demand for industrial production and manufactured goods, and **economic downturn**. An example of an urban project that failed due a downturn in the Turkish economy is that of Burg Al Babas. The project started in 2014 but went into bankruptcy in 2018; the chateau-style houses remained unoccupied in 2019 (see **FIGURE 3**).

Other examples of urban environments that have declined due to human-induced factors can be seen in **FIGURE 4**.

FIGURE 3 Burj Al Babas, Turkey, is a ghost town due to a downturn in the economy.

FIGURE 4 Cities abandoned due to changing human and physical factors

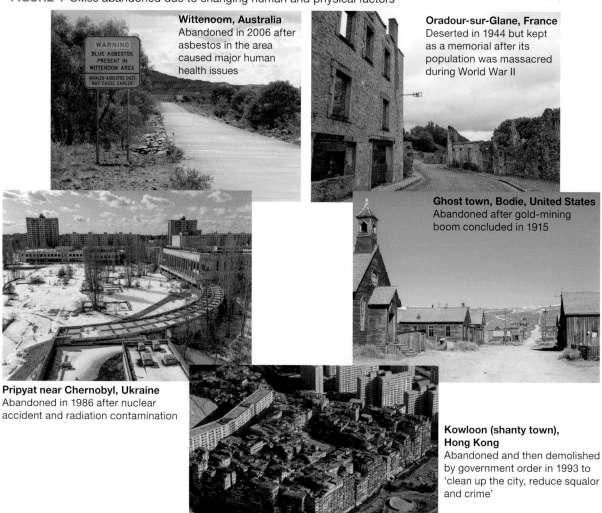

Wittenoom, Australia
Abandoned in 2006 after asbestos in the area caused major human health issues

Oradour-sur-Glane, France
Deserted in 1944 but kept as a memorial after its population was massacred during World War II

Ghost town, Bodie, United States
Abandoned after gold-mining boom concluded in 1915

Pripyat near Chernobyl, Ukraine
Abandoned in 1986 after nuclear accident and radiation contamination

Kowloon (shanty town), Hong Kong
Abandoned and then demolished by government order in 1993 to 'clean up the city, reduce squalor and crime'

Note: This photograph was taken prior to the demolition of Kowloon.

7.7.3 CASE STUDY: Venice, a sinking city

Urban centres that are built on low-lying coastal plains, where features such as **river deltas**, wetlands, lagoons, sand dunes, bars and barriers are found, are susceptible to the environmental impacts of the sea.

Storms and high tides, when added together, can lead to destructive surges that cause erosion and damage to cities. The prospect of rising sea levels as a result global warming-induced ice cap melting will require specialised management techniques such as the construction of coastal defence works to protect property and life.

Venice, Italy, is a city built on mud islands in a coastal **lagoon** at the head of the Adriatic Sea (see **FIGURES 5** and **6**). Although Venice has a population of only around 270 000, its **historical architecture**, life on the canals and cultural events such as Carnevale attract around 20 million tourists per year. Not surprisingly, the Venetians are keen to protect their heritage and manage the impacts of erosion and rising sea levels into the future.

FIGURE 5 Venice and surrounding areas

Source: © OpenStreetMap contributors.

Why is Venice sinking?

When Venice was established almost 2000 years ago, the sea level was two metres lower than current levels and buildings seemed secure from the impacts of the sea. Over time, the sea level has risen, and in more recent times this rate of increase has accelerated due to global warming. Also affecting the stability of buildings was the removal of fresh water from artesian wells near Venice in the 1950s. This practice, which fortunately has stopped, led to building subsidence.

FIGURE 6 Aerial view of Venice showing the built area in the lagoons and the canals

Floods or *aqua alta*

Venetians refer to floods as *aqua alta* or 'high water'. Flooding events occur each year, usually between autumn and spring, and their intensity varies. In October 2018, high sea waters with a depth of 1.56 metres above average sea levels submerged nearly two-thirds of the city. A combination of high tides and winds forced waters over the canal banks and into buildings and public areas (see **FIGURE 7**). These types of flooding events make life difficult for locals and tourists as buildings and public areas are flooded and transport is restricted, with some boats unable to fit under bridges. The actual number of tourists is another significant problem for Venice — in the busy season, tourist numbers can reach 60 000 per day. Management plans and limits to tourist numbers are now under consideration.

FIGURE 7 Flooding in Venice

Reducing the impact of floods

Completed in 2018, the MOSE (MOdulo Sperimentale Elettromeccanico) or Experimental Electromechanical Module Project aims to reduce the impact of floods on Venice. It consists of rows of mobile gates that are able to isolate the lagoon and canals from high tides above 110 centimetres (to a maximum of three metres). The project has been criticised by some who say that flushing of the canals would be reduced and the huge cost of the project cannot be justified as it may only be effective for a few years if sea levels continue to rise.

DISCUSS

'Let it sink. Venice is not worth saving.' Suggest an argument that supports and an argument that would challenge this viewpoint.

[Critical and Creative Thinking Capability]

7.7 INQUIRY ACTIVITIES

1. Use the internet to find out about the decline of the cities of Gary and Detroit in the United States, and make a list of reasons why these cities have declined and how they might be reinvigorated.

   Examining, analysing, interpreting

2. Use the internet to research and learn more about the MOSE project in Venice. Create a labelled diagram to show how it will work.

   Classifying, organising, constructing

3. Select a capital city in Australia and find out more about the impact of rising sea levels on suburbs close to the coast. How might the social and economic impacts of rising seas be managed?

   Examining, analysing, interpreting

7.7 EXERCISES

Geographical skills key: GS1 Remembering and understanding **GS2** Describing and explaining **GS3** Comparing and contrasting **GS4** Classifying, organising, constructing **GS5** Examining, analysing, interpreting **GS6** Evaluating, predicting, proposing

7.7 Exercise 1: Check your understanding

1. **GS1** What *environmental* hazards can lead to destruction or damage to urban *environments*?
2. **GS1** Why was the town of Pripyat near Chernobyl in the Ukraine abandoned?
3. **GS2** What reasons can you put forward to explain why cities decline over time?
4. **GS2** Why was water essential to the survival and decline of the city of Angkor?
5. **GS1** Where is Venice located?

7.8.3 What are the challenges for cities?

Many cities in the world face environmental, social and economic challenges. Issues such as extensive areas of slum housing and a general lack of infrastructure to support what may be called a socially and economically just lifestyle undermine the sustainability of these environments. Particularly in the poorest or least developed countries, there are significant environmental management issues associated with large cities. Some of these are detailed in **TABLE 1**. Note that there are issues even in cities that could be called wealthy or most developed.

TABLE 1 Urban challenges

Level	Challenges to be addressed to ensure a sustainable urban environment
Least developed countries	• Poverty and inequality • Rapid and chaotic development of slum housing • Increasing demand for housing, urban infrastructure, services and employment • Education and employment needs of the majority population of young people • Shortage of skills in the urban environment sector
Transition countries	• Slow (or even negative) population growth and ageing • Shrinking cities and deteriorating buildings and infrastructure • Urban sprawl and preservation of inner-city heritage buildings • Growing demand for housing and facilities by an emerging wealthy class • Severe environmental pollution from old industries • Rapid growth of vehicle ownership • Financing of local authorities to meet additional responsibilities
Developed countries	• Recent mortgage and housing markets crises • Unemployment and impoverishment due to changing availability of jobs • Large energy use of cities caused by car dependence, huge waste production and urban sprawl • Slow population growth, ageing and shrinking of some cities

7.8.4 How can we plan for the future?

By promoting sustainable urban environments at all levels of scale (local, regional, national and global), problems can be overcome.

Some management strategies that will foster socially, economically and environmentally sustainable urban environments include:

- building energy-efficient houses based on materials and energy sources that reduce the **ecological footprint** of cities
- reducing waste by recycling and reusing materials
- improving public transport systems to reduce reliance on cars
- redeveloping to include **medium-density housing** to reduce urban sprawl
- exchanging ideas between governments about planning and building policies and best and successful practice in design.

FIGURE 3 Vertical gardens can be used to add green spaces to medium- and high-density housing developments.

― Explore more with my**World** Atlas ―

Deepen your understanding of this topic with related case studies and questions.
* Investigate additional topics > Urbanisation > **Brisbane: an eco-city**

on Resources

Interactivity Where am I? (int-3303)

7.8 EXERCISES

Geographical skills key: GS1 Remembering and understanding **GS2** Describing and explaining **GS3** Comparing and contrasting **GS4** Classifying, organising, constructing **GS5** Examining, analysing, interpreting **GS6** Evaluating, predicting, proposing

7.8 Exercise 1: Check your understanding

1. **GS1** How many people in the world in 2030 will need new housing based on current predictions of urbanisation?
2. **GS1** What are the SDGs and why were they developed?
3. **GS2** Which of the SDGs do you think relate to the issue of *sustainable* urbanisation? Explain your view.
4. **GS5** From **TABLE 1**, identify the urban challenges that relate to life in Australia.
5. **GS2** List three management strategies for *sustaining* urban *environments* and explain the contribution that each of these would make.

7.8 Exercise 2: Apply your understanding

1. **GS6** How would migration help solve the problems of ageing populations in developed Western cities?
2. **GS2** Many cities of the world have programs such as 'urban infill' to overcome the problem of housing shortages. What is meant by this term and how can it help solve the problem of 'urban sprawl'?
3. **GS6** It has been said that if all nations had the same ecological footprint as the developed countries (e.g. the United States, Australia and most European nations), we would need four new worlds the size of planet Earth to accommodate the growth in resource consumption. In what ways can we achieve energy, food and water security with an aim of *sustainability* into the future?
4. **GS5** Which of the SDGs will directly improve social conditions in urban *environments*? Give reasons for your answer.
5. **GS5** Which of the SDGs will directly improve *environmental* conditions in urban *environments*? Give reasons for your answer.

Try these questions in learnON for instant, corrective feedback. Go to www.jacplus.com.au.

7.9 Thinking Big research project: Slum improvement proposal

SCENARIO

You have been employed by the local council in one of the world's megacities to carry out a study identifying issues associated with life in the city's slums, and to develop a plan for improving living conditions for slum residents.

Select your learnON format to access:

- the full project scenario
- details of the project task
- resources to guide your project work
- an assessment rubric.

 Resources

 ProjectsPLUS Thinking Big research project: Slum improvement plan (pro-0216)

7.10 Review

7.10.1 Key knowledge summary

Use this dot point summary to review the content covered in this topic.

7.10.2 Reflection

Reflect on your learning using the activities and resources provided.

 Resources

 eWorkbook Reflection (doc-31773)

Crossword (doc-31774)

Interactivity Sustaining urban environments crossword (int-7674)

KEY TERMS

alpha world city a city generally considered to be an important node in the global economic system

biophysical environment all elements or features of the natural or physical and the human or urban environment, including the interaction of these elements

conurbation an urban area formed when two or more towns or cities (e.g. Tokyo and Yokohama) spread into and merge with each other

desertification the transformation of land once suitable for agriculture into desert by processes such as climate change or human practices such as deforestation and overgrazing

developing nation a country whose economy is not well developed or diversified, although it may be showing growth in key areas such as agriculture, industries, tourism or telecommunications

ecological footprint a measure of human demand on the Earth's natural systems in general and ecosystems in particular; the amount of productive land required by each person in the world for food, water, transport, housing, waste management and other purposes

economic downturn a recession or downturn in economic activity that includes increased unemployment and decreased consumer spending

gross domestic product (GDP) the value of all goods and services produced within a country in a given period, usually discussed in terms of GDP per capita (total GDP divided by the population of the country)

historical architecture urban environment that has significant value due to its unique form and history of development

human–environment systems thinking using thinking skills such as analysis and evaluation to understand the interaction of the human and biophysical or natural parts of the Earth's environment

Industrial Revolution the period from the mid 1700s into the 1800s that saw major technological changes in agriculture, manufacturing, mining and transportation, with far-reaching social and economic impacts

infrastructure the basic physical and organisational structures and facilities (e.g. buildings, roads, power supplies) needed for the operation of a society

lagoon a shallow body of water separated from the sea by a sand barrier or coral reef

medium-density housing a form of residential development such as detached, semi-attached and multi-unit housing that can range from about 25 to 80 dwellings per hectare

megacity a settlement with 10 million or more inhabitants

river delta a landform composed of deposited sediments at the mouth of a river where it flows into the sea

rural–urban fringe the transition zone where rural (country) and urban (city) areas meet

slum rundown area of a city with substandard housing

urban environment the human-made or built structures and spaces in which people live, work and recreate on a day-to-day basis

urban infilling the division of larger house sites into multiple sites for new homes

urban renewal redevelopment of old urban areas including the modernisation of household interiors

urban sprawl the spreading of urban developments into areas on the city boundary

water rights refers to the right to use water from a water source such as a river, stream, pond or groundwater source

water security the reliable availability of acceptable quality water to sustain a population

UNIT 2
GEOGRAPHIES OF HUMAN WELLBEING

Organisations and governments devise programs that attempt to improve human wellbeing for their own and other countries, but significant variations in wellbeing exist from one place to another, both within countries and across the globe. What are the factors that affect human wellbeing, and how can we measure and compare wellbeing? Why are there such variations from one place to another and what can we do to address imbalances?

FIELDWORK INQUIRY: COMPARING WELLBEING IN THE LOCAL AREA   onlineonly
Task

Your task is to produce a fieldwork report you could present to your local council that outlines variations within your local area, reasons for the differences and strategies to improve the situation in the future. The aim of the fieldwork is for you to explore some of these variations by comparing two places at the local scale.

Select your learnON format to access:
- an overview of the project task
- details of the inquiry process
- resources to guide your inquiry
- an assessment rubric.

on Resources

ProjectsPLUS Fieldwork inquiry: Comparing wellbeing in the local area (pro-0151)

8 Measuring and improving wellbeing

8.1 Overview

Everyone wants a good life, but what does that mean for different people? Can wellbeing actually be measured and how can we improve it if it's not measuring up?

8.1.1 Introduction

We all want a better life for ourselves, our families and our children, no matter where we live. We care about the wellbeing and progress of our communities, our country and our world. But how can we measure these things? What does wellbeing really mean, and what do we count when we measure progress? How do we know if we are succeeding in our efforts to create a better life for everyone?

---

on Resources

☑ **eWorkbook**   Customisable worksheets for this topic

▦ **Video eLesson** The good life? (eles-1713)

LEARNING SEQUENCE

8.1 Overview
8.2 Understanding and measuring wellbeing
8.3 **SkillBuilder:** Constructing and interpreting a scattergraph   `online only`
8.4 Wealth and wellbeing
8.5 **SkillBuilder:** Interpreting a cartogram   `online only`
8.6 Improving wellbeing
8.7 The way forward
8.8 The importance of human rights
8.9 **Thinking Big research project:** SDG progress infographic   `online only`
8.10 **Review**   `online only`

To access a pre-test and starter questions and receive immediate, **corrective feedback** and **sample responses** to every question, select your learnON format at www.jacplus.com.au.

8.2 Understanding and measuring wellbeing

8.2.1 A good life

A new global movement has emerged seeking to produce measures of progress that go beyond a country's income. Driven by citizens, policy makers and statisticians around the world and endorsed by international organisations like the United Nations, the concept of **wellbeing** offers us a new perspective on what matters in our lives.

Wellbeing is experienced when people have what they need for life to be good. But how do we measure a good life? We can use **indicators** of wellbeing to help us. Indicators are important and useful tools for monitoring and evaluating progress, or lack of it. There are **quantitative indicators** and **qualitative indicators**.

Traditionally, **development** has been viewed as changing one's environment in order to enhance economic gain. Today, the concept of development is not only concerned with economic growth, but includes other aspects such as providing for people's basic needs, equity and social justice, sustainability, freedom and safety. We have built on this traditional concept for measuring progress by considering wellbeing, which emphasises what is positive and desirable rather than what is lacking. The most successful development programs address all areas of wellbeing, rather than simply focusing on economic, health or education statistics. There is a growing awareness that human beings and their happiness cannot simply be reduced to a number or percentage. We can measure development in a variety of ways, but the most common method remains to use economic indicators that measure economic progress using data such as **gross domestic product** (GDP).

Using indicators

Indicators can be classified into a range of broad categories (see **FIGURE 1**). Economic indicators measure aspects of the economy and allow us to analyse its performance. Social indicators include demographic, social and health measures. Environmental indicators assess resources that provide us with the means for social and economic development, and gauge the health of the environment in which we live. Political indicators look at how effective governments are in helping to improve people's **standard of living** by ensuring access to essential services. Wellbeing can also be influenced by technological indicators in such fields as transport, industry, agriculture, mining and communications.

FIGURE 1 Categories of wellbeing indicators

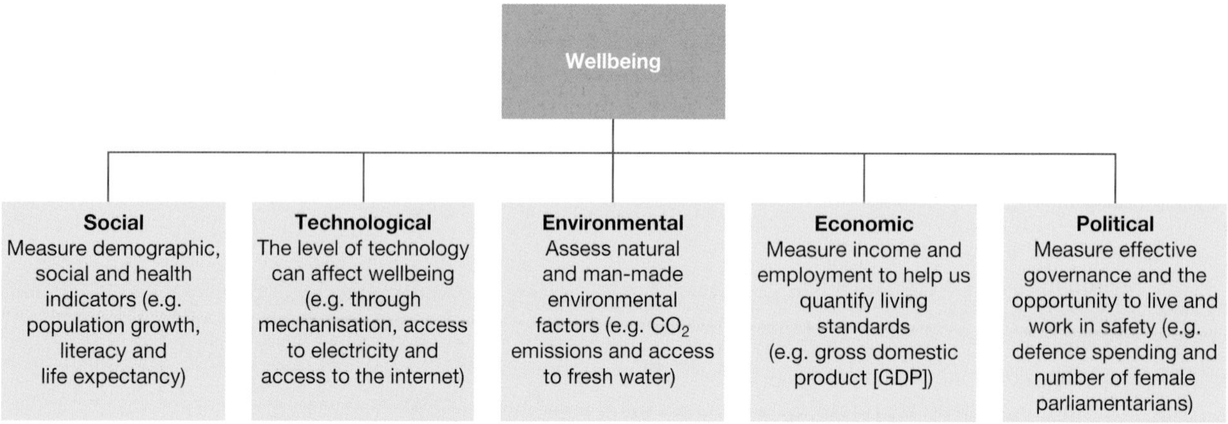

Wellbeing				
Social Measure demographic, social and health indicators (e.g. population growth, literacy and life expectancy)	**Technological** The level of technology can affect wellbeing (e.g. through mechanisation, access to electricity and access to the internet)	**Environmental** Assess natural and man-made environmental factors (e.g. CO_2 emissions and access to fresh water)	**Economic** Measure income and employment to help us quantify living standards (e.g. gross domestic product [GDP])	**Political** Measure effective governance and the opportunity to live and work in safety (e.g. defence spending and number of female parliamentarians)

Using patterns to describe wellbeing

Geographers use the spatial dimension, which helps us to identify patterns of where things are located over Earth's space and attempt to explain why these patterns exist. Identifying patterns across the globe may help to explain why the world is so unequal. Factors that affect equality across areas in a positive way may include the availability of natural resources or an educated workforce, whereas susceptibility to natural disasters or corruption may create more inequality.

Inequalities may exist between individuals, but also within and between countries, regions and continents (often referred to as 'spatial inequality'). Just as each person has their own unique strengths and weaknesses, places are either endowed with, or lack, various resources.

8.2.2 Describing development

Whichever method of measuring development or wellbeing we choose, it is important to understand the terms that have been used, the values that underpin it, and what perspective (often Western) we take. With an overwhelming amount of data available to us, the world is often divided simplistically into extremes such as 'rich' or 'poor'. Is this the best way? The annotated classifications in **FIGURE 2** have been used in the past century, but they are very general and as such have been questioned by geographers for their accuracy (and sometimes offensiveness).

FIGURE 2 World map showing various definitions of development

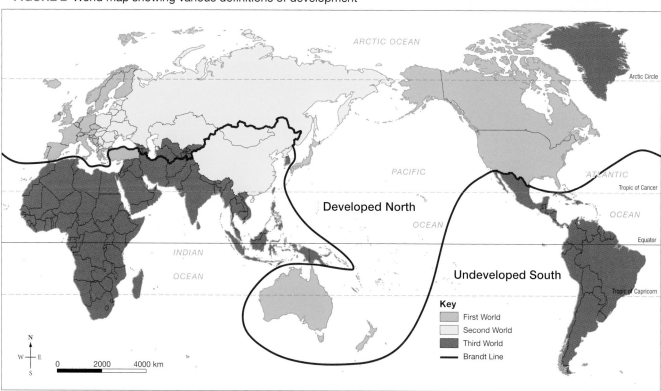

Source: United Nations Development Report.

DEVELOPED OR DEVELOPING?

One of the most common ways of talking about the level of development in various places is to label them as 'developed' or 'developing' (previously often referred to as 'undeveloped'). These terms assume that development is a linear process of growth, so each country can be placed on a continuum of development. Countries that are developing are still working towards achieving a higher level of living standard or economic growth, implying that the country could ultimately become 'developed'.

NORTH OR SOUTH?

In 1980, the Chancellor of West Germany, Willy Brandt, chaired a study into the inequality of living conditions across the world. The imaginary Brandt Line divided the rich and poor countries, roughly following the line of the equator. The North included the United States, Canada, Europe, the USSR, Australia and Japan. The South represented the rest of Asia, Central and South America, and all of Africa. Once again, these terms have become obsolete as countries have developed differently and ignored these imaginary boundaries.

Today, we use terminology such as 'more economically developed country' (MEDC) and 'less economically developed country' (LEDC) to describe levels of development — in the economic, social, environmental and political spheres. A newly **industrialised** country (NIC) is one that is modernising and changing quickly, undergoing rapid economic growth. Emerging economies (EEs) are places also experiencing rapid economic growth, but these are somewhat volatile in that there are significant political, monetary or social challenges.

FIGURE 3 World map showing MEDCs and LEDCs

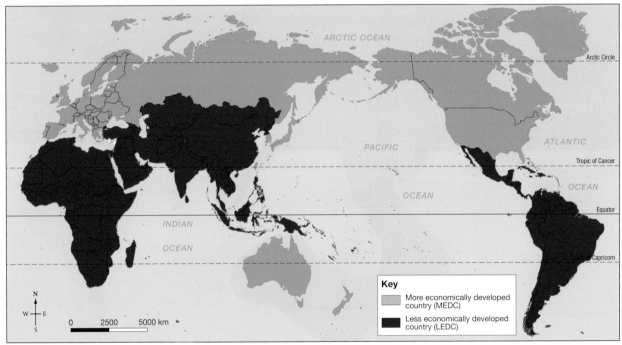

Source: CantGeoBlog.

8.2.3 Defining poverty

There is a strong interconnection between development and poverty. The United Nations defines poverty as

> a denial of choices and opportunities, a violation of human dignity ... It means not having enough to feed and clothe a family, not having a school or clinic to go to, not having the land on which to grow one's food or a job to earn one's living ... It means susceptibility to violence, and it often implies living in marginal or fragile environments, without access to clean water or sanitation.

However, poverty is most often measured using solely economic indicators. More than one billion people live in extreme poverty, as represented in **FIGURE 4**.

FIGURE 4 The proportion of the world's population living on less than US $1.90 per day, the World Bank's global poverty line indicator

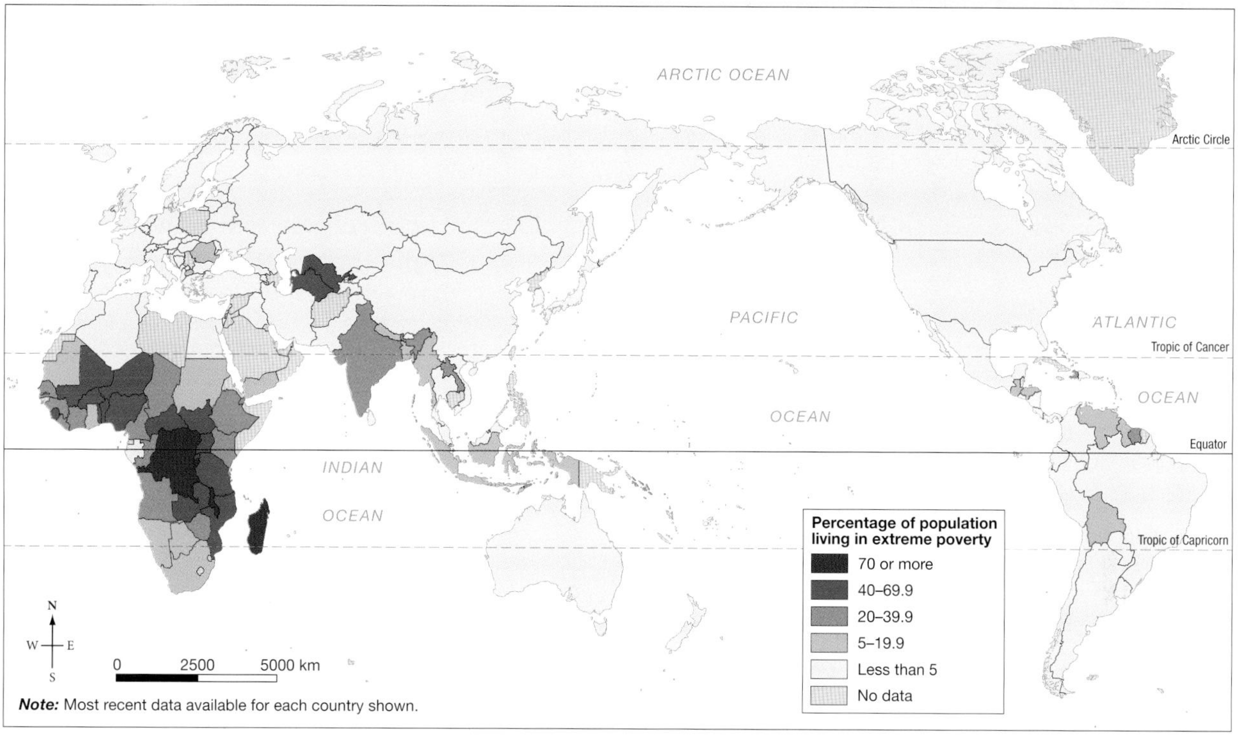

Note: Most recent data available for each country shown.

Percentage of population living in extreme poverty
- 70 or more
- 40–69.9
- 20–39.9
- 5–19.9
- Less than 5
- No data

Source: World Bank – World Development Indicators.

Explore more with myWorldAtlas

Deepen your understanding of this topic with related case studies and questions.
- Investigating Australian Curriculum topics > Year 10: Geographies of human wellbeing > **Wellbeing in Western Sydney**
- Investigating Australian Curriculum topics > Year 10: Geographies of human wellbeing > **Wellbeing in Sudan**

8.2 INQUIRY ACTIVITY

Select one of the indicator categories: social, economic or ***environmental***. In pairs or small groups, brainstorm the various indicators that you think might be used to measure the category. Create a short list of at least five before checking the World Statistics section of your atlas to see which indicators are commonly used.

Evaluating, predicting, proposing

8.2 EXERCISES

Geographical skills key: GS1 Remembering and understanding **GS2** Describing and explaining **GS3** Comparing and contrasting **GS4** Classifying, organising, constructing **GS5** Examining, analysing, interpreting **GS6** Evaluating, predicting, proposing

8.2 Exercise 1: Check your understanding

1. **GS1** Identify two examples of ***places*** that would have been classified as 'developed North' and two that would have been classified as 'undeveloped South'.
2. **GS2** What do you think about Australia being labelled a part of the 'developed North'? Explain.
3. **GS5** Look at **FIGURES 2** and **3**. List any differences you can see with the different forms of categorisation. ▶

---

4. **GS2** Although indicators measure different aspects of quality of life, they are also ***interconnected***. For example, if a country goes through an economic recession, other indicators will be affected. Explain with examples (a flow chart may be useful to step out your thinking).
5. **GS3** Complete **TABLE 1** to compare the differences between MEDCs and LEDCs (try to include your own explanations where possible).

TABLE 1 Comparison of MEDCs and LEDCs

	MEDC	LEDC
Birth rate		High — many children die so the birth rate increases to counteract fatalities
Death rate	Low — good medical care available	
Life expectancy	High — good medical care and quality of life	
Infant mortality rate		High — poor medical care and nutrition
Literacy rate	High — access to schooling, often free	
Housing type		Poor — often no access to fresh water, no sanitation, infrequent or no electricity

8.2 Exercise 2: Apply your understanding

1. **GS4** Refer to **FIGURE 1**.
   (a) Using the figure as a guide, create a table to classify each of the following as either a quantitative or qualitative indicator.
   - Motor vehicle ownership rates
   - Unemployment
   - Forest area
   - Incidence of obesity
   - Freedom of speech
   - Proportion of seats held by women in national parliaments
   - Electric power consumption
   - Quality of teaching at your school
   - How safe you feel walking in the city at night
   - How much you trust your neighbours
   - Access to public transport

   (b) Which indicators were difficult to classify, and why do you think this is the case?
2. **GS6** Look back over the indicators in question 1. Indicators can also suggest further information about a country's progress, rate of ***change*** or development. Could these indicators be clues to the factors affecting the development of a country? If so, what else do they tell you?
3. **GS2** Does your pet dog or cat have a good life? What indicators would you use to measure this? Write a selection of five quantitative and five qualitative indicators to help determine the wellbeing of your pet.
4. **GS6** The concept of wellbeing is relative to who you are and the ***place*** where you live. Consider the following statements. Does the term 'wellbeing' have any relevance to these people? Does wellbeing hold any relevance for people in the direst poverty? Write a paragraph to explain your view.
   Person A: 'We live in constant fear, starvation; there is a lack of government. Personal safety is crucial, so wellbeing is not there yet. Things are very difficult as people are living in despair.'
   Person B: 'Before, we always talked of improving living standards, which mostly meant material needs. Now we talk of the importance of relationships among people and between people and the environment.'
   Person C: 'The land looks after us. We have plenty to eat, but things are changing. There are no fish now, not like when my father was a boy.'

5. **GS6** How do you compare? As a teenager in Australia, you might think you have it tough. But, when we look at the indicators, is that really the case? Decide whether you are better off or worse off for each indicator in **TABLE 2** by evaluating the data. What reasons could account for these differences?

TABLE 2 Australia versus the world — a selection of quantitative indicators, 2017

Life expectancy (years)	Australia	83	Sierra Leone (Africa)	52
Mobile phones (subscriptions per 100 people)	Australia	131	Eritrea (Africa)	6
Adolescent fertility rate (births per 1000 women 15–19 years of age)	Australia	12.9	Denmark (Europe)	4
Proportion of seats held by women in national parliament (%)	Australia	32.7	Rwanda (Africa)	56
Gross National Income per capita (US$)	Australia	43 560	Qatar (West Asia)	116 818
Literacy rate (% of youth aged 15–24)	Australia	99	Mozambique (Africa)	Males 80 Females 57

Try these questions in learnON for instant, corrective feedback. Go to www.jacplus.com.au.

8.3 SkillBuilder: Constructing and interpreting a scattergraph

What is a scattergraph?

A scattergraph is a graph that shows how two or more sets of data, plotted as dots, are interconnected. This interconnection can be expressed as a level of correlation.

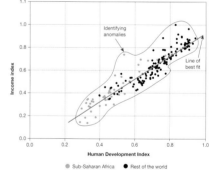

Select your learnON format to access:

- an overview of the skill and its application in Geography (Tell me)
- a video and a step-by-step process to explain the skill (Show me)
- an activity and interactivity for you to practise the skill (Let me do it)
- questions to consolidate your understanding of the skill.

on Resources

Video eLesson Constructing and interpreting a scattergraph (eles-1756)

Interactivity Constructing and interpreting a scattergraph (int-3374)

8.4 Wealth and wellbeing

8.4.1 The multiple component index

A wellbeing approach to development takes into account a variety of quantitative and qualitative indicators. Some of these are a little more difficult to measure, such as the idea of happiness. Before you read on, make a list of 10 indicators that you think would give an accurate measure of a teenager's happiness in their country of residence.

A single indicator gives us only a narrow picture of the development of a country. A country may have a very high GDP but, if we dig a little deeper and look at each individual's share in that country's income or their **life expectancy**, we may not find what we expected. Inequalities may be revealed.

GLOBAL WEALTH

The richest one per cent of adults worldwide owned 47 per cent of global assets in the year 2018, and the richest 10 per cent of adults accounted for 85 per cent of the world total. In contrast, the bottom half of the world adult population owned less than one per cent of global wealth.

Wealth is heavily concentrated in North America, western Europe and high-income Asia–Pacific countries (excluding China). People in these countries collectively hold 78 per cent of total world wealth.

Source: Credit Suisse Wealth Report, 2018

A combination of many indicators will create a more accurate picture of the level of wellbeing in a particular place. Much like using our five senses to try a new cuisine, a combination of indicators will give us better insight into a country's wellbeing. The **Human Development Index** (HDI) is one such index. It was developed in 1990 and measures wellbeing according to three key indicators (see **FIGURE 1**).

FIGURE 1 The HDI measures quality of life according to three key factors.

8.4.2 Other measures of wellbeing

Over thousands of years, different societies have measured progress in different ways. A GDP-led development model focuses solely on boundless economic growth on a planet with limited resources — and this is not a balanced equation. The HDI has become one of the most common ways to measure wellbeing, but it has also attracted criticism for its narrow approach. These measures do not recognise some of the greatest environmental, social and humanitarian challenges of the twenty-first century, such as pollution or stress levels.

Measuring twenty-first century wellbeing

The new Happy Planet Index (HPI) results map the extent to which 151 countries across the globe produce long, happy and sustainable lives for the people that live in them. The coloured shading in the **FIGURE 2** cartogram represents each country's HPI score, while the relative sizes and shapes of countries are determined by their population size (see the subtopic 8.5 SkillBuilder to learn more).

FIGURE 2 Cartogram showing relative global populations and Happy Planet Index scores, 2016

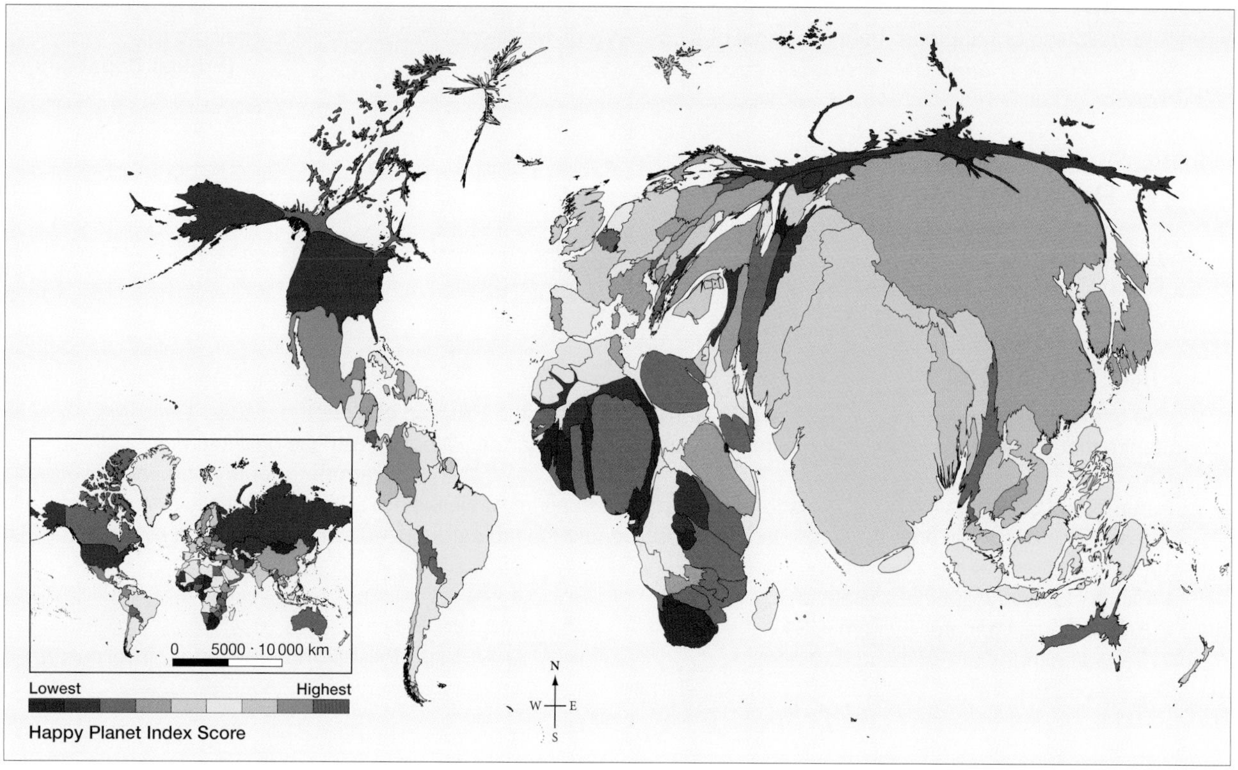

Happy Planet Index Score (Lowest — Highest)

Source: http://happyplanetindex.org/

Each of the three component measures — life expectancy, **experienced wellbeing** and **ecological footprint** — is given a traffic-light score based on thresholds for good (green), middling (amber) and bad (red) performance. These scores are combined to an expanded six-colour traffic light for the overall HPI score. To achieve bright green (the best of the six colours), a country would have to perform well on all three individual components.

$$\text{Happy Planet Index} = \frac{\text{experienced wellbeing} \times \text{life expectancy}}{\text{ecological footprint}}$$

Gross National Happiness

In 2011, the Prime Minister of Bhutan (Central Asia) demonstrated his country's commitment to its wellbeing by developing the world's first measure of national happiness, and he encouraged world nations to do the same. Former UN Secretary-General Ban Ki-moon supported this innovation: 'Gross national product (GNP) ... fails to take into account the social and environmental costs of so-called progress ... Social, economic and environmental wellbeing are indivisible. Together they define gross global happiness.'

Australia's assessment of wellbeing

The Australian National Development Index (ANDI), which was approved in 2015, incorporates 12 indicators measuring elements of progress including health, education, justice and Indigenous wellbeing. Measures such as this demonstrate a new direction in articulating wellbeing, recognising that happiness is not directly proportional to our bank balance or how long we expect to live. This new measure of wellbeing will reflect what is important to Australians to feel happy as individuals, as well as the happiness of our communities. It will allow Australians to measure the future we want.

DISCUSS

Should wellbeing or happiness be a core goal of a country's government? Debate this in a small group.

[Ethical Capability]

 Resources

 Weblink HPI

8.4 INQUIRY ACTIVITIES

1. Use the **HPI** weblink in the Resources tab to learn more about the HPI and explore the results. List three results that surprised you, and why. Compare your list with a partner. What similarities or differences did you find? **Examining, analysing, interpreting**
2. A number of countries have already adopted a national measure of wellbeing. Either individually or in pairs, research the history of one of the following indices, identify the indicators used to measure it and evaluate its success.
   - Gross National Happiness (Bhutan)
   - Key National Indicator System (USA)
   - Canadian Index of Wellbeing (Canada) **Examining, analysing, interpreting**

8.4 EXERCISES

Geographical skills key: GS1 Remembering and understanding **GS2** Describing and explaining **GS3** Comparing and contrasting **GS4** Classifying, organising, constructing **GS5** Examining, analysing, interpreting **GS6** Evaluating, predicting, proposing

8.4 Exercise 1: Check your understanding

1. **GS2** Define the wellbeing approach and show how it is a multiple component index.
2. **GS2** Explain why a multiple component index is significant.
3. **GS2** Provide a detailed explanation of each of the indicators used to calculate the HDI. Is the HDI the best indicator of a country's development? Give reasons for your answer.
4. **GS6** Without referring to **FIGURE 2**, name three *places* you would expect to appear high on the Happy Planet Index and three you would expect to appear low. Now, check your predictions on the map. Were you correct?
5. **GS6** The measurement of happiness has become important in the twenty-first century. Why do you think this is so?

8.4 Exercise 2: Apply your understanding

1. **GS5** What does the Credit Suisse Wealth Report of 2018 say about the inequality of wealth across the world?
2. **GS5** Using the Happy Planet Index and **FIGURE 2**, explain what wellbeing conditions you might find in the following countries:
   (a) South Africa
   (b) France
   (c) the United States.
3. **GS6** Suggest why a range of indices is being developed in the twenty-first century to measure wellbeing.
4. **GS6** Suggest two indicators that might be used in the ANDI.
5. **GS5** Comment on the distribution of the happiest and unhappiest countries across the world according to the data in **FIGURE 2**. What do you think would make a country unhappy?

Try these questions in learnON for instant, corrective feedback. Go to www.jacplus.com.au.

8.5 SkillBuilder: Interpreting a cartogram

What is a cartogram?

A cartogram is a diagrammatic map; that is, it looks like a map but is not a map as we usually know it. These maps use a single feature, such as population, to work out the shape and size of a country. Therefore, a country is shown in its relative location but its shape and size may be distorted. Cartograms are generally used to show information about populations and social and economic features.

Select your learnON format to access:

- an overview of the skill and its application in Geography (Tell me)
- a video and a step-by-step process to explain the skill (Show me)
- an activity and interactivity for you to practise the skill (Let me do it)
- questions to consolidate your understanding of the skill.

on Resources

Video eLesson   Interpreting a cartogram (eles-1757)

Interactivity   Interpreting a cartogram (int-3375)

8.6 Improving wellbeing

8.6.1 Giving aid to bridge the gap

We have much to be thankful for. We live in a world where we live much longer than our ancestors, we have better nutrition and education, and we generally have a better outlook for our lives. But in an age where some are globally connected, educated, fed, clothed and medicated, it is easy to forget that many of our fellow human beings go without, each and every day.

Have you ever given some loose change to a tin-shaker on the street or helped collect money for a fundraiser? If so, then you are already a part of the cycle of aid. Aid (also known as international aid, overseas aid or foreign aid) is the voluntary transfer of resources from one country to another, given at least partly with the aim of benefiting the receiving country.

Why do we give aid?

Aid may be given by government, private organisations or individuals. **Humanitarianism** is still the most significant motivation for the giving of aid, but it may be motivated by other functions as well:

- as a sign of friendship between two countries
- to strengthen a military ally
- to reward a government for actions approved by the donor
- to extend the donor's cultural influence
- to gain some kind of business or commercial access to a country.

8.6.2 What types of aid exist?

Bilateral aid is aid given by governments to donor countries. Multilateral aid is provided through international institutions such as UNICEF. **Non-government organisation** (NGO) or charity aid is

---

voluntary, private, individual donations collected by organisations such as the Red Cross. Aid takes many forms: money, food, medicine, equipment, expertise, scholarships, training, clothing or military assistance (to name just a few). Large-scale aid (top-down aid) is usually given to the government of a developing country so that it can spend it on the projects that it needs. Small-scale aid projects (bottom-up aid) target the people most in need of the aid and help them directly, without any government interference. Aid from NGOs tends to be bottom-up aid.

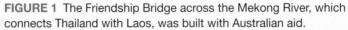

FIGURE 1 The Friendship Bridge across the Mekong River, which connects Thailand with Laos, was built with Australian aid.

There are positive and negative impacts of aid (see **TABLE 1**). Aid can increase the dependency of LEDCs on donor countries. Sometimes aid is not a gift but a loan, and poor countries may struggle to repay the money. Aid may also be used to put political or economic pressure on a country, which may leave its people feeling like they owe their donors a favour. There is always the threat that corruption among politicians and officials will prevent aid from reaching the people who need it most. If aid does not provide for and empower citizens, then wellbeing will not be improved.

TABLE 1 Advantages and disadvantages of different types of aid

	Advantages	**Disadvantages**
Bilateral aid	• Helps expand infrastructure: roads, railways, ports, power generation • Aid that directly supports economic, social or environmental policies can result in successful programs.	• 'Tied aid' obliges the country receiving aid to spend it on goods and services from the donor country (may be expensive). • Inappropriate technology may be given (e.g. tractors are of little use if there are no spare parts or fuel).
Multilateral aid	• The organisations have clear aims about what they are trying to achieve (e.g. WHO combats disease and promotes health). • Leading experts in their field work to achieve multilateral aid program objectives.	• Sometimes directed only towards specific areas or organisations, leaving many without benefit • May come with conditions to make big changes to structures, which can be difficult to manage once aid has 'finished'
NGO/charity aid	• Usually targeted at long-term development within a country • Raises awareness of specific situations in a country or region	• The greatest source of need may not be prioritised (e.g. the 2006 tsunami devastation received many donations, but areas in sub-Saharan Africa were just as much in need daily). • Up to 30 per cent of donations may be 'eaten up' by administration costs.

8.6.3 How does Australia help?

The Australian government's official development assistance (ODA) is designed to promote prosperity, reduce poverty and enhance stability in developing countries. In 2018–19, Australia provided $4.2 billion worth of official development assistance, 90 per cent of which was allocated to the Indo-Pacific region. Australia's ODA focuses on strengthening private sector development and enabling human development. Specifically, it contributes to investment in trade, infrastructure, agriculture, fisheries, water, health, education, gender equality and effective governance.

FIGURE 2 Australian aid has helped these primary school students in north-western Laos.

DISCUSS

Is aid ever inappropriate? Discuss this in a small group. **[Ethical Capability]**

 Resources

Interactivity Helping others (int-3305)

8.6 INQUIRY ACTIVITIES

1. Think of a charity that you or your family have supported in the past. Find out more about the charity. Where is your money going, and who are the beneficiaries? **Examining, analysing, interpreting**
2. Using the internet, review some of the Australian government programs currently in operation. In which **places** are most of these programs focused? **Examining, analysing, interpreting**
3. Discuss in a small group what limitations might exist in administering an aid program in (a) a developing country and (b) a country that has been devastated by a natural disaster (e.g. an earthquake). Suggest possible ways of overcoming the problems you identify. **Evaluating, predicting, proposing**

Geographical skills key: GS1 Remembering and understanding **GS2** Describing and explaining **GS3** Comparing and contrasting **GS4** Classifying, organising, constructing **GS5** Examining, analysing, interpreting **GS6** Evaluating, predicting, proposing

8.6 Exercise 1: Check your understanding

1. **GS2** What is the difference between the three types of aid?
2. **GS1** List the various forms of aid mentioned in this subtopic. Can you add any more types to this list?
3. **GS1** What is the difference between large-*scale* aid and small-*scale* aid? Provide an example of each to illustrate your answer.
4. **GS1** What are the motivating factors behind the giving of aid?
5. **GS2** What difference does aid as a gift make for a country, rather than aid as a loan?

8.6 Exercise 2: Apply your understanding

1. **GS2** Study **FIGURES 1** and **2**. What benefits would each of these aid projects bring to the recipients?
2. **GS6** Do you think the Australian government's focus will shift in 10 years' time? In 50 years' time? Which region do you think we might have to shift our focus to?
3. **GS5** Reflect on what you have studied so far in this topic. Why are some people 'poor' and some people 'rich'?
4. **GS6** Think about the challenges that might be faced by someone delivering emergency aid to an LEDC. How might they be affected by the physical and emotional conditions of their work?
5. **GS6** Do the positives of aid outweigh the negatives? Outline your view.

Try these questions in learnON for instant, corrective feedback. Go to www.jacplus.com.au.

8.7 The way forward

8.7.1 Defining poverty

Poverty is a living condition that is hard to measure and difficult to define. For example, people living in poverty in sub-Saharan Africa may struggle to find a daily meal or access to safe drinking water. This standard of living is known as **absolute poverty** and approximately 1.3 billion people experience this daily. In more developed countries, we may refer to **relative poverty**. In Australia, for example, we compare living standards to a benchmark called the **poverty line**. The majority of the population lives above the poverty line, but those who fall below it struggle to meet their day-to-day needs of adequate shelter, food and education.

Sub-Saharan Africa — the poorest region in the world

Sub-Saharan Africa is affected by many different forms of poverty. HDI scores in most countries of sub-Saharan Africa have worsened since 1990, making this region the poorest in the world. Poverty is more common in young families, who often have less assets and higher dependency ratios (people who are too young or too old to work). For others, the poverty is chronic rather than temporary. This means poverty is experienced for most of one's life and is often passed on to one's children, creating a generational cycle that is hard to break. The chronically poor experience hunger, undernutrition, illiteracy, lack of access to safe drinking water and basic health services, social discrimination, physical insecurity and political exclusion. Many will

FIGURE 1 The countries of sub-Saharan Africa are affected by many forms of poverty.

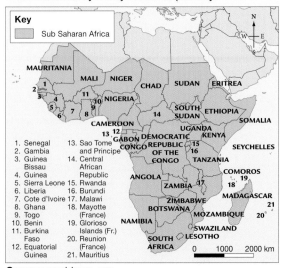

Key
Sub Saharan Africa

MAURITANIA
MALI NIGER CHAD SUDAN ERITREA
NIGERIA
SOUTH ETHIOPIA
SUDAN SOMALIA
CAMEROON
GABON DEMOCRATIC UGANDA KENYA
CONGO REPUBLIC
OF THE SEYCHELLES
CONGO TANZANIA
ANGOLA COMOROS
ZAMBIA MADAGASCAR
ZIMBABWE
BOTSWANA MOZAMBIQUE
NAMIBIA SWAZILAND
SOUTH LESOTHO
AFRICA

1. Senegal
2. Gambia
3. Guinea Bissau
4. Guinea
5. Sierra Leone
6. Liberia
7. Cote d'Ivoire
8. Ghana
9. Togo
10. Benin
11. Burkina Faso
12. Equatorial Guinea
13. Sao Tome and Principe
14. Central African Republic
15. Rwanda
16. Burundi
17. Malawi
18. Mayotte (France)
19. Glorioso Islands (Fr.)
20. Reunion (France)
21. Mauritius

0 1000 2000 km

Source: worldmap.org

die prematurely of easily preventable deaths. With appropriate support and resources, people living in chronic poverty have the ability to overcome the obstacles that trap them in poverty, and can create a better future for themselves and their children. There are numerous organisations and agencies that work with such communities to help them break the poverty cycle.

World hunger

One of the most pressing issues of poverty is hunger, a situation where people experience scarcity of food. Malnutrition is a general term that indicates a lack of some or all nutritional elements necessary for human health. There are two basic types of malnutrition. When we refer to world hunger, we are talking about a lack of food that provides energy (measured in calories; see **FIGURE 2**), obtained from all the basic food groups, and a lack of protein (from meat and other sources; see **FIGURE 3**). Another type of malnutrition is micronutrient (vitamin and mineral) deficiency, which may or may not occur with hunger. Recently there has also been a move to include obesity as a third form of malnutrition, expanding on the idea of poor nutrition.

FIGURE 2 Daily calorie intake, per person

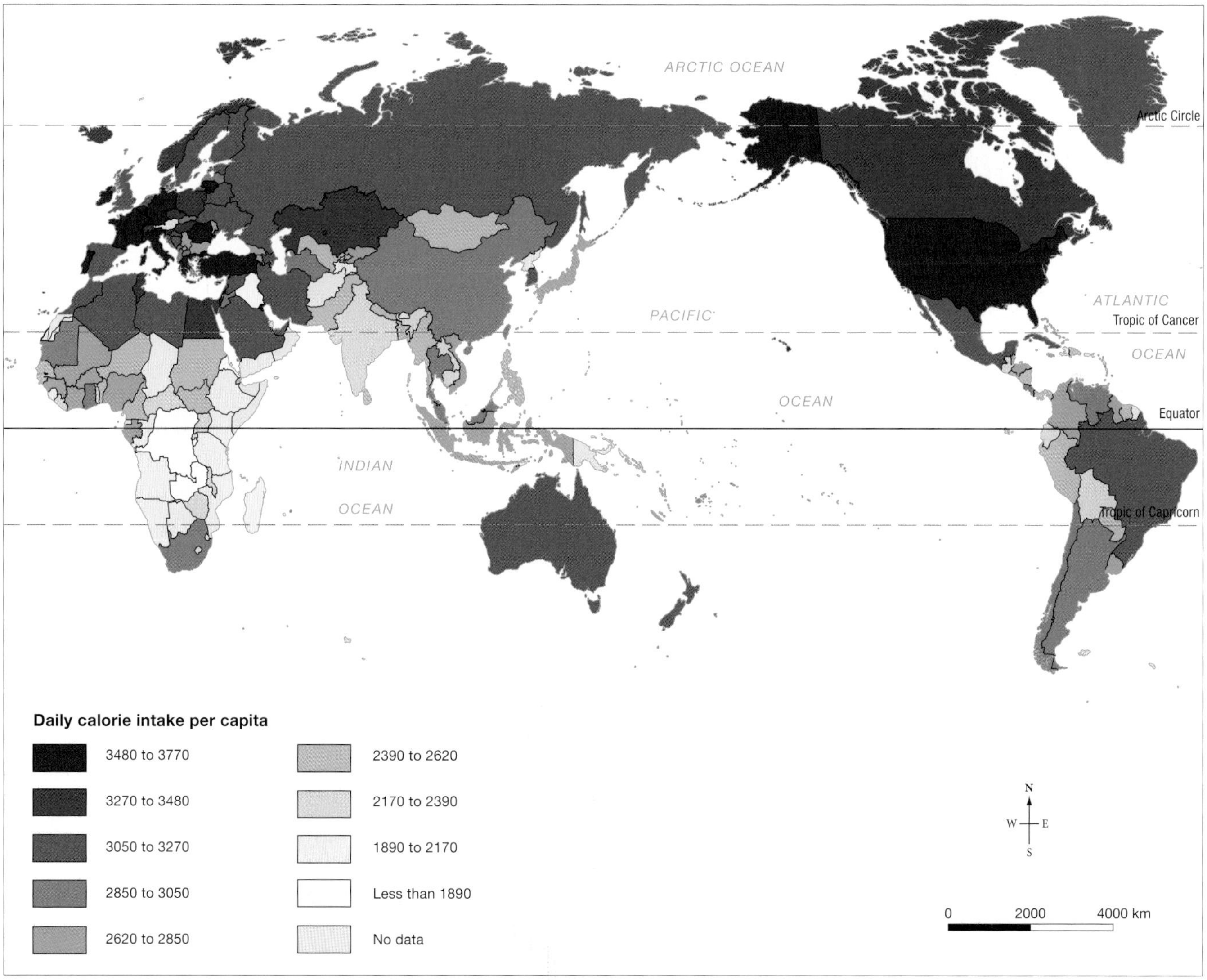

ARCTIC OCEAN
Arctic Circle
PACIFIC
OCEAN
ATLANTIC
OCEAN
Tropic of Cancer
Equator
INDIAN
OCEAN
Tropic of Capricorn
Daily calorie intake per capita
3480 to 3770
3270 to 3480
3050 to 3270
2850 to 3050
2620 to 2850
2390 to 2620
2170 to 2390
1890 to 2170
Less than 1890
No data
N
W
E
S
0
2000
4000 km

Source: FAO 2013 Made with Natural Earth. Map by Spatial Vision.
Note: Most recent data available

FIGURE 3 Prevalence of undernourishment in the world, 2017, by region

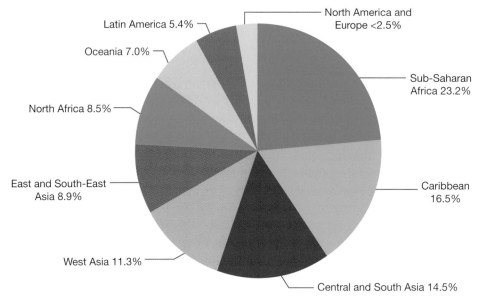

Latin America 5.4%

Oceania 7.0%

North Africa 8.5%

East and South-East Asia 8.9%

West Asia 11.3%

North America and Europe <2.5%

Sub-Saharan Africa 23.2%

Caribbean 16.5%

Central and South Asia 14.5%

Source: Data from *The State of Food Security and Nutrition in the World 2018*, report jointly prepared by FAO, IFAD, UNICEF, WFP and WHO.

FIGURE 4 While obesity levels are increasing every year in populations of countries such as Australia, billions of people across the world still suffer from starvation.

Source: Global Food Consumption — Richard & Slavomir Svitalsky/Cartoon Movement.

8.7.2 Sustainable Development Goals

The Sustainable Development Goals (SDGs) came into being on 25 September 2015 to replace the expired Millennium Development Goals (MDGs). The three overarching themes are to end poverty, protect the planet and ensure prosperity for everyone over the 15 years to 2030. Each of the 17 goals has a number of targets to be met. Indicators are used to assess each target. These SDGs apply to all countries. The United Nations provides an update on the progress towards 2030 in an annual report.

Former New Zealand Prime Minister and United Nations Development Programme Administrator Helen Clark commented, 'This agreement marks an important milestone in putting our world on an inclusive and sustainable course. If we all work together, we have a chance of meeting citizens' aspirations for peace, prosperity, and wellbeing and to preserve our planet'.

TABLE 1 A brief outline of the Sustainable Development Goals

Sustainable development goal	Targets
Goal 1: End poverty in all its forms everywhere	Major target: • By 2030 no-one should live on less than $1.25 per day.
Goal 2: End hunger, achieve food security and improved nutrition and promote sustainable agriculture	Significant targets include: • By 2030 ensure access by all people to safe, nutritious and sufficient food all year round. • By 2030 end all forms of malnutrition.
Goal 3: Ensure healthy lives and promote wellbeing for all at all ages	Targets for 2030 include: • Reduce the global maternity mortality ratio to less than 70 per 100 000 live births. • Attain an under-five mortality of, at most, 25 per 1000 live births. • End the epidemics of AIDS, tuberculosis and malaria.
Goal 4: Ensure inclusive and quality education for all and promote lifelong learning	Targets to achieve include: • By 2030 all boys and girls can complete free, equitable and quality primary and secondary schooling with effective outcomes. • All women and men have equal access to affordable and quality ongoing educational opportunities.
Goal 5: Achieve gender equality and empower all women and girls	Targets include: • End discrimination against all women and girls everywhere and eliminate violence towards them too.
Goal 6: Ensure access to water and sanitation for all	Targets by 2030 include: • Achieve universal and equitable access to safe and affordable drinking water for all. • Achieve access to adequate and equitable sanitation and hygiene for all.
Goal 7: Ensure access to affordable, reliable, sustainable and modern energy for all	Targets for 2030 include: • Provide access to affordable, reliable and modern energy services, especially renewable energies.

(continued)

TABLE 1 A brief outline of the Sustainable Development Goals *(continued)*

Sustainable development goal	Targets
Goal 8: Promote inclusive and sustainable economic growth, employment and decent work for all	Major targets: • Sustain economic growth and productivity aiming to achieve by 2030 full and productive employment and decent work for all. • By 2025 eliminate child labour in all its forms, including forced labour, modern slavery, human trafficking and child soldiers.
Goal 9: Build resilient infrastructure, promote sustainable industrialisation and foster innovation	General targets include: • Develop quality, reliable, sustainable and resilient infrastructure to support economic development and human wellbeing. • Promote inclusive and sustainable industries that raise industries' share of employment and GDP. • Provide universal and affordable internet access to least developed countries by 2020.
Goal 10: Reduce inequality within and among countries	Key target: • By 2030 achieve and sustain income growth of the bottom 40 per cent of the population.
Goal 11: Make cities inclusive, safe, resilient and sustainable	Targets include: • By 2030 ensure adequate, safe, affordable and sustainable housing and transport for all. • Protect and safeguard the world's cultural and natural heritage.
Goal 12: Ensure sustainable consumption and production patterns	2030 targets include: • Achieve the sustainable management and efficient use of natural resources • Halve per capita global food waste by consumers and during production. • Ensure all people have the relevant information and awareness for sustainable development and lifestyles in harmony with nature.
Goal 13: Take urgent action to combat climate change and its impacts	Targets include: • Strengthen resilience and adaptive capacity to climate-related hazards and natural disasters in all countries. • Integrate climate change measures into national policies, strategies and planning.
Goal 14: Conserve and sustainably use oceans, seas and marine resources	Significant targets: • By 2025 prevent and reduce marine pollution of all kinds. • By 2020 sustainably manage and protect marine and coastal ecosystems. • By 2020 regulate and end overfishing.

	Goal 15: Sustainably manage forests, combat desertification, halt and reverse land degradation, halt biodiversity loss	Targets: • By 2020 protect inland freshwater ecosystems and all types of forests. • By 2020 prevent the introduction of invasive alien species and prevent the extinction of threatened species. • By 2030 combat desertification and protect mountain ecosystems.
	Goal 16: Promote just, peaceful and inclusive societies	Targets: • End abuse, exploitation, trafficking and all forms of violence against children. • Reduce bribery and corruption in all forms. • By 2030 provide legal identity for all, including birth registration.
	Goal 17: Strengthen the means of implementation and revitalise the global partnership for sustainable development	Targets: • Address finance, technology, capacity building, trade and systemic issues to support sustainable development goals..

8.7.3 What can we do?

The fact that over a billion people experience severe hardship every day can appear extremely daunting. Compassion is the first step, but action is what is required. It is important to remember that those experiencing hardship do not live exclusively in less developed countries. Some of them live in more developed countries, like Australia. Many of them live in places where wellbeing for some is improving dramatically, but there are many who are being left behind.

International aid provides just one avenue for change. It is **grassroots movements** that will provide the greatest and most effective change for those who are most disadvantaged or without a voice. Education can provide a means for people to change their own destiny. Lowering population growth will reduce population pressures on a country's resources and government, and good governance can build a strong economy and provide opportunities for residents.

8.7 INQUIRY ACTIVITIES

1. Within your class, divide into groups and assign the SDGs across the class. Using the internet, research the targets of each goal and provide a tick or a cross depending on whether you think the world will meet each target as set down. (A number of targets were to have been met by 2020 – have these been achieved?) Be prepared to argue your point of view in a class debate on 'Are there too many goals and targets to be met by all countries in the world?'   **Evaluating, predicting, proposing**

2. In pairs or small groups, design a country-specific program that you think would help alleviate chronic poverty in one of sub-Saharan Africa's most poverty-stricken countries.   **Evaluating, predicting, proposing**

3. Australians living below the poverty line are assisted by charitable organisations and individuals who commit to improving the wellbeing of others. Research and list the services that are available in your local area. Consider other possible initiatives. For example, the Young Australian of the Year Award 2016 went to two individuals who provided a mobile laundry for the homeless. What other services could be provided to assist those in need?   **Examining, analysing, interpreting**

4. Poverty can be caused by a number of factors, both natural and human. Create a mind map that explores a natural or human-made cause of poverty; for example, exactly how would conflict or *environmental* crisis create conditions of poverty?   **Classifying, organising, constructing**

8.7 EXERCISES

Geographical skills key: GS1 Remembering and understanding **GS2** Describing and explaining **GS3** Comparing and contrasting **GS4** Classifying, organising, constructing **GS5** Examining, analysing, interpreting **GS6** Evaluating, predicting, proposing

8.7 Exercise 1: Check your understanding

1. **GS1** What is the poverty line?
2. **GS2** Explain the difference between absolute poverty and relative poverty.
3. **GS1** What are the three overarching themes of the SDGs?
4. **GS1** By when do the SDGs aim to be achieved?
5. **GS5** Refer to **FIGURE 3**.
   (a) Which two regions combined make up nearly 40 per cent of the world's undernourished?
   (b) Which two regions have the lowest prevalence of undernourishment?

8.7 Exercise 2: Apply your understanding

1. **GS4** Using the three overarching themes of the SDGs, draw up a table to show where each of the 17 goals is aligned.
2. **GS2** Refer to **FIGURE 2**. Identify the region of sub-Saharan Africa. If we compare the general results for this region with other regions across the world, how does it rate in terms of calorie intake? Give examples of specific *places* where appropriate.
3. **GS6** Given sub-Saharan Africa's slow progress in improving the wellbeing of its people, choose three SDGs that are most likely to have an impact on improving the region. Justify your answer.
4. **GS5** Analyse the **FIGURE 4** cartoon from the perspective of each person.
5. **GS6** How ambitious is Helen Clark's quote (in section 8.7.2)? Do you think this is achievable by 2030? Which aspects do you think may be more achievable than others? Write an extended response outlining your view, and providing reasons for your opinion.

Try these questions in learnON for instant, corrective feedback. Go to www.jacplus.com.au.

8.8 The importance of human rights

8.8.1 The basis of human rights

Human rights are so much a part of our daily lives here in Australia that we tend to take them for granted. Many principles that have been adopted in international human rights practices have their roots in traditions and religions that are thousands of years old. Different countries, societies and cultures have come up with their own definitions over time to suit their particular environment or context.

In some societies, human rights may be enshrined in law and legislation, whereas in others they may simply exist as guidelines that reflect the values of that particular community. In short, the concept of human rights stems from the belief that there is an instinctive human ability to distinguish right from wrong.

FIGURE 1 Origins of human rights

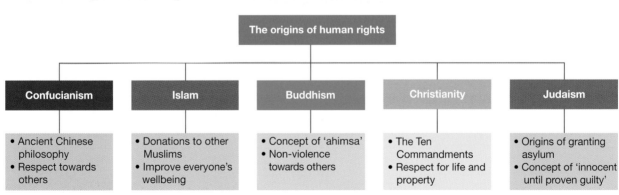

The origins of human rights

Confucianism	Islam	Buddhism	Christianity	Judaism
• Ancient Chinese philosophy • Respect towards others	• Donations to other Muslims • Improve everyone's wellbeing	• Concept of 'ahimsa' • Non-violence towards others	• The Ten Commandments • Respect for life and property	• Origins of granting asylum • Concept of 'innocent until proven guilty'

Human rights can be defined in different ways. The Australian Human Rights Commission notes that definitions may include:

- the recognition and respect of people's dignity
- a set of moral and legal guidelines that promote and protect a recognition of our values
- our identity and ability to ensure an adequate standard of living
- the basic standards by which we can identify and measure inequality and fairness
- those rights associated with the Universal Declaration of Human Rights.

8.8.2 The role of the United Nations

The UN was formed in the aftermath of World War II on 24 October 1945 by countries committed to preserving peace through international cooperation and security. Today, nearly every nation (currently 193 countries) in the world belongs to the UN. One of the main aims of the UN Charter is to promote respect for human rights. The **Universal Declaration of Human Rights**, proclaimed by the UN General Assembly in 1948, sets out basic rights and freedoms to which all women and men are entitled, including:

- the right to life, liberty and nationality
- the right to freedom of thought, conscience and religion
- the right to work and to be educated
- the right to food and housing
- the right to take part in government.

These rights are legally binding by virtue of two **International Covenants**, to which most states are parties. One covenant deals with economic, social and cultural rights, and the other deals with civil and political rights. Together with the Declaration, they constitute the **International Bill of Human Rights**. **FIGURE 2** shows where political rights exist or are lacking throughout the world.

FIGURE 2 Political rights around the world, 2018. A free country is one where political rights are available and protected. A country that is not free is one where basic political rights are absent, and basic civil liberties are widely and systematically denied.

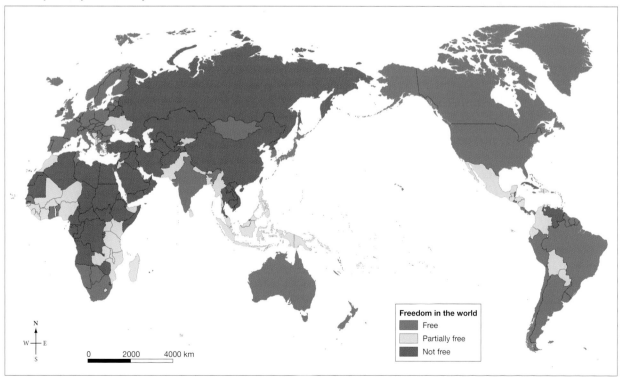

Source: Freedom House. Map drawn by Spatial Vision.

8.8.3 Who protects our human rights?

Although Australia has agreed to be bound by these major international human rights treaties, they do not form part of Australia's domestic law unless they have been specifically written into Australian law through legislation. The Australian Human Rights Commission is the national organisation that advocates for promotion and protection of human rights. In addition to monitoring economic, social, cultural, civil and political rights, other areas of human rights include peacekeeping, eradication of poverty and the humanitarian tribunals (for example, the International Criminal Court that deals with mass human rights violations, such as genocide). Amnesty International is a global organisation that works to uphold human rights. One area on which it focuses its human rights advocacy is the death penalty (capital punishment), a contentious issue on the global political stage (see **FIGURE 3**).

FIGURE 3 The death penalty violates the right to life as proclaimed in the Universal Declaration of Human Rights.

At least 21 919 people worldwide were under sentence of death at the end of 2017.	
Top five countries where most executions happened in 2017	**Top five countries where most people were sentenced to death in 2017**
China (unknown, but Amnesty International estimates executions to have been in the thousands)	China (unknown)
Iran 507+	Nigeria 621+
Saudi Arabia 146	Egypt 402+
Iraq 125+	Bangladesh 273+
Pakistan 60+	Sri Lanka 218+
USA 23	Pakistan 200+

According to Amnesty International, of 142 countries worldwide, more than 70 per cent of all the world's countries are abolitionist in law or practice.

8.8.4 Protecting the vulnerable

International human rights organisations recognise that children have special human rights because of their vulnerability to exploitation and abuse. The United Nations General Assembly adopted the Convention on the Rights of the Child (the CRC) in November 1989. How are your rights protected? And what are some of the big issues for children's rights today?

Some of the rights and protections that a **child** is entitled to according to the CRC include:
- the right to life
- the right to a name and a nationality
- the right to live with their parents
- the right to freedom of thought, conscience and religion
- the right to privacy
- protection from abuse and neglect
- the right to education
- the right to participate in leisure, recreation and cultural activities
- protection from economic exploitation
- protection from or prevention of abduction, sale or trafficking.

Two key areas that are currently a focus for rights are the use of children in conflict and the use of children for labour.

Child soldiers and child labour

The issue of children in armed conflict has become a pressing one over the past few decades because of the serious risks of involving children in war or conflict situations. Approximately 300 000 children are believed to be combatants in conflicts worldwide. **Child soldiers** have gone to battle in a range of countries, including Afghanistan, the Central African Republic, the Democratic Republic of the Congo, Myanmar, Somalia, South Sudan, Sudan, Syria and Yemen.

SDG 8 has a target specifically relating to child labour (see **TABLE 1** in subtopic 8.7). From 2012 to 2016 child labour declined, but the rate of decline is slowing. Sub-Saharan Africa has seen an increase in child labour. Places in conflict, experiencing disasters, and of low income have a heightened risk of child labour.

FIGURE 4 Children's involvement in child labour and hazardous work, by region, 2016

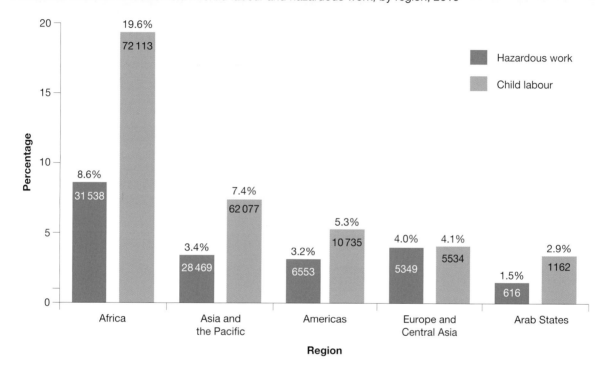

International Labour Organization (ILO) figures for 2016 show that:
- globally, one in ten children work (152 million children between the ages of 5 and 17)
- 73 million children work in hazardous conditions
- the highest number of child labourers are in Africa; almost half the children in labour globally are found there (71.2 million, or one in five children)
- the highest proportion of child labourers is in sub-Saharan Africa, where 27 per cent of children (59 million) work.

In many countries, poor girls are put to work as domestic servants for richer families. In many places, children (especially girls) perform unpaid work for their families. In all cases, children are exploited, and in many cases, they are excluded from attending school (denying them their right to education).

DISCUSS

How might the values and beliefs differ between countries with a higher incidence of child labour and hazardous work compared to those countries with a lower incidence of these issues?   **[Intercultural Capability]**

Resources

✅ **eWorkbook** Child labour (doc-32100)
🔗 **Weblink** Human Rights Watch

8.8 INQUIRY ACTIVITIES

1. Use the internet to find out when Human Rights Day occurs each year and why the date was chosen.
   Examining, analysing, interpreting
2. Refer to **FIGURE 2**. Identify one of the countries that is not free and conduct additional research. What violations of this area of human rights have contributed to this rating? You may wish to use the **Human Rights Watch** weblink in the Resources tab as one source of information.
   Examining, analysing, interpreting
3. Some of the basic human rights are outlined in this subtopic. In pairs or small groups, develop a 'Teenagers' Bill of Rights' (include at least 10 rights) that you believe would provide for a better existence for all teenagers.
   Classifying, organising, constructing
4. Complete the **Child labour** worksheet to investigate and learn more about child labour throughout the world.
   Examining, analysing, interpreting

8.8 EXERCISES

Geographical skills key: GS1 Remembering and understanding **GS2** Describing and explaining **GS3** Comparing and contrasting **GS4** Classifying, organising, constructing **GS5** Examining, analysing, interpreting **GS6** Evaluating, predicting, proposing

8.8 Exercise 1: Check your understanding

1. **GS1** What is the International Bill of Human Rights?
2. **GS2** Define 'human rights' in your own words.
3. **GS2** Who does the Universal Declaration of Human Rights apply to?
4. **GS1** What is the name of the document that sets out the rights of children?
5. **GS1** How is a *child* defined?
6. **GS2** Why do children need a separate declaration outlining their rights?

8.8 Exercise 2: Apply your understanding

1. **GS2** Study **FIGURE 1**. Which philosophies have influenced your understanding of human rights?
2. **GS5** Consider **FIGURE 2**.
   (a) What does this map illustrate?
   (b) Which *places* around the world are 'free', and which are 'not free'?
3. **GS6** Only a small selection of the rights outlined by the Convention on the Rights of the Child (CRC) is provided in this subtopic.
   (a) How would you rank the 10 rights listed in this section in order of importance (1 being the most important)? Justify your choices.
   (b) Do you think someone in sub-Saharan Africa would agree with your choices? Explain.
4. **GS6** Since the 2008 global financial crisis, the situation for child labourers has *changed* for the worse. Why do you think this might be the case? Justify your explanation.
5. **GS6** If you had to stay home and babysit your younger siblings, then your right to an education may be compromised. How might simple daily events prevent you from achieving your rights or protections as outlined by the CRC?

Try these questions in learnON for instant, corrective feedback. Go to www.jacplus.com.au.

8.9 Thinking Big research project: SDG progress infographic

SCENARIO

Help the UN make its annual SDG Report more understandable! Working in small groups, you will create an engaging infographic detailing the global progress towards achieving one of the SDGs and outlining one particular country's response to that goal.

Select your learnON format to access:

- the full project scenario
- details of the project task
- resources to guide your project work
- an assessment rubric.

 Resources

 ProjectsPLUS Thinking Big research project: SDG progress infographic (pro-0217)

8.10 Review

8.10.1 Key knowledge summary
Use this dot point summary to review the content covered in this topic.

8.10.2 Reflection
Reflect on your learning using the activities and resources provided.

 Resources

 eWorkbook Reflection (doc-31775)

Crossword (doc-31776)

 Interactivity Measuring and improving wellbeing crossword (int-7675)

KEY TERMS

absolute poverty experienced when income levels are inadequate to enjoy a minimum standard of living (also known as extreme poverty)

child any person below 18 years of age

child soldier a child who is, or who has been, recruited or used by an armed force or armed group in any capacity, including but not limited to children, boys and girls, used as fighters, cooks, porters, messengers, spies or for sexual purposes. This term does not refer only to a child who is taking or has taken a direct part in hostilities.

development According to the United Nations, development is defined as 'to lead long and healthy lives, to be knowledgeable, to have access to the resources needed for a decent standard of living and to be able to participate in the life of the community'.

▶

ecological footprint a measure of human demand on the Earth's natural systems in general and ecosystems in particular; the amount of productive land required by each person for food, water, transport, housing, waste management and other purposes

experienced wellbeing an individual's subjective perception of personal wellbeing

extreme poverty a state of living below the poverty line (US$1.90 per day), and lacking resources to meet basic life necessities (also known as absolute poverty)

grassroots movement action by ordinary citizens, as compared with the government, aid or a social organisation

gross domestic product (GDP) the value of all goods and services produced within a country in a given period, usually discussed in terms of GDP per capita (total GDP divided by the population of the country)

Human Development Index (HDI) measures the standard of living and wellbeing by measuring life expectancy, education and income

humanitarianism concern for the welfare of other human beings

indicator a value that informs us of a condition or progress. It can be defined as something that helps us to understand where we are, where we are going and how far we are from the goal

industrialised having developed a wide range of industries or having highly developed industries

International Bill of Human Rights the informal name given to the Universal Declaration of Human Rights and the two International Covenants

International Covenants a multilateral treaty adopted by the United Nations General Assembly, in force from 1976. It commits those who have signed the Covenant to respect the civil and political rights of individuals and their economic, social and cultural rights.

life expectancy the number of years a person can expect to live, based on the average living conditions within a country

non-government organisation (NGO) an organisation that operates independently of government, usually to deliver resources or serve some social or political purpose

poverty line an official measure used by governments to define those living below this income level as living in poverty

qualitative indicators subjective measures that cannot easily be calculated or measured; e.g. indices that measure a particular aspect of quality of life or that describe living conditions, such as freedom or security

quantitative indicators objective indices that are easily measured and can be stated numerically, such as annual income or the number of doctors in a country

relative poverty where income levels are relatively too low to enjoy a reasonable standard of living in that society

standard of living a level of material comfort in terms of goods and services available. This is often measured on a continuum; for example, a 'high' or 'excellent' standard of living compared to a 'low' or 'poor' standard of living.

Universal Declaration of Human Rights the first specific global expression of rights to which all human beings are inherently entitled

wellbeing a good or satisfactory condition of existence; a state characterised by health, happiness, prosperity and welfare

9 Global variations in human wellbeing

9.1 Overview

The world's population is constantly increasing. Can we fit so many people in the space we have without affecting our quality of life?

9.1.1 Introduction

As the world's living standards have improved, so too our population has grown. In April 2019, the world's population reached 7.7 billion. It is expected that the number of people on the planet will continue to grow, with experts estimating a population of 9.8 billion in 2050 and 11.2 billion by 2100. It is not just a matter of how many people we can fit in a particular place, but also the manner in which we live (our ecological footprint) that affects our wellbeing. Our wellbeing is clearly interconnected with our population characteristics.

---

on Resources

☑ **eWorkbook**　Customisable worksheets for this topic

▦ **Video eLesson**　A long life (eles-1714)

To access a pre-test and starter questions and receive immediate, **corrective feedback** and **sample responses** to every question, select your learnON format at www.jacplus.com.au.

9.2 Global population distribution

9.2.1 Describing global spatial variations

Whether you have travelled within Australia or to another country, you would be aware that there is considerable variation in the number of people found in one place compared to another. The 7.7 billion people on our planet are not spread evenly across space (continents, countries, and rural or urban areas). We may feel crowded in or feel we have plenty of space; we may feel isolated within or embraced as part of a community. This has a major impact on our wellbeing.

Although there is an average of 47.7 persons per square kilometre across the globe, as shown in **FIGURE 1**, the **population density** varies considerably. Places well below this figure include large regions of most continents, particularly in the inland sections, such as central Asia and Australia, with well below 10 persons per square kilometre. Regions of highest density are clustered in Europe, East and South-East Asia and in the eastern half of the United States of America. For example, Germany has a population density of 230 persons per square kilometre, and Japan has 336 persons per square kilometre.

FIGURE 1 also shows that most regions of high density are dominated by large cities. The majority of the world's population lives in urban environments. Within the largest cities, population density may be considerably higher than the country average. Dhaka, the capital city of Bangladesh (see **FIGURE 2**), is considered to be the most densely populated place in the world, with an estimated 47 400 persons per square kilometre.

FIGURE 1 Global population density and the world's largest cities

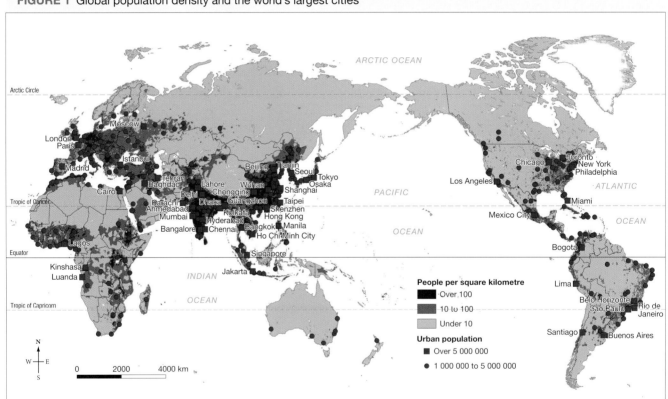

Source: Spatial Vision.

FIGURE 2 View of Dhaka, Bangladesh — an area of high density

9.2.2 Why does population distribution vary so much?

Physical factors play a large part in determining global **population distribution**. Characteristics of the natural environment that favour human settlement include the availability of freshwater resources, fertile soil, moderate climate and sea ports. Inhospitable features such as mountains, jungles and deserts tend to deter high population densities.

Of course, human factors also influence population distribution. Urban places around the world are attracting an increasing number and percentage of people due to the availability of employment, particularly in the manufacturing and service sectors (for example, the industrial areas of Mumbai, India and Yokohama, Japan, and the coastal ports of Rotterdam in the Netherlands and Rio de Janiero in Brazil). Population density is closely interconnected with energy demands, as reflected in the pattern of lights visible in **FIGURE 3**. Not all regions of high density are urban places; rural environments, such as in parts of central Europe and South-East Asia, may contain large numbers of people per square kilometre.

Government policies may also affect population distribution. Examples include migration of people from one country to another, such as from Mexico to the United States of America where the demand for unskilled labour is high close to the border. (This is set to become more difficult, however, if the United States' construction of its border wall progresses as planned.) Chinese government policy led to the movement of Han Chinese into Tibet. At home in Australia, the establishment of service centres based around resource development, such as in the Pilbara in Western Australia, has affected population distribution, as has the use of migration to combat population shrinkage and labour shortages in other regional communities such as Nhill in Victoria and Dalwallinu in Western Australia.

FIGURE 3 Population distribution on display at night

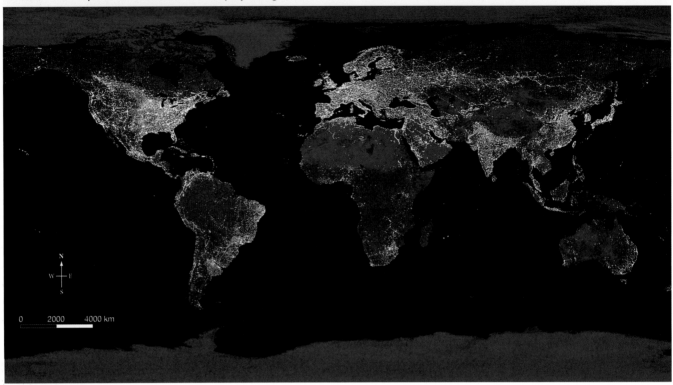

Source: NASA.

 Resources

 **Weblink** Population concepts

9.2 INQUIRY ACTIVITIES

1. Use the **Population concepts** weblink in the Resources tab to distinguish between population density and population distribution. **Comparing and contrasting**
2. Refer to the Australian Bureau of Statistics website or the website for your local government area. Find out the population density of your local government area, or calculate it by dividing the size of the area by the number of people. How does it compare to Dhaka's population density? **Comparing and contrasting**

9.2 EXERCISES

Geographical skills key: GS1 Remembering and understanding **GS2** Describing and explaining **GS3** Comparing and contrasting **GS4** Classifying, organising, constructing **GS5** Examining, analysing, interpreting **GS6** Evaluating, predicting, proposing

9.2 Exercise 1: Check your understanding

1. **GS1** What is the global average population density?
2. **GS1** Name the continents with the highest and lowest population density.
3. **GS2** Account for (give reasons for) the uneven distribution of global population.
4. **GS1** What are the main natural factors that favour human settlement?
5. **GS2** Outline two human factors that influence population distribution.

9.2 Exercise 2: Apply your understanding

1. **GS6** How might global population distribution *change* in the next 20 years? Justify your answer.
2. **GS6** Why would some governments want to redistribute their populations away from existing large cities?
3. **GS5** What disadvantages might there be for people moving from an area of dense population?

9.3 Life expectancy and wellbeing

9.3.1 Life expectancy

How long we can expect to live when we are born is referred to as our **life expectancy** and is calculated according to the conditions in a particular country in that year. Life expectancy is one of the major indicators of wellbeing. Globally, on average, people are expected to live longer than at any previous time in history. Sayings such as '60 is the new 50' reflect our changing expectations in Australia as to how long we expect to lead active lives. However, with the variation in living conditions around the world, the answer to the question 'how long can we expect to live?' also varies considerably.

Worldwide, a child born in 2018 can expect to live, on average, 72 years. If this child is born in Japan, they can expect to live 84 years, while one born in the African country of Somalia can expect to live only 56 years. **FIGURE 1** shows variations in life expectancy worldwide.

FIGURE 1 Global life expectancy

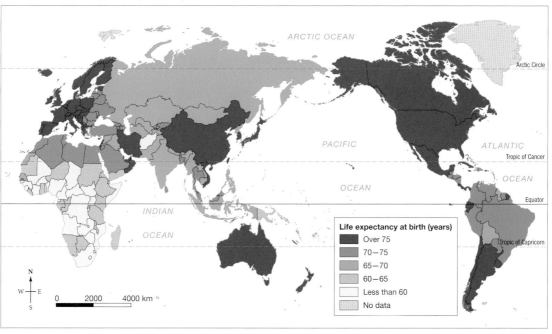

Life expectancy at birth (years)
- Over 75
- 70–75
- 65–70
- 60–65
- Less than 60
- No data

Source: © United Nations Publications.

Life expectancy around the world started to increase in the mid 1700s due to improvements in farming techniques, working conditions, nutrition, medicine and hygiene. There is a clear interconnection between wealth and life expectancy: wealthier people in all countries can expect to live longer than poorer people. In general, women outlive men. A higher income enables people to have better access to education, food, clean water and health care.

However, improvements that have led to increased life expectancy have not been uniform across the world. Regions where life expectancy continues to be low have a higher prevalence of infectious diseases. For example, 72 per cent of people worldwide living with HIV and AIDS are in sub-Saharan Africa, and the number of deaths in this region from AIDS is the highest in the world. In 2017, 25 per cent of new infections were in South Africa and Nigeria, with the latter also recording the highest rate of new infections among children under 14.

9.3.2 Child mortality

Life expectancy is closely interconnected with child mortality: countries with high death rates for children under five years of age have low life expectancy. The highest rates are recorded in countries in sub-Saharan Africa; in Somalia, for example, 127 deaths per 1000 were recorded in 2018, while 100 were recorded in Nigeria and 122 in the Central African Republic. Afghanistan, with 68 recorded deaths per 1000, has the highest child mortality rate outside of this region. These rates are well above the United Nations Sustainable Development Goal 3 target of 25 deaths per 1000 by 2030 (see **FIGURE 2**). Young children are particularly vulnerable to infectious diseases due to their lower levels of immunity. Major causes of death include pneumonia, diarrhoea, measles and malnutrition. In wealthier households, child deaths are lower as these children are more likely to have better nutrition and to be immunised, and parents are more likely to be educated and aware of how to prevent disease.

FIGURE 2 Under-five mortality rate (deaths per 1000 live births) and percentage change, 1990–2016

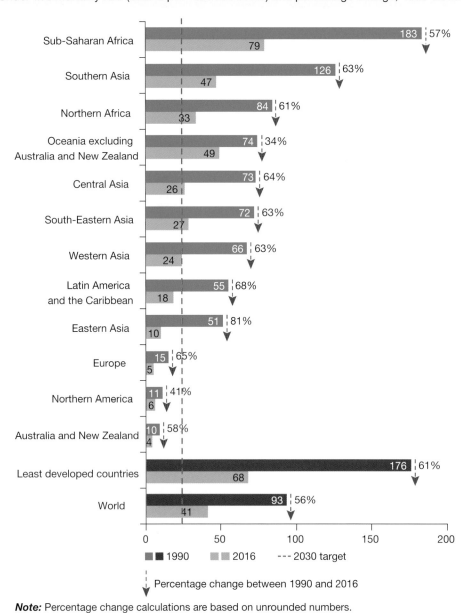

Note: Percentage change calculations are based on unrounded numbers.

Under the United Nations' Millennium Development Goals program, which operated from 1990 to 2015, child mortality was reduced considerably. The number of deaths of children under the age of five declined from 12.7 million in 1990 to 6 million in 2015 — the equivalent of nearly 17 000 fewer children dying each day. The greatest success occurred in northern Africa and eastern Asia.

Substantial improvements have been made in the area of preventable childhood diseases. For example, before the introduction of a measles vaccine in 1963, major epidemics occurred every two to three years, resulting in 2.6 million deaths each year. In 2017, 85 per cent of the world's children received the measles vaccine before their first birthday, 13 per cent more than in 2000. The World Health Organization reports an 80 per cent drop in measles-related childhood deaths, thus preventing 21.1 million deaths. Despite this reduction, in 2017 there were still 110 000 deaths globally due to measles.

Life expectancy and child mortality allow us to measure and compare human wellbeing in different places.

9.3.3 Births and deaths

Every minute there are an estimated 250 births and 105 deaths worldwide. This natural increase equates to an extra 145 people at a global scale every minute. However, the rate of population change

FIGURE 3 Her future

varies considerably across the world, with some places experiencing a decline rather than an increase in numbers of people. Rates of population change have an impact on wellbeing, both now and in the future.

FIGURE 4 shows the global distribution of birth rates. The continent of Africa clearly stands out here with the highest figures. In 2010, Africa had a population of one billion; the United Nations projects that by 2050 the African population will be 2.5 billion — three times that of Europe. The majority of this growth will occur in sub-Saharan Africa, where the average **fertility rate** in 2018 was five children per woman. Countries such as Niger and Somalia have fertility rates as high as 7.2 and 6.2 respectively, while Tunisia has the lowest fertility rate in Africa with 2.2 births per woman. Europe has very low birth rates with a fertility rate of 1.6. Ireland and France are slightly higher than this average with fertility rates of 2.0, while the poorest country in Europe, Moldova, has the lowest fertility rate of 1.3. Taiwan recorded the lowest fertility rate in the world in 2018 with 1.2 births per woman. Australia's fertility rate is 1.8.

For death rates, as **FIGURE 5** illustrates, sub-Saharan Africa is again at the high end of the spectrum, with many places in that region experiencing death rates above 10 per 1000. However, high death rates are more dispersed, with many European countries, such as Bulgaria and Ukraine, included. Low death rates are widely distributed across the regions of the Americas, much of Asia and Oceania.

Whether a population increases or decreases is largely dependent on variations in births and deaths producing a **natural increase**. Where a fertility level is well above the **replacement rate** of 2.1 children, population growth will occur. Conversely, fewer births over a period of time will ultimately result in a declining population. **FIGURE 6** indicates the rate of natural population change, which ranges from over 3 per cent growth primarily in African nations (this would result in a doubling of population in approximately 23 years) to negative growth primarily in Europe. It should be noted that on a national scale, population change is also affected by migration.

FIGURE 4 Global distribution of birth rates

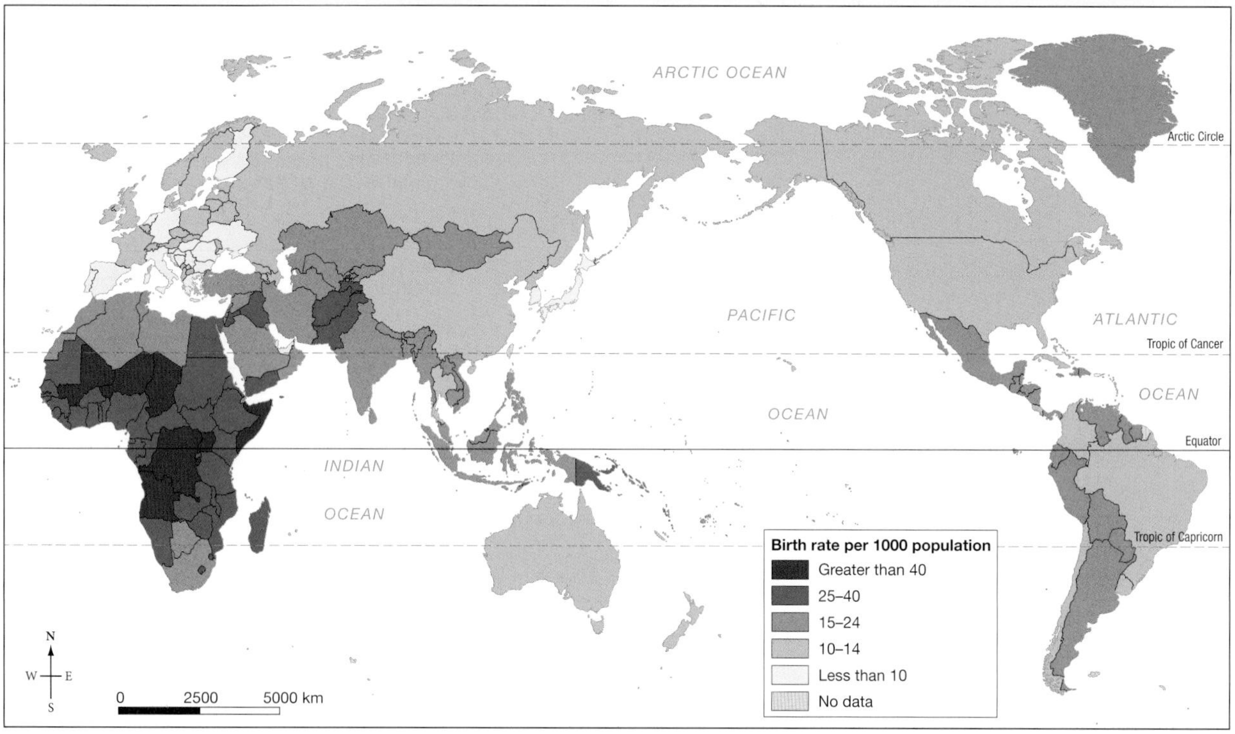

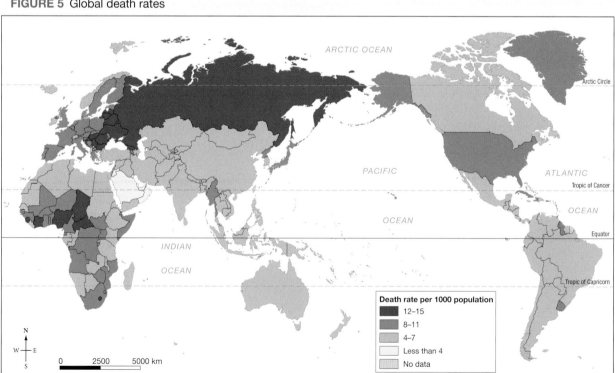

Source: © United Nations Publications; The World Bank.

FIGURE 5 Global death rates

Source: © United Nations Publications; The World Bank.

FIGURE 6 Population change

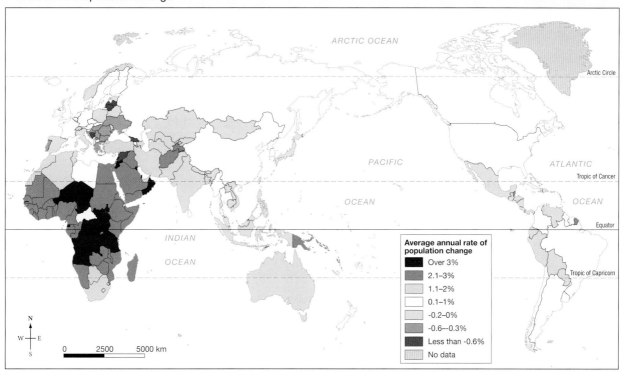

Source: © United Nations Publications.

9.3 INQUIRY ACTIVITY

Create your own version of 'Her future' (**FIGURE 3**), depicting what you imagine your life will be like when you are 35 years old.   **Evaluating, predicting, proposing**

9.3 EXERCISES

Geographical skills key: GS1 Remembering and understanding **GS2** Describing and explaining **GS3** Comparing and contrasting **GS4** Classifying, organising, constructing **GS5** Examining, analysing, interpreting **GS6** Evaluating, predicting, proposing

9.3 Exercise 1: Check your understanding

1. **GS2** Describe the distribution of life expectancy shown in **FIGURE 1**.
2. **GS1** What age groups are likely to have the highest mortality rates? Why?
3. **GS5** Using **FIGURES 4 and 5** and an atlas, identify countries that are exceptions to these patterns:
    (a) high birth rates in Africa
    (b) low birth rates in Europe
    (c) low death rates in Asia
    (d) low death rates in Europe.
4. **GS4** Using **FIGURE 6**, identify two countries for each category of natural population *change*.
5. **GS2** What are the major contributors to population *change* at a national *scale*?
6. **GS2** Explain the significance of a national fertility level of 2.1 children per woman.

9.3 Exercise 2: Apply your understanding

1. **GS6** Predict how life expectancies in sub-Saharan Africa may *change* if a cure for HIV/AIDS is discovered.
2. **GS6** Will increased incomes always lead to increased life expectancy? Justify your answer.
3. **GS5** What implications does an increase in life expectancy have on the provision of health care?
4. **GS5** What are the implications for a country if its fertility rate is below the replacement rate?

  ▶

9.4 The link between population growth and wellbeing

9.4.1 Population growth over time

From our examination of birth and death rates, we have seen that subsequent population change varies considerably across the world. To generalise, the more developed countries of the world tend to have lower birth and death rates and lower population growth, while developing nations experience higher rates of births and deaths and higher growth. What are the reasons for such a large variation? What impacts does this have on wellbeing?

Global population growth has been rapid. From approximately one billion people in the year 1800, our planet now supports nearly eight billion people. Annually, global population is growing by some 82 million, although the rate of growth has now slowed. These changes have been due to gains in wellbeing. Improvements in food production, education, medicine and hygiene have resulted in rapidly decreasing death rates, especially in infants and young children, and increased life expectancy. The **demographic transition model** (**FIGURE 1**) attempts to explain changes in population growth by examining the interconnection between population characteristics and changes in wellbeing.

Most of the global population growth is taking place in developing countries, particularly the poorest regions (see **FIGURE 2**). By 2050, with an estimated 9.7 billion people in the world, some eight billion people (82 per cent) will be in developing nations; the United Nations predicts that more than half of the world's population growth between 2015 and 2050 will occur in Africa. Despite continued global population growth, global fertility rates are falling (see **FIGURE 3**). Declines in fertility have coincided with improvements in living conditions, greater access to education (particularly for women), improved health care and access to contraception. It is anticipated that fertility rates in developing regions will continue to fall, particularly with increasing rural–urban migration. In the cities, a child is more likely to be an economic burden than an asset, and there is better access to health services and family planning programs.

FIGURE 1 The demographic transition model

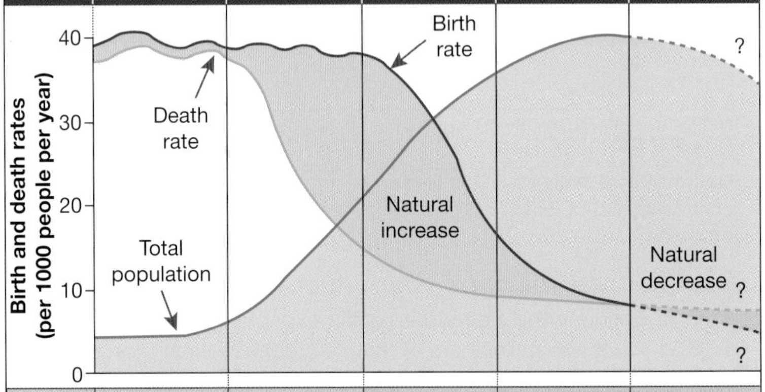

Stage	1 High stationary	2 Early expanding	3 Late expanding	4 Low stationary	5 Declining ?
Examples	A few remote groups	Egypt, Kenya, India	Brazil	USA, Japan, France, UK	Germany
Birth rate	High	High	Falling	Low	Very low
Death rate	High	Falls rapidly	Falls more slowly	Low	Low
Natural increase	Stable or slow increase	Very rapid increase	Increase slows down	Stable or slow increase	Slow increase
Reasons for changes in birth rate	Many children needed for farming. Many children die at an early age. Religious/social encouragement. No family planning.		Improved medical care and diet. Fewer children needed.	Family planning. Good health. Improving status of women. Later marriages.	
Reasons for changes in death rate	Disease, famine. Poor medical knowledge so many children die.	Improvements in medical care, water supply and sanitation. Fewer children die.		Good health care. Reliable food supply.	

One example of the link between rising urbanisation and decreasing birth rates is South Korea. From 1970 to 2018, its urban population increased from 40 to 82 per cent, and its fertility rate fell from 4.53 to 0.98.

FIGURE 2 World population growth per region

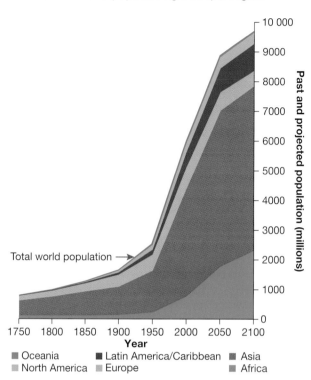

Total world population →

| Oceania | Latin America/Caribbean | Asia |
| North America | Europe | Africa |

FIGURE 3 Changing global fertility rates

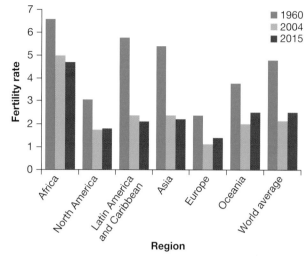

FIGURE 4 Seoul, South Korea — as urbanisation has increased here, fertility rates have dropped significantly.

9.4.2 Population structure and wellbeing

The level of wellbeing of a population in terms of its health and life expectancy is reflected in its **population structure**. Increases in life expectancy and a decrease in the number of children being born has resulted in an increasing proportion of people in the older age groups. This ageing is expected to occur on a global scale in both developed and developing countries, with the rate of change being faster in the latter.

The proportion of the population in the **dependent population** affects the wellbeing of a country. A youthful population, as in Niger and Kenya, has implications in terms of future provision of infrastructure, education and employment. In addition, a high proportion of youth means that the population has momentum to cause large future growth, placing stress on a country's resources. However, if those young people are healthy and well educated, they provide a potential skilled workforce in future years. An **ageing population** and a high percentage of elderly population, as found in Germany and Japan, has implications in terms of a decreasing workforce and tax base and increased demands on health services. On the positive side, the aged population does make a significant economic contribution, often in terms of voluntary labour such as caring for grandchildren and assisting with community projects.

on Resources

🔗 **Weblinks**   Population growth
                  HDI over time

9.4 INQUIRY ACTIVITIES

1. Refer to the **Population growth** weblink in the Resources tab. What is the relationship between population growth and wellbeing, and how is this likely to *change* in the future?   **Examining, analysing, interpreting**
2. Refer to the **HDI over time** weblink in the Resources tab. How has wellbeing, as measured by the Human Development Index, *changed* over time?   **Examining, analysing, interpreting**

9.4 EXERCISES

Geographical skills key: GS1 Remembering and understanding **GS2** Describing and explaining **GS3** Comparing and contrasting **GS4** Classifying, organising, constructing **GS5** Examining, analysing, interpreting **GS6** Evaluating, predicting, proposing

9.4 Exercise 1: Check your understanding

1. **GS1** Which regions of the world have the highest and lowest fertility rates?
2. **GS1** Describe how world population growth has *changed* over time.
3. **GS2** Explain why fertility is likely to decrease with an increasing proportion of people living in cities.
4. **GS2** Explain what is meant by the term *dependent population*.
5. **GS6** What do you think is meant by the statement 'Children can be an economic burden or an economic asset'?

9.4 Exercise 2: Apply your understanding

1. **GS6** In which stage of the demographic transition model (refer to **FIGURE 1**) is Kenya likely to be in 10 years? Justify your answer.
2. **GS6** If a country reaches stage 5 of the demographic transition model, is it likely to remain there? Justify your answer.
3. **GS6** At what stage of the demographic transition model (**FIGURE 1**) is Australia most likely to be? Justify your answer.
4. **GS5** A dependent population can be described as 'youthful' or 'ageing'. What are the implications for a country if its population falls into either of these categories?
5. **GS6** Suggest reasons for fertility rates in Africa remaining almost twice that of the world average.

Try these questions in learnON for instant, corrective feedback. Go to www.jacplus.com.au.

9.5 Government responses to population and wellbeing issues

The number of children in a family directly affects their wellbeing. In Australia, parents may comment on the cost of raising children. In other countries, children may make a contribution to family income by completing simple jobs such as collecting firewood. At a national scale, the numbers of children also affect the wellbeing of the country as a whole. Concerns about too few or too many children have resulted in a variety of government responses. Two are outlined in the following sections.

TABLE 1 Selected demographic characteristics for Japan and Kenya

Demographic characteristic	Kenya	Japan
Population mid 2018	51.0 million	126.5 million
Life expectancy at birth	67 years	84 years
Fertility rate	3.9	1.4
Natural increase	2.6%	−0.3%
Infant mortality	36 per 1000	2 per 1000
Projected population 2050	95.5 million	101.8 million
Population under 15 years	41%	12%
Population 60+ years	3%	28%
Percentage urban	32%	92%
Gross national income per capita (US$)	3250	45 470

Source: PRB, Data Sheet, 2018.

9.5.1 Kenya: response to a youthful population

Although Kenya's fertility rate has fallen substantially (from a high of 8.1 children per woman in 1967, down to 3.9 children per woman in 2015), the country still has a relatively high rate of population growth. Its population structure has a high proportion of young people (see **TABLE 1** and **FIGURE 2(a)** and **(b)**), so by 2030 it is estimated that there will be over 65 million people, with further growth up to 95.5 million by 2050. This increase will put pressure on Kenya's resources in terms of providing food, services and employment. With a predominantly rural population, the amount of arable land per person is falling.

FIGURE 1 The Kenyan population has a high proportion of young people.

FIGURE 2 Population pyramid for Kenya (a) 2019 and (b) 2050

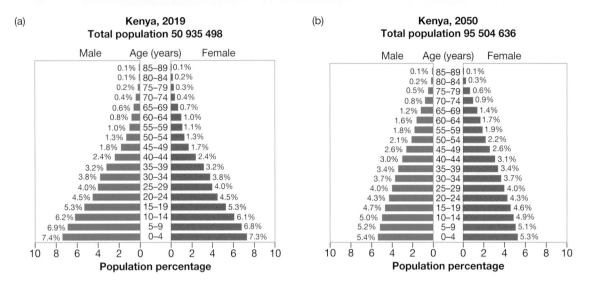

(a) **Kenya, 2019**
Total population 50 935 498

Male   Age (years)   Female

Male	Age (years)	Female
0.1%	85–89	0.1%
0.1%	80–84	0.2%
0.2%	75–79	0.3%
0.4%	70–74	0.4%
0.6%	65–69	0.7%
0.8%	60–64	1.0%
1.0%	55–59	1.1%
1.3%	50–54	1.3%
1.8%	45–49	1.7%
2.4%	40–44	2.4%
3.2%	35–39	3.2%
3.8%	30–34	3.8%
4.0%	25–29	4.0%
4.5%	20–24	4.5%
5.3%	15–19	5.3%
6.2%	10–14	6.1%
6.9%	5–9	6.8%
7.4%	0–4	7.3%

Population percentage

(b) **Kenya, 2050**
Total population 95 504 636

Male   Age (years)   Female

Male	Age (years)	Female
0.1%	85–89	0.1%
0.2%	80–84	0.3%
0.5%	75–79	0.6%
0.8%	70–74	0.9%
1.2%	65–69	1.4%
1.6%	60–64	1.7%
1.8%	55–59	1.9%
2.1%	50–54	2.2%
2.6%	45–49	2.6%
3.0%	40–44	3.1%
3.4%	35–39	3.4%
3.7%	30–34	3.7%
4.0%	25–29	4.0%
4.3%	20–24	4.3%
4.7%	15–19	4.6%
5.0%	10–14	4.9%
5.2%	5–9	5.1%
5.4%	0–4	5.3%

Population percentage

Under Kenya's Vision 2030, a national framework for development, population management is an essential component of achieving wellbeing goals for health, poverty reduction, gender equality and environmental sustainability. The United Nations Population Fund (UNFPA) has been working with the Kenyan government since the 1970s to help improve wellbeing in the country. Through the provision of financial assistance, a range of services have been enabled, including family planning with free contraceptives provided, increased availability of maternal and newborn health services, services to prevent the contraction of HIV and sexually transmitted infections, and advocacy for the education of girls and elimination of gender-based violence. Unfortunately, despite this work, there is still a huge unmet need for family planning in Kenya, particularly among the poorest women, where almost half report they have unplanned pregnancies.

9.5.2 Japan: response to an ageing population

Japan has one of the highest life expectancies in the world, and this, combined with a very low fertility rate, has led to an ageing population, with almost one-third of Japan's population in the 60-plus age group (see **TABLE 1** and **FIGURES 4(a)** and **(b)**). Fertility in Japan has been consistently below replacement level since the 1970s. A high standard of living, increased participation of women in the workforce, high costs of raising children and lack of supporting childcare facilities have all contributed to this. Japan's total population is expected to decline from 126 million in 2019 to 120 million in 2030, 109.5 million in 2050 and 83 million by 2100.

The workforce is expected to fall 15 per cent over the next 20 years and halve in the next 50 years. This means that in 2025, three working people will have to support two retirees. The Japanese government also faces rising pension and healthcare costs. These economic concerns led to the Japanese government implementing a number of measures in 1994 such as subsidised child care and bonus payments for childbirth via a policy known as the Angel Plan (revised in 1999). In 2009, the government introduced the 'plus one' policy, offering further incentives for families by offering free education, more childcare places and providing fathers with up to 12 months subsidised paternity leave. These policies have been largely ineffective: although the fertility rate rose slightly initially, it has remained well below replacement level.

FIGURE 3 High life expectancy and a low fertility rate have led to an ageing Japanese population.

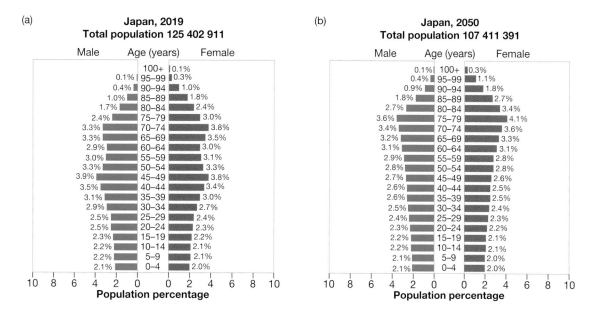

FIGURE 4 Population pyramid for Japan (a) 2019 and (b) 2050

The Japanese government has historically been reluctant to use immigration to fill labour shortages, and although this may change slowly, improving female workforce participation rates, particularly after marriage, may be a more viable option. In 2018, Japan recorded its highest level of natural decline, with the population falling by 449 000. The difference between the 921 000 recorded births and 1.37 million deaths.

9.5 INQUIRY ACTIVITY

Research the Japanese government's Angel Plan, New Angel Plan and Plus One Policy, designed to help increase the country's falling birth rate.
a. What aspects of the government's beliefs and values led to the development of these plans?
b. What aspects of Japanese culture, values and beliefs have prevented the plans from achieving success?

Examining, analysing, interpreting

9.5 EXERCISES

Geographical skills key: GS1 Remembering and understanding **GS2** Describing and explaining **GS3** Comparing and contrasting **GS4** Classifying, organising, constructing **GS5** Examining, analysing, interpreting **GS6** Evaluating, predicting, proposing

9.5 Exercise 1: Check your understanding

1. **GS1** How has an improvement in living conditions led to a *change* in population structure?
2. **GS2** Account for (give reasons for) the variation in shape of the population pyramids for Japan and Kenya in 2019 and 2050.
3. **GS3** Calculate the difference in life expectancy between Kenya and Japan.
4. **GS1** By how much is Kenya's population expected to grow between 2030 and 2050?
5. **GS1** What factors have contributed to low fertility rates in Japan since the 1970s?

9.5 Exercise 2: Apply your understanding

1. **GS2** Describe the *changing* percentage of aged population between 2019 and 2050 in Kenya and Japan.
2. **GS6** What problems does the Kenyan government face with a large proportion of young population?
3. **GS6** What problems does Japan face with a large proportion of aged population?
4. **GS6** How do these issues affect the wellbeing of people in those countries?
5. **GS6** Of the problems you identified in questions 2 and 3, which do you consider more serious? Why?

Try these questions in learnON for instant, corrective feedback. Go to www.jacplus.com.au.

9.6 SkillBuilder: Using Excel to construct population profiles

Why do we use Excel to construct population profiles?

When constructing population profiles, there is a large amount of data and large numbers to handle. The use of an Excel spreadsheet simplifies the process.

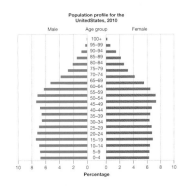

Select your learnON format to access:

- an overview of the skill and its application in Geography (Tell me)
- a video and a step-by-step process to explain the skill (Show me)
- an activity and interactivity for you to practise the skill (Let me do it)
- questions to consolidate your understanding of the skill.

on Resources

Video eLesson Using Excel to construct population profiles (eles-1758)

Interactivity Using Excel to construct population profiles (int-3376)

9.7 Variations in wellbeing in India

9.7.1 How and why is India's population changing?

China has the biggest population in the world, with a population of 1.4 billion in 2019; however, its population is expected to drop to 1.35 billion by 2050. The population of India, on the other hand, is expected to surpass that of China by 2025, rising from its current level of 1.37 to 1.46 billion. With a predicted population of 1.7 billion by 2050, what happens to India's population will have major implications in terms of the wellbeing of the people of that country.

Although the number of children per woman in India has declined substantially from 5 in the 1970s to 2.3 in 2018, there is considerable regional variation in this rate. Overall, India's population is growing at a rate of 1.1 per cent per year. Improvements in water supply, a decrease in infectious diseases and an increase in education levels have resulted in a reduced death rate since the 1950s, while the birth rate has not declined to the same extent. Infant mortality remains high as over two-thirds of the population are rural dwellers who may not have ready access to health and reproductive services. Children remain a vital part of the family's labour force both on farms (as shown in **FIGURE 1**) and for old age support, so it is essential for families to have more children to improve the chance of them surviving to adulthood. Of the entire population, 27 per cent is under 15 years of age, creating huge momentum for future growth (see **FIGURE 2**).

FIGURE 1 Indian children assisting with rice planting

FIGURE 2 Population pyramid for India (a) 2019 and (b) 2050

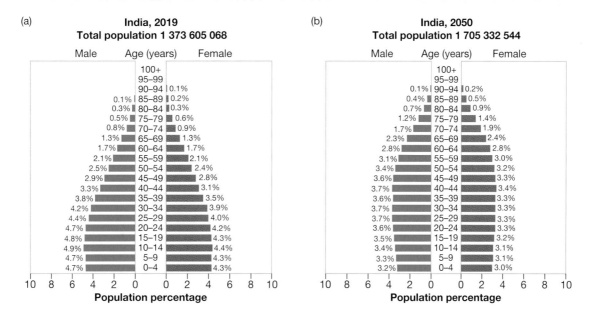

FIGURE 3 Proportion of children 0–6 years to total population, India, 2011

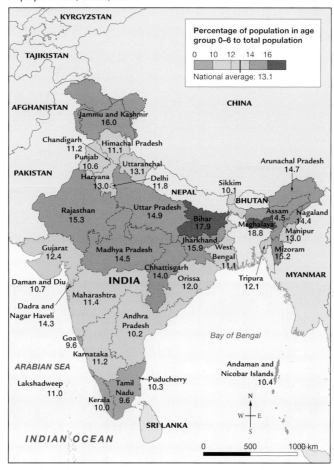

Note: Data drawn from 2011 census; next census due in 2021
Source: Spatial Vision.

9.7.2 Regional variation in wellbeing

In addition to variations in population structure, the states of India show differences in other important wellbeing-related demographics. The levels of literacy and poverty shown in **FIGURES 4** and **5** reflect a varying distribution of wellbeing in India.

FIGURE 4 Poverty levels in India

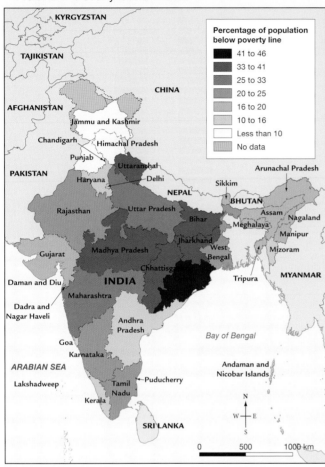

Percentage of population below poverty line

- 41 to 46
- 33 to 41
- 25 to 33
- 20 to 25
- 16 to 20
- 10 to 16
- Less than 10
- No data

Note: Data drawn from 2011 census; next census due in 2021
Source: Spatial Vision.

FIGURE 5 Literacy rates (percentage) in India, 2011

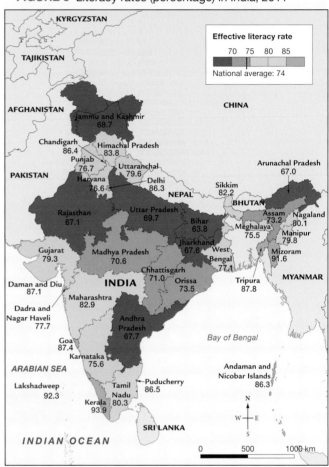

Effective literacy rate

70 75 80 85

National average: 74

Note: Data drawn from 2011 census; next census due in 2021
Source: Spatial Vision.

FIGURE 6 Poverty and literacy levels vary throughout India. States with both high levels of literacy and lower levels of poverty, such as Kerala and Mizoram, are rare.

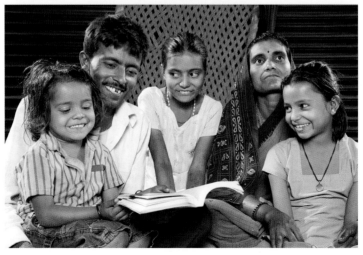

9.7 EXERCISES

Geographical skills key: GS1 Remembering and understanding **GS2** Describing and explaining **GS3** Comparing and contrasting **GS4** Classifying, organising, constructing **GS5** Examining, analysing, interpreting **GS6** Evaluating, predicting, proposing

9.7 Exercise 1: Check your understanding

1. **GS1** With reference to the **FIGURE 2** population pyramids, account for India's *changing* population growth.
2. **GS2** Explain why India is set to overtake China in terms of total population.
3. **GS2** From **FIGURE 3**, which three regions of India have the lowest proportion of children aged 0–6 years?
4. **GS2** Study **FIGURE 4**. Describe the distribution of the states with the highest percentage of the population living below the poverty line.
5. **GS2** Look at **FIGURE 5**. Name five Indian states that have a literacy rate above the national average.

9.7 Exercise 2: Apply your understanding

1. **GS5** Use **FIGURES 3**, **4** and **5** to describe the average characteristics of the population in Uttar Pradesh.
2. **GS5** Describe the characteristics of the state of Orissa in India.
3. **GS5** Name two states of India that have below the national average for percentage of population aged 0–6, less than 16 per cent of the population living below the poverty line and literacy rates above 85 per cent.
4. **GS5** Using the data provided throughout this subtopic, describe and account for the variation in wellbeing in India.
5. **GS6** In which state (or states) of India do you think wellbeing would be highest? Explain your response.

Try these questions in learnON for instant, corrective feedback. Go to www.jacplus.com.au.

9.8 Population characteristics of Australia

9.8.1 Australia's population

According to the Australian Bureau of Statistics, Australia's population reached 25 million in August 2018. Statistically speaking, a typical Australian in that year would be female, born in Australia, aged 37 years and living in a household consisting of a couple and children (although the average household size was only 2.6 people). Of course, Australia's demographic characteristics are much more diverse than this. To what extent do you fit the 'typical' profile?

Most of Australia's population is concentrated in coastal regions in the south-east and east and, to a lesser extent, in the south-west. The population within these regions is concentrated in urban centres, particularly the capital cities (see **FIGURE 1**). Seventy-one per cent of Australians live in major cities with a population of over 100 000; 29 per cent live in rural and remote areas, of whom around 2 per cent live in small towns with a population of less than 1000 people.

FIGURE 1 Australia's population distribution

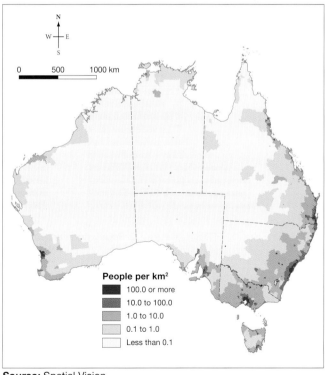

People per km²
- 100.0 or more
- 10.0 to 100.0
- 1.0 to 10.0
- 0.1 to 1.0
- Less than 0.1

Source: Spatial Vision.

FIGURE 2 Brisbane, a typical Australian urban environment

FIGURE 3 Innamincka: an outback town

Aboriginal and Torres Strait Islander peoples make up 3.3 per cent of the Australian population, with a median age of 23 years. With 65 per cent of the Indigenous population under 30 and only 3.4 per cent over the age of 65, they are a young population. The non-Indigenous population is ageing; with a median age of 37 years; 39 per cent of the population is under 13 and 14.1 per cent is over the age of 65.

FIGURE 4 Population pyramid for Indigenous and non-Indigenous Australians

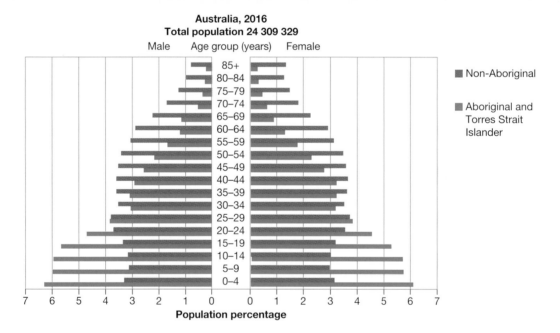

Australia's population has increased considerably over time and is continuing to grow. Between 2000 and 2019, the population increased by over six million people, at an average rate of 1.7 per cent per year (see **FIGURE 5**).

Our population growth is due to immigration rather than natural increase. The level of migration is set annually by the Federal Government and is currently about 160 000 per year.

Our rate of fertility has declined steadily since the 1970s and is now well below replacement rate. Despite attempts to increase the number of children via a Federal Government baby bonus of approximately $5000 per baby, which was in place between 2003 and 2013, our fertility rate was only 1.8 in 2018.

FIGURE 5 Australia's changing population growth

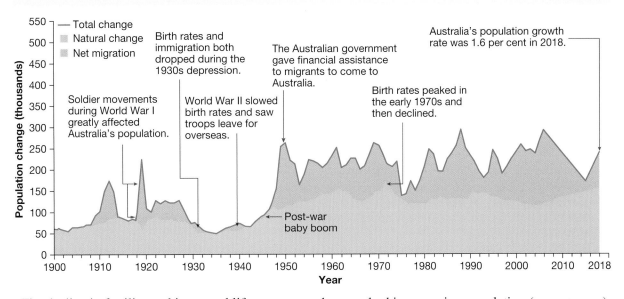

The decline in fertility and increased life expectancy has resulted in an ageing population (see **FIGURE 6**). The proportion of the population aged 65 years and over increased from 11.3 per cent to 16 per cent between 30 June 1991 and 30 June 2018.

FIGURE 6 Australia's changing population structure: age distribution of Australian population and migrants

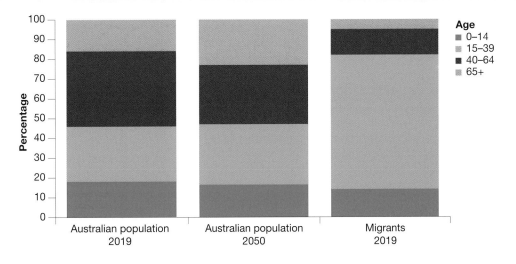

DISCUSS

Australia is one of the world's most multicultural nations. Identify some of the challenges and benefits of living and working in our multicultural society. **[Intercultural Capability]**

9.8 INQUIRY ACTIVITY

a. Use the Australian Bureau of Statistics website to access statistics on four demographic characteristics of your Local Government Area.
b. How do the statistics for your Local Government Area compare to those for Australia as a whole and those of your state? **Examining, analysing, interpreting**

9.8 EXERCISES

Geographical skills key: GS1 Remembering and understanding **GS2** Describing and explaining **GS3** Comparing and contrasting **GS4** Classifying, organising, constructing **GS5** Examining, analysing, interpreting **GS6** Evaluating, predicting, proposing

9.8 Exercise 1: Check your understanding

1. **GS1** What factors have accounted for Australia's *changing* population growth over time?
2. **GS1** How is Australia's population structure expected to *change* in the future?
3. **GS2** Account for the variation in Australia's population distribution.
4. **GS6** Predict the impact of Australia's ageing population on our demand for different facilities.
5. **GS2** Explain why migration is an important part of ensuring Australia has a skilled workforce.

9.8 Exercise 2: Apply your understanding

1. **GS6** Sketch the shape of how you think Australia's population pyramid will look in 50 years' time. Justify your drawing.
2. **GS6** What other methods could the Australian government use in order to encourage population growth in Australia?
3. **GS5** What are the advantages and disadvantages of a 'big Australia' and a projected future population of 35 million?
4. **GS6** Study the population pyramids for Indigenous and non-Indigenous Australians. Suggest reasons for the differences you observe.
5. **GS6** Predict the *changes* that might occur in Australia's population if migration levels were doubled.

Try these questions in learnON for instant, corrective feedback. Go to www.jacplus.com.au.

9.9 SkillBuilder: How to develop a structured and ethical approach to research

onlineonly

What is a structured and ethical approach to research?

A structured and ethical approach to research involves organising your work clearly and meeting research standards without pressuring anyone into providing material and without destroying environments while gathering the data. Your work must also be your own, and anything that is someone else's work must be referenced in the text and included in the reference list.

Select your learnON format to access:

- an overview of the skill and its application in Geography (Tell me)
- a video and a step-by-step process to explain the skill (Show me)
- an activity and interactivity for you to practise the skill (Let me do it)
- questions to consolidate your understanding of the skill.

 Resources

   Video eLesson How to develop a structured and ethical approach to research (eles-1759)

   Interactivity How to develop a structured and ethical approach to research (int-3377)

9.10 Health and wellbeing

9.10.1 SDG targets for health and wellbeing

Health is a key indicator of wellbeing. Goal 3 of the Sustainable Development Goals (SDGs) is to 'ensure healthy lives and promote wellbeing for all at all ages'. Key among the 2030 targets for this goal is the eradication of tuberculosis, AIDS and malaria epidemics. Developing nations, where the standard of living and health care services are generally low, are more likely to be affected by such epidemics because they lack the resources to deal with these health issues. In this subtopic, we explore the threats to wellbeing associated with malaria and HIV/AIDs and what is being done to combat them globally.

9.10.2 What is malaria?

Have you ever been kept awake on a warm summer's night by an annoying mosquito buzzing around your head? For us, such an occurrence is merely a nuisance. However, for many people, particularly children under five years of age in Africa, a bite from a malaria-carrying mosquito can be life threatening. Malaria therefore has a major impact on their wellbeing.

Fortunately, we do not have that potentially deadly mosquito in Australia.

Malaria is a preventable and curable disease caused by plasmodium parasites transmitted by the female Anopheles mosquito. These parasites destroy red blood cells and initially cause headaches and fever but may also cause death if sufficient organs are affected. There are four species of these parasites, which are found in tropical and subtropical places of the world, with the species commonly found in Africa being particularly severe in impact. There is therefore a significant interconnection between such tropical conditions and incidence of this disease.

FIGURE 1 Malaria-carrying mosquito

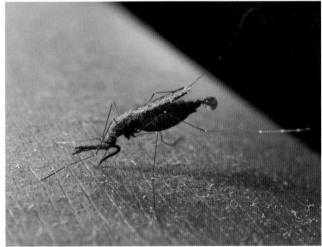

9.10.3 Why should we worry about malaria?

According to the World Health Organization (WHO), in 2017 about 3.75 billion people (half the world's population) located in 87 countries were at risk of malaria. There were around 219 million cases of malaria in these 87 countries; however, 49 per cent of malaria cases occurred in just five countries, and four of these are located in sub-Saharan Africa — 92 per cent of recorded cases and 93 per cent of all deaths from malaria were in Africa. Approximately 435 000 people died as a result of contracting the disease.

The most at-risk group is children under the age of five; in 2017 approximately 61 per cent (266 000) of the recorded deaths were in this age group. This is the equivalent of one child dying every two minutes. Pregnant women are also particularly vulnerable to malaria and the disease can cause complications such as still and premature births, and low birth weights. Twenty-five million pregnant women are at risk of malaria each year.

Malaria costs Africa approximately $12 billion per year. People affected are not able to work, so this affects both an individual's and a country's income. In addition, malaria accounts for 40 per cent of public health expenditure and between one-third and one-half of hospital admissions.

FIGURE 2 Distribution of malaria

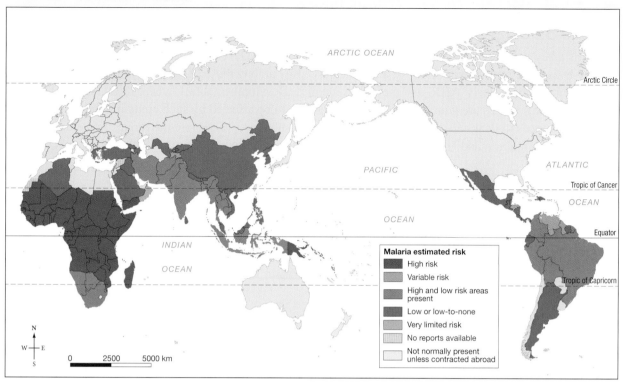

Source: CDC, July 2018, https://www.cdc.gov/malaria/travelers/country_table/a.html

9.10.4 Malaria control and elimination

To address the aims of malaria control and elimination, the WHO developed the Global Technical Strategy for Malaria 2016–2030, setting a target of reducing global malaria cases and mortality rates by at least 40 per cent by 2020 and 90 per cent by 2030. Before this, under the Millennium Development Goals action, both the number of cases of malaria and the number of deaths due to this disease declined as a result of increased prevention and control measures. Between 2000 and 2015, there was a 37 per cent reduction in global malaria incidence and a fall of 60 per cent in malaria mortality. Progress has not been uniform, however, and has stalled to some extent since 2015. Although 21 countries are set to be declared malaria-free, of the 11 countries that record the most cases of malaria, only India has recorded any significant progress, with a 24 per cent reduction in cases since 2016. Some countries in sub-Saharan Africa have actually recorded an increase in the number of cases of malaria.

FIGURE 3 Insecticide-treated nets, Kenya

Many government and non-government organisations have been involved in combating malaria, for example the WHO Global Malaria Programme, Roll Back Malaria, the President's Malaria Initiative,

Malaria No More and the Bill and Melinda Gates Foundation. Such organisations have focused on one or more of the following strategies:

- provision of insecticide-treated nets to protect people at night, when mosquitoes most like to bite. More than half the households in sub-Saharan Africa now have at least one bed net.
- use of insecticide sprays both to treat inside houses and in mosquito habitat areas
- improved access to diagnostic testing and treatment.

In addition, the Bill and Melinda Gates Foundation is funding the development of a malaria vaccine. In conjunction with the WHO, the vaccine had favourable initial trials and pilot studies are continuing.

FIGURE 4 Proportion of children under age five sleeping under insecticide-treated nets in sub-Saharan African countries, 2001 and 2013

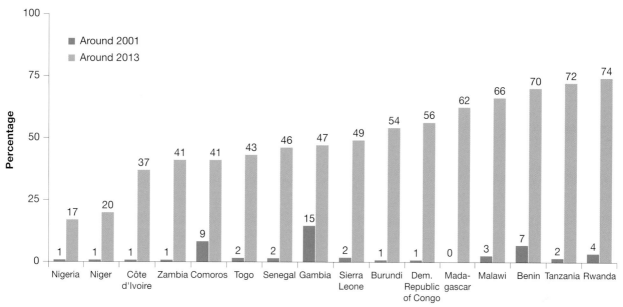

Note: 'Around 2001' refers to a survey conducted during 1999–2003. 'Around 2013' refers to a survey conducted during 2012–2014. Most recent data available.

Source: © UNDP.

One problem for the continued fight against malaria is insufficient funding. Although US$3.1 billion was available in 2017, an estimated US$8.7 billion is needed annually to achieve global malaria targets by 2030. Another concern is drug resistance, particularly along the Cambodia–Thailand border. The WHO launched a global plan to address this issue in 2010. In 2017, a new program 'High burden to high impact — a targeted response' was launched. This is a country-led approach to targeting those countries where malaria rates are increasing.

9.10.5 HIV and AIDS

When HIV (Human Immunodeficiency Virus) was first identified in the early 1980s, it was seen as a death sentence. Those infected succumbed to Acquired Immune Deficiency Syndrome (AIDS) and there was little to be done but wait. However, advances in treatments and extensive use of drug combinations mean that today, depending on the infected person's location, having HIV no longer means AIDS is inevitable. Despite this, the disease has a major impact on wellbeing, not only on a local scale but also at a national and regional level.

AIDS is caused by HIV. The virus reduces the body's ability to fight infections. According to the WHO, since 1981, more than 70 million people have been infected and approximately 35 million people have been killed by this pandemic. UNAIDS reported that at the end of 2017, 36.9 million people were living with

HIV, of whom three million were children and adolescents under the age of 20. UNAIDS estimates that in 2017 there were 1.8 million new infections (about 5000 every day); of these, 180 000 were children under the age of 15. It is estimated that in 2017, 940 000 people died of AIDS-related illnesses.

The virus is spread via infected bodily fluids entering a person's bloodstream. This may occur via unprotected sex, intravenous drug use, or from mother to child in the womb or through breastfeeding. Women and young people (aged 15–24 years) are disproportionately affected.

As **FIGURE 5** shows, the prevalence of HIV infections in children aged 0–14 is concentrated in places that lie within sub-Saharan Africa, which, as a region, is home to nearly two-thirds of all people living with HIV. The most affected countries are Swaziland (27.4 per cent), Botswana (22.8 per cent) and South Africa (18.8 per cent). New infections fell by 8 per cent in western and central Africa and 30 per cent in eastern and southern Africa between 2010 and 2017. Globally, the incidence rate of new infections has fallen by 18 per cent in the adult population and by 35 per cent for children since 2010. In 2017, 80 per cent of pregnant women had access to medications to prevent the transmission of HIV to their unborn babies. This is an increase over 2010 levels, where only 40 per cent of pregnant women had access to such medication. However, while 62 per cent of women take antiviral medication while pregnant, only 49 per cent continue to take this medication while breastfeeding. Globally, almost 330 children die every day due to an illness attributed to AIDS.

FIGURE 5 Distribution of children under 15 who are living with HIV, 2017

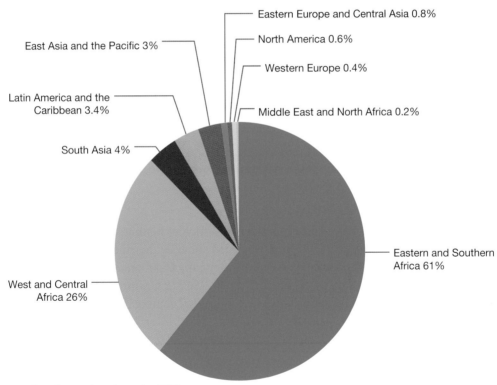

Note: Due to rounding, figures do not sum to 100%.

HIV has major impacts across the communities affected, and is reversing decades of improvements in living conditions. The pressures of illness and caring for sick family members can result in loss of income, and children may be taken out of school. Those in poverty may feel they have no choice but to engage in high-risk behaviours to earn income, such as women taking up sex work. In addition, loss of adults to AIDS has had devastating effects on children: globally, an estimated 12.2 million children in 2017 had lost at least one parent to the disease (80 per cent of these live in sub-Saharan Africa), compounding the likelihood they will not attend school and maintaining the **poverty cycle**.

9.10.6 HIV and AIDS successes

There is currently no cure for HIV or AIDS or a vaccination to prevent people contracting the virus. However, since the mid 1990s, antiretroviral treatment (ART) has been available to persons with HIV. These drugs reduce the level of the virus in the infected person, not only extending their life but also reducing the risk of them transmitting the disease to an uninfected partner. In 2017, around 21.7 million (57 per cent of people living with AIDS) had access to this treatment. The number of AIDS-related deaths is now declining, as is the rate of new infections. Approximately US$22−24 billion is needed annually to meet the UNAIDS vision; at the end of 2017 around US$21 billion in funding was available to fund the fight against AIDS in low-income and middle-income countries. An estimated US$29.3 billion will be needed annually by 2030. **FIGURE 6** shows how some countries in southern Africa experienced a major decline in life expectancy with the onset of AIDS; however, with improvements in treatment, life expectancy is increasing again.

FIGURE 6 Impact of AIDS on life expectancy, 1950 to 2015

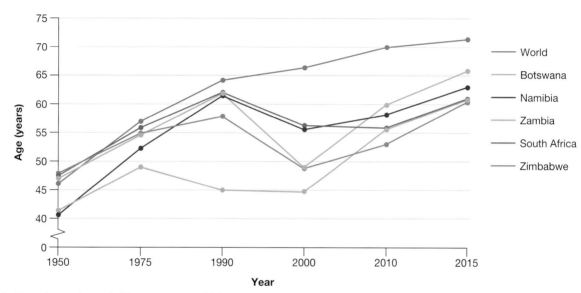

Note: Data shows change in life expectancy at birth in selected countries in southern Africa

AusAID work in Papua New Guinea

The Australian government, via the Department of Foreign Affairs and Trade, committed $100 million in the 2017–18 foreign aid budget to support Papua New Guinea in strengthening its health services. Of this, $12.6 million was directly allocated to fund services linked to sexual and reproductive services. Papua New Guinea faces particular issues in dealing with HIV and AIDS because of its predominantly rural population and because access to many locations is difficult due to rugged topography. The main response activities include education, condom promotion and distribution, **STI** and HIV testing, and treatment delivered mostly through a variety of organisations such as the National Catholic AIDS Office, Save the Children Fund and UNICEF. As a result of this work, more than 2700 people with sexually transmitted diseases such as AIDS were able to access antiretroviral medication.

on Resources

☑ **eWorkbook** Malaria and HIV (doc-32097)

9.10 EXERCISES

Geographical skills key: GS1 Remembering and understanding **GS2** Describing and explaining **GS3** Comparing and contrasting **GS4** Classifying, organising, constructing **GS5** Examining, analysing, interpreting **GS6** Evaluating, predicting, proposing

9.10 Exercise 1: Check your understanding

1. **GS1** Which locations are particularly vulnerable to malaria? Why?
2. **GS2** Explain why children would be more likely to develop malaria than adults.
3. **GS1** Which *places* in the world are most affected by HIV/AIDS?
4. **GS1** How is HIV transmitted?
5. **GS5** From **FIGURE 6**, during what period does HIV/AIDS appear to have had the greatest impact on life expectancy?

9.10 Exercise 2: Apply your understanding

1. **GS6** Predict how the global distribution of malaria may *change* with an increase in temperatures due to global warming.
2. **GS6** Predict how the distribution of malaria may *change* if the problem of drug resistance is not overcome.
3. **GS2** In what way is malaria both a disease of poverty and a cause of poverty?
4. **GS6** Imagine a vaccine to prevent malaria is available but in limited supplies. Which countries and which groups of people would you target to receive this vaccination? Justify your answer.
5. **GS6** How might *changing* global economic conditions affect the response to HIV/AIDS by aid organisations or donor companies?
6. **GS6** What could be done to help poor families affected by HIV/AIDS break the poverty cycle?

Try these questions in learnON for instant, corrective feedback. Go to www.jacplus.com.au.

9.11 Thinking Big research project: UN report — Global wellbeing comparison

SCENARIO

Life expectancy, child mortality, and the prevalence of disease will all come under your microscope as you investigate and prepare a report for the UN on changes and variations in human wellbeing across one developed and one developing country.

Select your learnON format to access:

- the full project scenario
- details of the project task
- resources to guide your project work
- an assessment rubric.

Resources

ProjectsPLUS Thinking Big research project: UN report — Global wellbeing comparison (pro-0218)

9.12 Review

9.12.1 Key knowledge summary

Use this dot point summary to review the content covered in this topic.

9.12.2 Reflection

Reflect on your learning using the activities and resources provided.

KEY TERMS

ageing population an increase in the number and percentage of people in the older age groups (usually 60 years and over)

demographic transition model a graph attempting to explain how a country's population characteristics change as the level of wellbeing in a country improves over time

dependent population those in the under 15 years and over 60 years age groups. People in these age groups are generally dependent on those in the working age groups, either directly or indirectly for support.

fertility rate the average number of children born per woman

life expectancy the number of years a person can expect to live, based on the average living conditions within a country

natural increase the difference between the birth rate (births per thousand) and the death rate (deaths per thousand). This does not include changes due to migration.

population density the number of people within a given area, usually per square kilometre

population distribution the spread of people across the globe

population structure the number or percentage of males and females in a particular age group

poverty cycle circumstances whereby poor families become trapped in poverty from one generation to the next

replacement rate the number of children each woman would need to have in order to ensure a stable population level — that is, to 'replace' the children's parents. This fertility rate is 2.1 children.

STI sexually transmitted infection

10 Factors affecting human wellbeing

10.1 Overview

Some people in the world have a better life than others. What are the reasons for this inequality?

10.1.1 Introduction

Measurements of wellbeing can be subjective judgements of how people perceive their own quality of life or they can be objective indicators such as life expectancy, education, income and population growth rates. Analysing these indicators across our society, we come to see that the distribution of human wellbeing is unequal. This inequality can be due to factors such as geographical location, access to natural resources, the experience of conflict, and the prevalence of poverty and disease. In this topic, we will explore some of the variation that exists in human wellbeing and the factors that underpin these differences.

---

on Resources

 eWorkbook   Customisable worksheets for this topic

 Video eLesson The other side of life (eles-1715)

LEARNING SEQUENCE

10.1 Overview
10.2 Gender as a factor in wellbeing
10.3 Poverty as a factor in wellbeing
10.4 Rural–urban wellbeing variation in Australia
10.5 Indigenous wellbeing in Australia
10.6 **SkillBuilder:** Understanding policies and strategies `online only`
10.7 **SkillBuilder:** Using multiple data formats `online only`
10.8 **Thinking Big research project:** Improving wellbeing in a low-HDI ranked country `online only`
10.9 **Review** `online only`

To access a pre-test and starter questions and receive immediate, **corrective feedback** and **sample responses** to every question, select your learnON format at www.jacplus.com.au.

10.2 Gender as a factor in wellbeing

10.2.1 Maternal mortality

Rarely would women in Australia consider that pregnancy and giving birth could threaten their lives. For most women in Australia and other developed countries, childbirth is something that occurs without significant health complications for either the mother or the baby. Unfortunately, for a huge number of women around the world this is not the case, and child-bearing can have a negative impact on their health and wellbeing.

Less economically developed countries (LEDCs) generally experience worse human wellbeing than more economically developed countries (MEDCs). For example, women in low-income countries have a 1 in 36 lifetime risk of **maternal mortality**, whereas this risk for women in high income countries is 1 in 3300. It is also true that within LEDCs, the health and wellbeing of women is generally worse than that of their male counterparts across most age groups. For this reason, ending gender discrimination, achieving gender equality and empowering women and girls are key elements of the United Nations' **Sustainable Development Goals** (SDGs). SDGs 3 and 5 clearly outline targets related to the health and wellbeing of women. Reducing maternal mortality rates is key among these aims.

Every day approximately 830 women die from complications related to pregnancy or childbirth. Most of these deaths are from preventable complications: severe bleeding, infections and complications from unsafe abortions. The incidence of maternal mortality and related illness is interconnected with poverty and lack of accessible, affordable quality health care.

FIGURE 1 shows the global distribution of maternal mortality. Eighty-six per cent of maternal deaths are in sub-Saharan Africa and southern Asia; sub-Saharan Africa accounts for two-thirds of all these deaths. Highest maternal mortality rates are recorded in Sierra Leone and Chad, where mothers have, respectively, a 1 in 17 and 1 in 18 risk of dying. At a national scale, two countries account for one-third of total global maternal deaths: Nigeria at 19 per cent (58 000), followed by India at 15 per cent (45 000).

FIGURE 1 Global scale distribution of maternal mortality

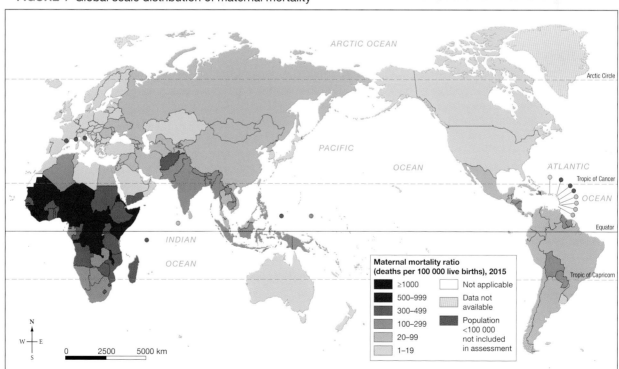

Source: World Health Organization.

The Millennium Development Goals (MDGs) — the forerunner to the SDGs — set a target of reducing the maternal mortality rate between 1990 and 2015 by three-quarters, and the achievement of universal access to reproductive health by 2015. While maternal mortality fell by 45 per cent during the MDG period, globally the 75 per cent target was not met, particularly in countries in sub-Saharan Africa.

FIGURE 3 indicates progress in terms of access to reproductive health. Most indicators fell well short of universal access (considered to be 80 per cent). While use of contraception has increased, wealthier women continue to have the best access to contraception. The unmet need for contraception among poorer women remains at levels similar to 1990. In sub-Saharan Africa for example, this figure was 28 per cent in 1990; in 2015 it remained high, at 24 per cent. Providing access to contraception is a means of empowering women to make choices about family size.

FIGURE 2 Child-bearing: a threat to wellbeing

FIGURE 3 Global reproductive health indicators, 1990–2015

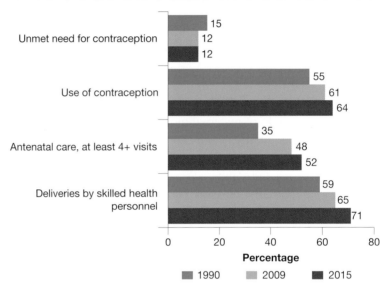

Source: Data from *The Millennium Development Goals Report 2015*, United Nations. pp. 39, 41.

The SDGs have set a target of a maternal mortality rate below 70 per 100 000 by 2030. This will require an annual drop of 7.5 per cent — more than three times the reduction that occurred under the MDGs. However, Cambodia, Rwanda and Timor-Leste all achieved this type of reduction rate in the 15 years from 2000 to 2015, proving that this goal is achievable.

10.2.2 Maternal mortality in India

As noted previously, India accounts for a large percentage of global maternal deaths. Although, on average, maternal mortality rates in India declined 68 per cent between 1990 and 2015, there is substantial variation within India, as **FIGURE 4** shows.

Maternal mortality is strongly interconnected with poverty in both rural areas and urban slums: places with poor **sanitation** and a lack of affordable health services are associated with high levels of maternal mortality. In addition, women are likely to be less well-nourished than males in a household. According to the 2011 Indian Census, women also have much lower literacy levels — a 65 per cent literacy rate compared to 82 per cent for men — so they are less likely to be able to access information on health and contraception. Unlike Australia, where census data is collected every five years, India only conducts its census every ten years. While it is important to use the most recent data available, we are often forced to use and analyse data that may not fully reflect current trends and patterns. The release of the next census data in 2021 will allow greater analysis of progress made in this important area of wellbeing.

The government of India launched the National Rural Health Mission in 2005, with a specific focus on maternal health. This was reinforced in their 2013 Call to Action. Efforts have been focused on those districts that account for 70 per cent of all infant and maternal deaths. Under this program, community workers have been trained to deliver babies, and 10 million women have been provided with a cash incentive to enable them to give birth in clinics rather than at home. Maternal mortality has fallen, but Human Rights Watch reports that many women are being charged for services as they are unaware of these entitlements.

FIGURE 4 Maternal mortality rates in India

Note: Data drawn from 2011 census; next census due in 2021
Source: Published and issued by Office of the Registrar General, India, Ministry of Home Affairs http://www.censusindia.gov.in/ vital_statistics/SRS_Bulletins/MMR_Bulletin-2010-12.pdf

A related issue for pregnant women in India is the pressure to produce a son. Census data in 2011 revealed the number of female children (0–6 years) has decreased from 927 to 914 girls per 1000 boys in the past decade, despite some overall improvement in the **sex ratio** across all age groups (see **FIGURE 5**). By comparison, the natural human sex ratio at birth would see around 952 girls born per every 1000 boys. Males are traditionally preferred over female children: sons are seen as the breadwinners who carry the family name, while daughters are often perceived as an economic burden. Although **female infanticide** is illegal, use of ultrasound for sex-determination tests has led to sex-selective abortions, with an estimated 500 000 girls aborted each year (although sex selective abortion is also illegal). The pressure to produce a son means that many Indian women have multiple pregnancies, thereby increasing their risk of maternal mortality over their reproductive years. **FIGURES 4** and **5** suggest an interconnection between the places of high maternal mortality and those with a large imbalance in the sex ratio.

FIGURE 5 Variation in sex ratio within India

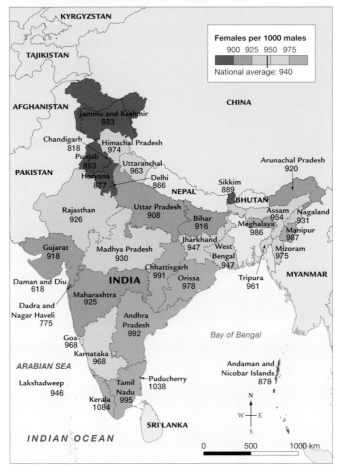

Note: Data drawn from 2011 census; next census due in 2021

Source: Government of India, Ministry of Home Affairs, Office of Registrar General. Map by Spatial Vision.

FIGURE 6 Son preference has resulted in an imbalance in India's sex ratio.

10.2.3 Comparing maternal mortality in India and Australia

Stark differences appear when the maternal mortality rates of India are compared with those of Australia. In 2015, India's maternal mortality rate (deaths per 100 000 live births) was 174 (45 000 Indian women died), compared to Australia's rate of just 5.3 (16 Australian women died). The major reason behind this difference is poverty. India's **poverty rate** in 2015 was 21.7 per cent compared to Australia's poverty rate of 13.2 per cent. Poverty limits the access and affordability of maternal health care, leading to a higher rate of maternal mortality. Australia's smaller population size, greater wealth and higher levels of geographical accessibility to medical services all contribute to its lower maternal mortality rate.

DISCUSS

How have cultural norms, religion and world views within Indian culture contributed to a lack of fairness and equality for females in Indian society? **[Ethical Capability]**

FIGURE 7 Access to affordable healthcare is a significant factor in maternal mortality rates.

10.2 EXERCISES

Geographical skills key: GS1 Remembering and understanding **GS2** Describing and explaining **GS3** Comparing and contrasting **GS4** Classifying, organising, constructing **GS5** Examining, analysing, interpreting **GS6** Evaluating, predicting, proposing

10.2 Exercise 1: Check your understanding

1. **GS2** Refer to **FIGURE 1**. Describe the variation in maternal mortality on a global *scale*.
2. **GS1** Explain how a lack of education may contribute to increased rates of maternal mortality.
3. **GS2** Why do LEDCs generally experience lower levels of human wellbeing than MEDCs?
4. **GS2** Explain why women living in developing *places* are more likely to have lower levels of wellbeing than their male counterparts.
5. **GS5** Identify three Indian states that have a sex ratio below the national average and a maternal mortality ratio above 150.

10.2 Exercise 2: Apply your understanding

1. **GS6** Predict the shape of India's population pyramid if the trends in India's sex ratio continue.
2. **GS6** Suggest measures that could be introduced by the Indian government to help Indian parents see the value of female babies as equal to that of boys.
3. **GS5** Refer to **FIGURE 4** in subtopic 9.7 'Variations in wellbeing in India', showing the distribution of poverty in India. To what extent is the *interconnection* between poverty and maternal mortality (as shown in **FIGURE 4** in section 10.2.2) evident?
4. **GS2** Describe and explain the difference in maternal mortality rates in India and Australia.
5. **GS6** Female infanticide occurs in some parts of India. Discuss the issues associated with this practice and its impact on the Indian population now and into the future.

Try these questions in learnON for instant, corrective feedback. Go to www.jacplus.com.au.

10.3 Poverty as a factor in wellbeing

10.3.1 The haves and have-nots of Rio

How would you like to live with spectacular views over one of the world's most beautiful coastlines? The only problem is that you could be living in a slum without running water and your only access in or out is via hundreds of stairs and laneways. This is what life is like in a typical **favela** in Rio de Janeiro.

According to World Bank statistics, Brazil is the eighth-largest economy in the world. This ranking places Brazil ahead of Russia, South Korea and Australia. Yet, despite the strength of Brazil's economy, the benefits of economic growth have not trickled down to the poor, resulting in large differences in wellbeing across the nation.

In Brazil, some of the most overt inequality of wealth distribution in the world is evident. Almost 55 million Brazilians — a quarter of the population — live in poverty. Incredibly, the wealthiest 5 per cent of Brazil's population earns the same income as the remaining 95 per cent.

FIGURE 1 shows the variation in wealth experienced across Brazil's different regions, as measured by **gross domestic product** (GDP).

FIGURE 1 Distribution of wealth per state of Brazil

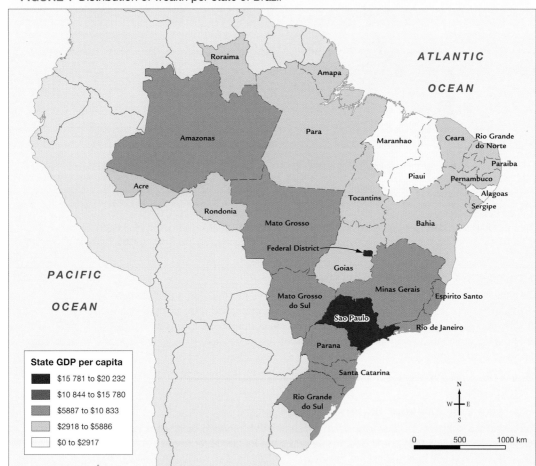

State GDP per capita
- $15 781 to $20 232
- $10 844 to $15 780
- $5887 to $10 833
- $2918 to $5886
- $0 to $2917

Source: Instituto Brasileiro de Geografia e Estatística. Made with Natural Earth. Map by Spatial Vision.

There is considerable **spatial variation** in wellbeing between regions in Brazil. The majority of industrial development in Brazil has occurred in the south and south-east regions, generating more wealth there. This contrasts markedly with the agriculturally based north-east region, which has higher rates of poverty and infant mortality and lower rates of nutrition.

Impacts of Rio de Janeiro's development on wellbeing

Rio de Janeiro is a well-known tourist destination in Brazil, famous for its beautiful beaches, spectacular scenery and carnivals. However, for many local people, these elements are far removed from their daily lives. Even within one of the wealthiest cities in the wealthiest region in Brazil, there is considerable variation in wellbeing and living conditions.

The city has experienced rapid growth, starting in the eighteenth century when freed slaves who had worked on plantations came into the city in search of employment. This rural–urban migration still exists today, with thousands flocking to the city in search of opportunity and a new life. New settlers faced the dual problems of low wages and high housing costs, thus forcing them to construct illegal shanties on wasteland or vacant land. Over time, these have developed into entire suburbs, known as favelas. Typically, these slums are located on steep slopes on the edges of the city, although, as the city has expanded, it has wrapped itself around the favelas.

Ironically, the poorest citizens live on unstable slopes with spectacular million-dollar views (see **FIGURE 2**), while the wealthier tend to live on the more stable flatter land closer to the city centre.

Brazil conducts its census every ten years. According to the census in 2010, 22 per cent of Rio de Janeiro's population of over 6.35 million people lived across some 763 favelas. Rocinha (shown in **FIGURE 3**) is considered Rio's largest favela, with its population estimated at between 150 000 and 300 000 people. It is located in the southern zone of the city, in close proximity to the famous beaches of Rio's Ipanema and Copacabana districts.

FIGURE 2 A favela located on a steep slope in Rio de Janeiro

FIGURE 3 Street view of Rocinha favela, Rio de Janeiro

The effect of favela-living on wellbeing

Living conditions in the favelas are extremely difficult as these areas have developed without any type of planning or government regulation and the housing is generally substandard. The resulting issues have an impact on the ongoing development of the city as well as the wellbeing of its citizens.

Issues affecting wellbeing include:

- *lack of infrastructure such as sanitation and piped water.* For example, almost one-third of favela households lack sanitation, leading to higher rates of disease. Garbage has to be put in sectioned-off dumping sites.
- *vulnerability to weather extremes.* For example, heavy rainfall creates landslides and floods on steep slopes. Timber shacks are more vulnerable to collapse than houses built of concrete bricks.
- *lack of access.* There is often only one main road, so movement around the favelas is via narrow lanes and steep staircases (see **FIGURE 3**).
- *long commuting times.* The average time to travel into the city centre of Rio is 1.5 hours by bus. The cost of public transport also takes a sizeable proportion of the average worker's salary. This in turn limits both educational and employment opportunities.
- *lower household income.* The average household income for people living in the favelas is approximately half that of those people living in the inner suburbs.

- *high crime rates*. The incidence of homicide and other crimes is high. This is linked to the influence of drug trafficking and criminal gangs who have established themselves within the relative safety of the favelas.
- *a sense of insecurity felt by residents*. Most people do not have legal title to their land or dwellings and can be moved by the government at any time.

Improvements to wellbeing in the favelas

In order to reduce crime and drug traffickers' control over the favelas, and to improve safety, in 2008 the government introduced Pacifying Police Units (known in Portuguese as the *Unidade de Polícia Pacificadora*, or UPP). By 2014, 37 UPPs were in place. These have had mixed success, with ongoing violence still very much an issue in many areas.

To help improve access for favela residents, the government has installed cable cars to transport people up and down the steep hillsides quickly and effectively, with local residents entitled to one free round trip per day. It is also hoped that the cable cars will allow for expansion of tourism. However, one favela community criticised the government's priorities, maintaining that locals were not properly consulted and that basic services such as sewerage and education should have come first.

As Brazil hosted the 2014 World Cup soccer tournament and the 2016 Olympic Games, the city expanded infrastructure and built new facilities. **FIGURE 4** highlights the major issue that many of the planned Olympic Zones were located on existing favela sites. Many residents were very unhappy at the prospect of being relocated to make way for new sporting venues. Over 3000 families were forcibly relocated. Favela residents claimed that the financial compensation offered was insufficient for a new home and that communities that had existed for generations were being destroyed.

FIGURE 4 Location of Rio's favelas and Olympic venues

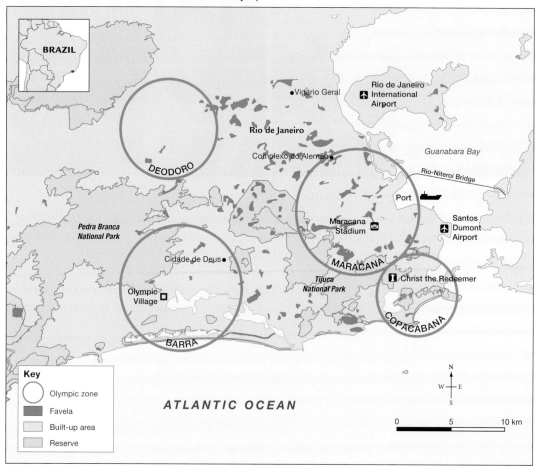

Source: UNEP-WCMC 2012. Made with Natural Earth.

Unfortunately, Brazil's ambitious plan to improve the wellbeing and living conditions of the favelas has been limited by recent economic uncertainty. After the hope provided by the Olympic Games upgrades, the futures of people living in the favelas are much more uncertain.

Preventing the continued growth of favelas by providing adequate low-income housing is the most cost-effective means of improving wellbeing. The cost of upgrading a favela with basic infrastructure is estimated to be two to three times as much as the cost of providing new high-rise housing estates. However, with 65 per cent of Rio's population growth coming from rural–urban migration, it is difficult for authorities to keep up with demand for housing and space.

Nationally, the government aimed to eliminate extreme poverty by 2014 with its Brazil Without Misery Plan. It involved the expansion of cash transfer payments to low-income families in exchange for them keeping their children in school and following a health and vaccination program. Improved infrastructure, vocational training and **micro-credit** were also part of this plan. Although the plan reduced the numbers of those living in extreme poverty from 10 per cent in 2004 to 4 per cent by 2012, it ultimately fell well short of its goal; extreme poverty remains, with rates back up to 4.8 per cent in 2017. The challenge of improving the lives of Rio's poor continues.

10.3.2 The interconnection between poverty, water supply, sanitation and wellbeing

In Australia most of us take it for granted that we can turn on a tap and get clean, drinkable water. We are confident that when we drink our tap water we will not get sick from it. We assume that our waste water and sewage will be treated and disposed of without posing a threat to our health. However, for many people around the world, lack of access to clean water and lack of adequate sanitation has had a major impact on their health and therefore their wellbeing.

Approximately 840 million people do not have access to clean, safe drinking water, and at least two billion people use a drinking water source contaminated with faeces. Over two billion lack basic sanitation. Safe drinking water and basic sanitation are of crucial importance to human health, especially for children. Water-related diseases are the most common cause of death among the poor in less developed countries — they kill an estimated 842 000 people each year, nearly half of whom are children under the age of five. Diseases such as cholera, typhoid, dysentery and worm infestations are directly attributable to contaminated water supplies.

Global progress

Remarkable progress was made under Millennium Development Goal 7 — to halve the proportion of people without access to safe drinking water by 2015. The target was met more than five years ahead of schedule, with 91 per cent of the global population using an improved drinking water source in 2015 compared with 76 per cent in 1990. Fifty-eight per cent of the world's people now have access to piped drinking water. Rural and urban coverage does vary, however, as shown in **FIGURES 5** and **6**.

Progress towards the MDG sanitation target of 75 per cent was much slower and was not met overall, although 95 countries did succeed. Only 68 per cent of the world's population has access to improved sanitation. New targets for improved access to clean water and sanitation have been set for 2030 under Sustainable Development Goal 6.

FIGURE 5 Trends in drinking water coverage (%), by rural and urban residency

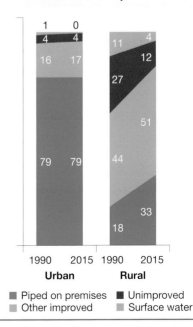

FIGURE 6 Proportion of the population using improved drinking water sources, 2015

Proportion of the population using improved drinking water sources, 2015
- 91—100%
- 76—90%
- 50—75%
- Less than 50%
- No data

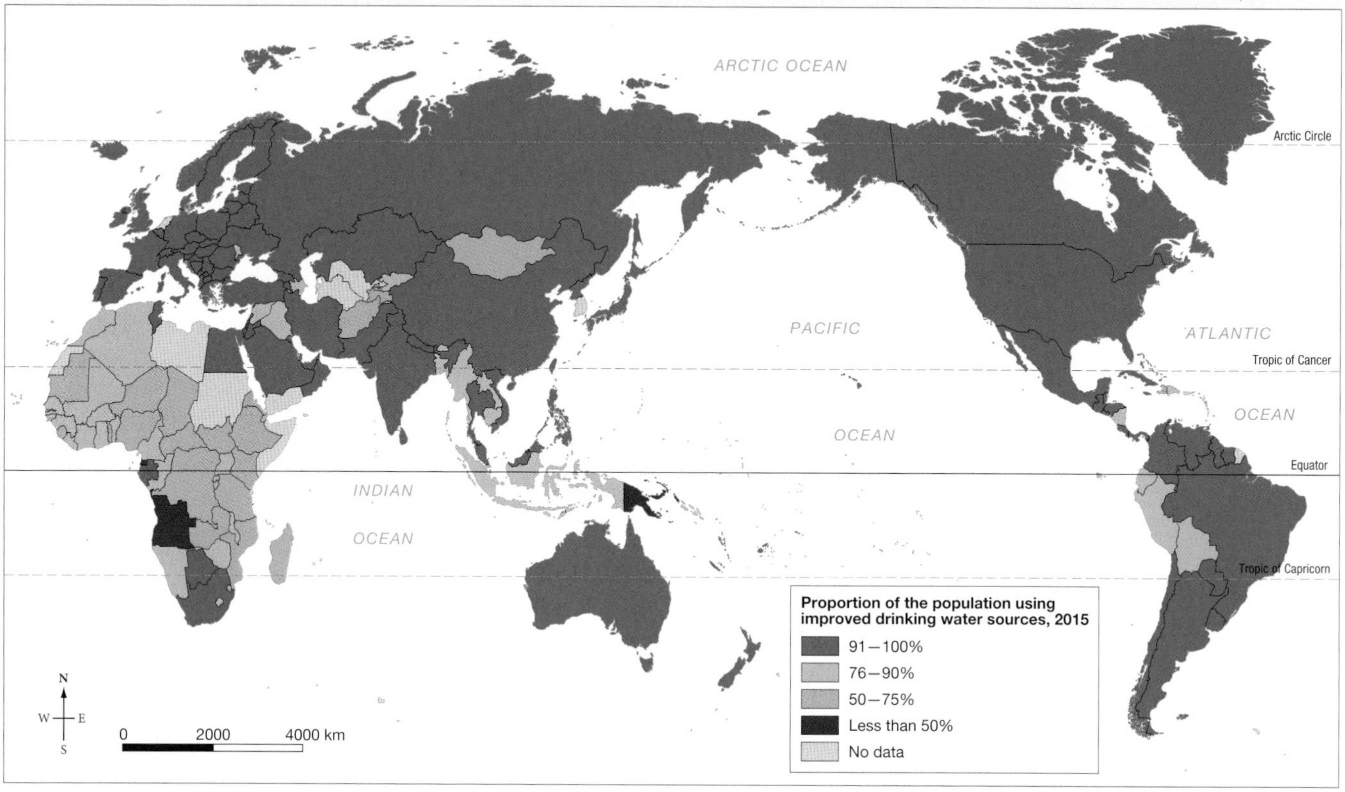

Source: © UNICEF.

There are huge regional variations across the globe as well as within countries. For example, 90 per cent or more of people in Latin America, Northern Africa and much of Asia have improved water supply in contrast with only 68 per cent in sub-Saharan Africa. In terms of sanitation, most people who lack access are again rural dwellers, even within locations where water supply has improved.

Success stories

Many non-government organisations (NGOs) have been involved in successful projects to improve access to water supply and sanitation. The success of these projects hinges not just on provision of clean water and toilets, but also on community involvement and education. Agencies such as Oxfam and World Vision work within communities to provide access to water filters, safe and clean toilets, water pumps and rainwater harvesting systems for collecting and storing water. In addition to the health benefits that they provide, these facilities also free up women and girls from hours of work spent carrying water or finding additional firewood in order to boil unsafe drinking water. Mothers now have more time to work on their farms, potentially improving food availability and income, and girls have more time free from essential chores, so they can instead attend school.

DISCUSS
'Access to adequate sanitation is increasingly a problem in urban places in the developing world.'
Evaluate this statement in small groups.   **[Critical and Creative Thinking Capability]**

 Resources

 **Weblink** Favelas

---

10.3 INQUIRY ACTIVITIES

1. Undertake research on one of the following to find out the role unclean water plays in its spread: cholera, typhoid, dysentery, schistosomiasis, worm infestations. **Examining, analysing, interpreting**
2. Go to the website for either Oxfam or World Vision. Gather the following information on one of their current projects involving improvements to the provision of clean water and sanitation: location of project, *scale* of project, problems the project aims to fix, what the project involves. **Examining, analysing, interpreting**

10.3 EXERCISES

Geographical skills key: GS1 Remembering and understanding **GS2** Describing and explaining **GS3** Comparing and contrasting **GS4** Classifying, organising, constructing **GS5** Examining, analysing, interpreting **GS6** Evaluating, predicting, proposing

10.3 Exercise 1: Check your understanding

1. **GS1** Outline the characteristics of a favela.
2. **GS1** Refer to **FIGURE 1**.
   (a) What is the average GDP for the state of Rio de Janeiro?
   (b) Describe the distribution of Brazilian states with an average GDP per capita of more than $12 000.
   (c) What reason can you give for this pattern?
3. **GS1** Refer to **FIGURE 2**. Describe how topography has influenced the development of favelas in Rio de Janeiro.
4. **GS1** With reference to **FIGURE 4**, describe the *spatial* distribution of favelas in Brazil.
5. **GS1** How has the development of Rio de Janeiro had an impact on people's wellbeing?
6. **GS2** Study **FIGURE 3**.
   (a) What difficulties would exist for people of different ages living in this street?
   (b) Why would the government find it cheaper to build new high-rise housing rather than upgrading existing favelas?
   (c) Predict the *changes* people would face if they are moved from living in a favela like this to living in a high-rise housing estate.
7. **GS2** What impact does poor sanitation have on health and wellbeing?
8. **GS5** Refer to **FIGURES 5** and **6**. How has the distribution of *places* with clean water *changed* over time?
9. **GS2** Explain the *interconnection* between poor sanitation, unclean water, health and wellbeing.

10.3 Exercise 2: Apply your understanding

1. **GS2** Explain the *interconnection* between the development of favelas and movement of people from rural areas into Rio de Janeiro.
2. **GS2** Discuss the government action to improve wellbeing in favelas on two different *scales*.
3. **GS6** Visiting favelas is increasingly popular among tourists. Suggest some positive and negative impacts of tourist tours of favelas. Formulate your own opinion: is such tourism exploiting or helping locals?
4. **GS6** Why would the estimated population of Rocinha (150 000 to 300 000) vary to such an extent?
5. **GS6** How might hosting the Olympic Games and World Cup have affected the wellbeing of Rio's residents?
6. **GS2** What impact does poor sanitation have on the natural *environment*?
7. **GS6** Which countries and regions do you expect to make greatest progress in terms of improving access to clean water and sanitation? Justify your answer.
8. **GS6** What is the likely impact of improved water provision on literacy for girls? Explain.
9. **GS6** If you were travelling to a *place* that does not have clean water or the level of sanitation you are used to, what steps could you take to ensure you did not become ill during your visit?

Try these questions in learnON for instant, corrective feedback. Go to www.jacplus.com.au.

10.4 Rural–urban wellbeing variation in Australia

10.4.1 Risk factor variations across regions

Geographical location is a significant factor in human health and wellbeing. In Australia, as in many other countries, the level of health is lower in **regional and remote areas** than in major cities. According to the Australian Institute of Health and Welfare's (AIHW) 2018 **rural** and remote health report, people living in rural and remote places tend to have shorter lives and higher levels of injury and disease and less access to health services than those living in major cities. **FIGURE 1** shows the remoteness classifications of regions across Australia.

FIGURE 1 Australia's population by remoteness classification

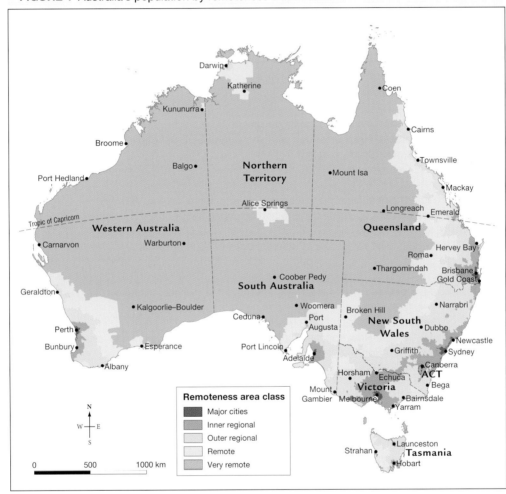

Source: Australian Bureau of Statistics.

According to the Australian Bureau of Statistics (ABS) the vast majority of Australians (71 per cent) live in major cities, 18 per cent live in inner regional areas, 9 per cent live in outer regional areas and only 2 per cent of the population live in remote and/or very remote locations.

People from regional and remote areas tend to be more likely than their major city counterparts to smoke and to drink alcohol beyond the recommended 'safe' levels. **TABLE 1** outlines differences in various health risk factors between people living in cities, inner regional areas and outer regional and remote areas.

TABLE 1 Rates of different health behaviours and risk factors across different living areas

Health risk factor	Major cities	Inner regional	Outer regional/remote
Daily smoker	13%	18%	22%
Overweight or obese	61%	67%	68%
No/low levels of exercise	64%	69%	72%
Exceed lifetime alcohol risk guideline	16%	18%	24%
High blood pressure	22%	24%	22%

Notes:
1. '%' represents prevalence of risk factor in each region (excluding *Very remote* areas of Australia).
2. Proportions were age standardised to the 2001 Australian Standard Population.

Shorter life expectancy and higher rates of injury and illness may be linked to differences in access to services, increased risk factors and the regional/remote environment. More physically dangerous occupations in rural areas lead to higher accident rates. Factors associated with driving, such as long distances, greater speed and animals on roads, contribute to higher road accident rates in country areas.

10.4.2 Variations in service accessibility

In general, people living in rural Australia do not always have the same opportunities for good health as those living in major cities. Residents of more inaccessible areas of Australia are generally disadvantaged in their access to health facilities staffed by skilled personnel.

FIGURE 2 Health services are less accessible for people living in Australia's regional and remote areas than for those in urban areas.

FIGURE 3 shows access to services is at least partially affected by the number of available health workers per population. Medical personnel in rural areas have a higher average age and face longer work hours than their city counterparts. Recruitment difficulties in rural areas also affect the sustainability of such services. **FIGURE 4** shows differences between regional and remote areas and cities in relation to access to and experience of medical services.

FIGURE 3 Health practitioners per 100 000 people, by remoteness area

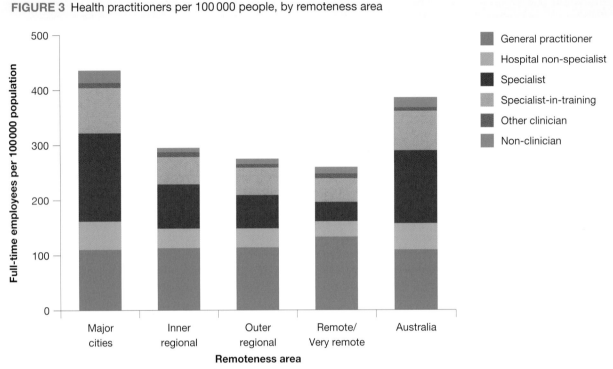

Source: Australian Institute of Health and Welfare.

Where health services are provided, regional and remote residents also tend to face higher out-of-pocket expenses. People with disability living outside major cities are significantly less likely to access disability support services.

In addition to access to health services, access to healthy food options also affects health and wellbeing. A government survey found that absence of competition in remote areas led to mark-ups of up to 500 per cent on some foods, particularly fresh fruit and vegetables, which could take up to two weeks to reach their destination. A typical packet of pasta could cost approximately five times more than in metropolitan stores.

Fewer educational and employment opportunities are other wellbeing challenges faced by those in regional and remote places.

On the positive side, despite the poorer health-related wellbeing typically associated with regional and remote areas, according to the Household, Income and Labour Dynamics in Australia (HILDA) survey, Australians living in non-urban areas tend to report higher levels of life satisfaction. This may be due to a greater sense of social cohesiveness, through community-based activities and higher rates of participation in volunteer organisations, as well as general feelings of safety in their community. The Country Women's Association is one such volunteer organisation that forms a vital part of rural community life (see **FIGURE 6**).

DISCUSS

'The benefits to wellbeing of living in regional areas outweigh the disadvantages.' Debate this issue in small groups or as a class. **[Critical and Creative Thinking Capability]**

FIGURE 4 Access to and experience of medical services in the past 12 months, by remoteness, 2016

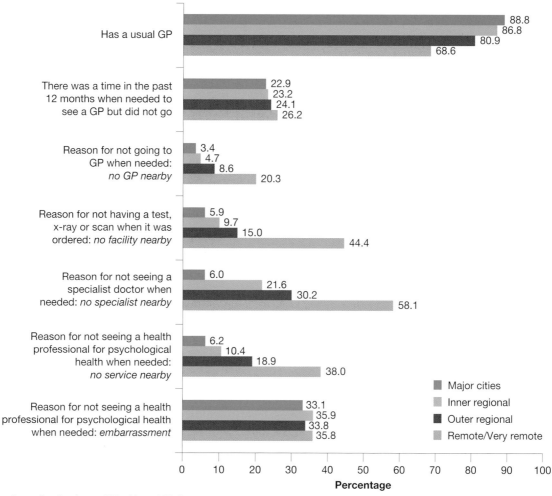

Source: Australian Institute of Health and Welfare

FIGURE 5 The Royal Flying Doctor Service provides vital health care for remote areas of Australia.

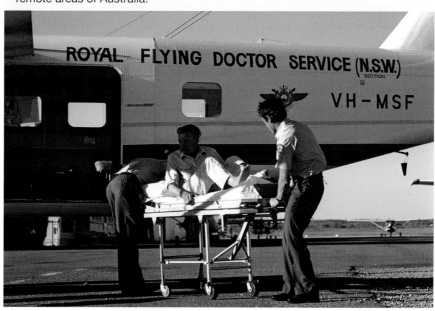

FIGURE 6 Members of the Canberra Evening Branch of the Country Women's Association held a charity bake-off for the Women's Refuge of the ACT.

Variation in wellbeing within cities

Although people in urban places generally have a higher standard of wellbeing according to many measures than those living in rural areas, levels of wellbeing are not uniform across towns and cities. If you live in a town or city yourself, you would be aware that not all parts of that location have the same access to facilities or the same types of housing. Elements such as public transport links, access to schools, shops and sporting and entertainment facilities have an impact on a place's liveability and therefore the wellbeing of its residents. Variations in wellbeing occur on a local scale as well as at national and global scales.

The cost of housing is a major expense for people so its affordability directly affects people's living standards. When more of an individual's income needs to be spent on mortgage or rent payments, they have less disposable income available for other activities. Thus, spending on wellbeing-enhancing activities, such as social events with friends and family, gym membership or sports participation and holidays, will be less, which may lead to a lower sense of general wellbeing. **TABLES 2** and **3** list the top ten most affordable suburbs in Melbourne for house and unit purchases, and the top ten most affordable suburbs within ten kilometres of the city centre. As the tables demonstrate, in general, greater proximity to the city centre leads to increased housing prices.

TABLE 2 Melbourne's top ten most affordable suburbs, 2019

Most affordable suburbs, houses	Median house price	Most affordable suburbs, units	Median unit price
Melton	$389 190	Junction Village	$303 505
Melton South	$404 812	Bacchus Marsh	$318 393
Millgrove	$419 830	Albion	$320 234
Cobblebank	$434 888	Darley	$325 919
Kurunjang	$436 026	Melton South	$329 168
East Warburton	$439 360	Carlton	$329 228
Melton West	$445 522	Melton	$330 490
Warburton	$447 262	Notting Hill	$332 055
Coolaroo	$453 602	Harkness	$344 225
Weir Views	$462 542	Werribee South	$345 024

TABLE 3 Melbourne's top ten most affordable suburbs within 10 km of city centre, 2019

Most affordable suburbs: houses	Median house price	Most affordable suburbs: units	Median unit price
Maidstone	$738 813	Carlton	$329 228
Coburg North	$750 857	Travancore	$355 798
Bellfield	$766 227	Flemington	$376 069
West Footscray	$800 115	Footscray	$415 862
Footscray	$804 367	Kingsville	$420 856
Kensington	$869 416	West Footscray	$432 156
Coburg	$882 328	Melbourne	$439 738
Collingwood	$887 392	Prahran	$459 113
Flemington	$897 218	Maribyrnong	$469 152
Preston	$903 583	Windsor	$469 815

A comparison of urban wellbeing — Mumbai, India

Located in the city of Mumbai, India, the slum of Dharavi (see **FIGURE 7**) is home to more than one million people in an area of around 220 hectares. Although this location has minimal formal infrastructure and has poor drainage, its cheap rent, manufacturing activities such as leather tanning and its location between two major suburban railway lines mean that it continues to grow. In contrast, **FIGURE 8** shows a new residential development in the eastern part of the same city. Mumbai's population of more than 22 million people includes India's richest person, Mukesh Ambani, whose US$54 billion fortune, made via the textile industry, also ranks him among the richest people in the world. The contrasts in wellbeing in India are clearly seen by these two vastly different environments.

While in Australia we may experience variation in wellbeing across the various areas within our cities, they are nothing like the scale of difference that is evident in these contrasting images of Mumbai.

FIGURE 7 Dharavi, a large slum in the city of Mumbai, India

FIGURE 8 New residential development for the wealthy in Mumbai, India

10.4 INQUIRY ACTIVITY

Find an outline map of Melbourne in your atlas or online and shade each of the suburbs in **TABLES 2** and **3** using a graded colour scheme to represent increasing price brackets (e.g. $300 000–$349 999, $350 000–399 999, $400 000–449 000). Write a paragraph to describe the distribution of the most affordable suburbs for house and unit purchases. What reasons can you propose for this? **Classifying, organising, constructing**

10.4 EXERCISES

Geographical skills key: GS1 Remembering and understanding **GS2** Describing and explaining **GS3** Comparing and contrasting **GS4** Classifying, organising, constructing **GS5** Examining, analysing, interpreting **GS6** Evaluating, predicting, proposing

10.4 Exercise 1: Check your understanding

1. **GS3** Draw a table to show the advantages and disadvantages of rural versus urban areas in terms of wellbeing in Australia.
2. **GS2** Refer to **FIGURES 3 and 4** to explain why people living in rural Australia do not always have the same opportunities for good health as those living in major cities.
3. **GS2** Explain why the information shown in **TABLES 2** and **3** may be considered a measure of wellbeing.
4. **GS5** Study **FIGURE 3**. For which type of health practitioner is access most noticeably different between major cities and regional and remote areas?
5. **GS3** Study **FIGURE 4**.
   (a) For which measures is the least difference evident? Suggest a reason why this is so.
   (b) For which measures are the greatest differences evident? Explain why this is the case.

10.4 Exercise 2: Apply your understanding

1. **GS6** What is the long-term potential outcome of the contrast in wellbeing shown in **FIGURES 7 and 8**?
2. **GS6** Why would it be difficult to measure wellbeing in a slum area such as Dharavi in Mumbai?
3. **GS6** Suggest a strategy for improving the accessibility of health services for people living in rural Australia.
4. **GS6** Suggest an alternative measure of wellbeing not mentioned in this subtopic that could highlight the variation in wellbeing within an urban area.
5. **GS6** Suggest the positive and negative impacts of the mining boom on levels of wellbeing in remote communities in Australia.

Try these questions in learnON for instant, corrective feedback. Go to www.jacplus.com.au.

10.5 Indigenous wellbeing in Australia

10.5.1 Are all Australians equal?

In a just society like Australia, we would expect that everyone is able to experience a similar standard of living. It would be unfair for one sector of a community to experience significant disadvantage when the rest of the community enjoys the privileges of a 'good life'. However, many Indigenous Australians consistently experience lower levels of health, education, employment and economic independence than those experienced by most non-Indigenous Australians. These **socioeconomic** factors inhibit the ability of Indigenous Australians, who make up 3.3 per cent of the Australian population, to contribute to and benefit from all that Australia has to offer.

Why does disadvantage exist?

The inequalities may be attributed to three main causes:
- the dispossession of land
- the displacement of people
- discrimination.

Many generations of Indigenous people have experienced difficulties in accessing the same services and opportunities as other Australians. Disadvantage in one area, for example, poor access to health services, may affect a student's ability to attend school, which may in turn alter their employment prospects. Compared with other Australians, Indigenous people (as a group) remain disadvantaged (see **FIGURE 1**).

FIGURE 1 Indicators of Indigenous wellbeing

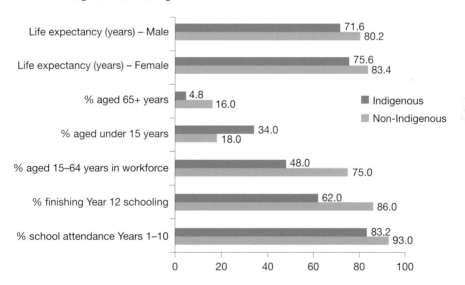

10.5.2 Improving wellbeing — closing the gap

In 2008, the Close the Gap program was launched with the aim of eliminating health and wellbeing differences between Indigenous and non-Indigenous Australians. It set out six key targets:
1. to close the life expectancy gap within a generation
2. to halve the gap in mortality rates for Indigenous children under five within a decade
3. to ensure access to early childhood education for all Indigenous four-year-olds in remote communities within five years
4. to halve the gap in reading, writing and numeracy achievements for children within a decade
5. to halve the gap for Indigenous students in Year 12 attainment rates by 2020
6. to halve the gap in employment outcomes between Indigenous and non-Indigenous Australians within a decade.

An additional target relating to school attendance was added in 2014. In the ten-year review of the program in 2018, it was noted that only two of the targets — early childhood education and Year 12 attainment — were on track to be met. The remaining target areas, although showing some improvements, were falling short of the desired levels. In the Closing the Gap Refresh program outlined in 2018, further targets were outlined in the key areas of:

- families, children and youth
- housing
- justice, including youth justice
- health
- economic development
- culture and language
- education
- healing
- eliminating racism and systemic discrimination.

FIGURE 2 Closing the gap for Indigenous Australians will take generations of commitment.

How can we measure Indigenous wellbeing?

Indigenous peoples are culturally and linguistically diverse, but Indigenous culture differs markedly from non-Indigenous Australian culture. Concepts of family structure and community obligation, language, obligations to country and the passing down of traditional knowledge are all viewed and practised very differently by Indigenous cultures in comparison to non-Indigenous cultures. These are important factors that contribute to both identity and wellbeing, yet as indicators, they may be difficult to measure.

The National Aboriginal and Torres Strait Islander Social Survey (NATSISS) is a six-yearly survey conducted by the federal government. It aims to measure the health and wellbeing of Aboriginal and Torres Strait Islander communities. Some of the data gathered from the survey are highlighted in **FIGURE 3**.

The data reveals some of the key wellbeing issues facing Aboriginal and Torres Strait Islander people, but also highlights the strong cultural and community connections felt by many Indigenous people and the importance of these connections in relation to overall sense of life satisfaction and wellbeing. **FIGURE 4** summarises the extent of this participation in family and community life.

FIGURE 3 Some findings from the National Aboriginal and Torres Strait Islander Social Survey (NATSISS) 2014–15

FIGURE 4 Indigenous Australians' participation in aspects of family, community and cultural life

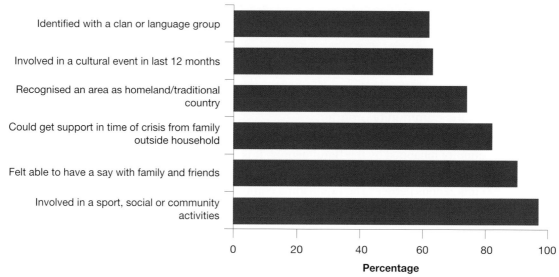

Source: ABS 2016; Table S6.1.2.

Programs to close the gap

Recognising the divides that exist at home, Australian governments and other agencies such as Oxfam are continuing to push initiatives aimed at tackling some of the problems that many Indigenous communities face. Ultimately, all Australians benefit from a united effort to address Aboriginal and Torres Strait Islander disadvantage. When disadvantage is overcome, the need for government expenditure is decreased. At the same time, Aboriginal and Torres Strait Islander peoples will be better placed to fulfil their cultural, social and economic aspirations.

The following initiatives are examples of how both government and non-government agencies are working to improve the wellbeing of Australia's Indigenous peoples.

- *The National Partnership Agreement on Closing the Gap in Indigenous Health Outcomes.* For example, the Many Rivers Aboriginal Medical Service Alliance in northern New South Wales brings together ten Aboriginal-controlled health organisations that share resources and programs servicing 35 000 people.
- *The Australian Government licensing scheme for community stores in the Northern Territory.* This scheme requires store managers to offer a range of healthy food and drinks and to make these attractive to customers. Prior to this, people in remote Indigenous communities often had little choice. Goods and food were of poor quality and basic consumer protection was lacking. More than 90 Northern Territory stores, such as the one pictured in **FIGURE 5**, are now licensed, with reported improvements in management, hygiene and employment of Indigenous staff.
- The *Wylak Project* builds the leadership capacity of Indigenous youth by offering grants in three key areas — cultural projects, advocacy/campaigning and learning events.

FIGURE 5 New food store at Ngukurr, Northern Territory

Lombadina Indigenous community program

Indigenous communities themselves are also working hard to improve their wellbeing. Lombadina is an Indigenous community inhabited by the Bardi people. It is located on the north-western coast of Western Australia (see **FIGURE 6**). Lombadina and the neighbouring Djarindjin community are home to approximately 200 Indigenous people. The Lombadina community is working towards self-sufficiency through ventures that include tourism operations, a general store, an artefact and craft shop, a bakery and a garage. The tourist ventures centre on sharing knowledge of an Indigenous lifestyle. In addition to providing serviced accommodation, many tours are offered, including cultural tours, fishing charters, kayaking and bushwalking. Lombadina has received a number of tourism awards. The considerable success of these businesses has contributed substantially to the wellbeing of this community.

Lombadina is also involved in the EON Thriving Communities Project. EON is a non-government organisation operating by invitation in Indigenous communities in Western Australia. It aims to close the gap in terms of health; for example, via the provision of practical knowledge about growing and preparing healthy food in schools and communities. The project has community ownership and is designed to be sustainable, thus improving wellbeing in the long term.

FIGURE 6 Location of Lombadina, Western Australia

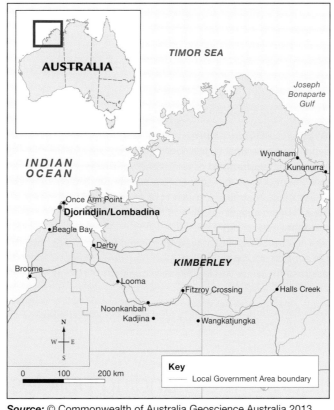

Source: © Commonwealth of Australia Geoscience Australia 2013. © Commonwealth of Australia Australian Bureau of Statistics 2013. Map drawn by Spatial Vision.

 Resources

🔗 **Weblink** Lombadina

10.5 INQUIRY ACTIVITIES

1. Conduct internet research into the Closing the Gap Refresh. Create an infographic poster outlining the new targets relating to each of the key focus areas. **Classifying, organising, constructing**

2. Another factor contributing to disadvantage may be the remoteness of Indigenous communities. Investigate the proportions of the Indigenous population who live in urban and regional areas compared with those who live in remote and very remote areas. What differences in wellbeing may result from these differences in location? **Examining, analysing, interpreting**

3. National Close the Gap Day is held in March each year to improve community awareness of the issue of Indigenous disadvantage and to publicise federal government action. Use the internet to find out what activities are taking place in your state and/or local area for the next Close the Gap Day.
Examining, analysing, interpreting

4. Using the internet, research one of the following organisations that have experienced success in combating some of the health, social or educational disadvantages experienced by Indigenous Australians. Why have they been successful? What outcomes will be *changed* for Indigenous people?
   • Aboriginal Women Against Violence (NSW)
   • MPower — Family Income Management Plan (Qld)
   • Indigenous Enabling Program at Monash University (Vic.) **Examining, analysing, interpreting**

10.5 EXERCISES

Geographical skills key: GS1 Remembering and understanding **GS2** Describing and explaining **GS3** Comparing and contrasting **GS4** Classifying, organising, constructing **GS5** Examining, analysing, interpreting **GS6** Evaluating, predicting, proposing

10.5 Exercise 1: Check your understanding

1. **GS1** What are some of the reasons that disadvantages exist for Indigenous Australians?
2. **GS1** Refer to **FIGURE 1**. What is the average life expectancy for Indigenous Australians and non-Indigenous Australians? What is the difference (in years) between these average life expectancies?
3. **GS2** Explain how the National Aboriginal and Torres Strait Islander Health Survey might give us more insight into the wellbeing of Indigenous Australians.
4. **GS1** What areas are being addressed by the federal government's Close the Gap program?
5. **GS2** Why is the Close the Gap program necessary?

10.5 Exercise 2: Apply your understanding

1. **GS2** Did any of the statistics about Australia's Indigenous people surprise you? Explain your reaction to them, and how they may have either *changed* or reinforced your own opinions or beliefs.
2. **GS6** 'Social justice' means fair and equitable access to a community's resources. Do you think Indigenous people experience social justice in Australia? Explain your answer.
3. **GS6** How might Indigenous tourism initiatives such as those run by the Lombadina community improve the wellbeing of people beyond that community?
4. **GS6** What do you consider the most significant wellbeing issue facing Indigenous Australians today? Use data from this subtopic to support your view.
5. **GS6** Suggest a way to address the issue you identified in question 4. How could wellbeing in this area be improved?

Try these questions in learnON for instant, corrective feedback. Go to www.jacplus.com.au.

10.6 SkillBuilder: Understanding policies and strategies

What are policies and strategies?

Policies are principles and guidelines that allow organisations to shape their behaviour and decisions, and to clarify future directions. Strategies ensure that the key components of a plan are implemented. Policies and strategies are particularly useful in large organisations, where information needs to be spread to all employees.

Select your learnON format to access:

- an overview of the skill and its application in Geography (Tell me)
- a video and a step-by-step process to explain the skill (Show me)
- an activity and interactivity for you to practise the skill (Let me do it)
- questions to consolidate your understanding of the skill.

 Resources

🎞 **Video eLesson** Understanding policies and strategies (eles-1760)

🔧 **Interactivity** Understanding policies and strategies (int-3378)

10.7 SkillBuilder: Using multiple data formats

What are multiple data formats?

Multiple data formats are varied forms of data presentation, used when a range of data needs to be shown. All the information must be read before the data can be interpreted.

Select your learnON format to access:

- an overview of the skill and its application in Geography (Tell me)
- a video and a step-by-step process to explain the skill (Show me)
- an activity and interactivity for you to practise the skill (Let me do it)
- questions to consolidate your understanding of the skill.

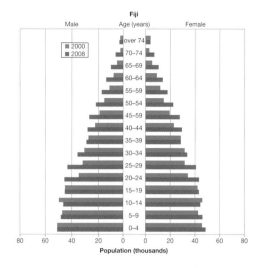

Resources

Video eLesson Using multiple data formats (eles-1761)

Interactivity Using multiple data formats (int-3379)

10.8 Thinking Big research project: Improving wellbeing in a low-HDI ranked country

SCENARIO

You will research and prepare a report on a country with one of the lowest Human Development Index rankings. What is the country's current development status, what are its top three most pressing problems, and what can the Australian government do to alleviate the situation?

Select your learnON format to access:

- the full project scenario
- details of the project task
- resources to guide your project work
- an assessment rubric.

Resources

ProjectsPLUS Thinking Big research project: Improving wellbeing in a low-HDI ranked country (pro-0219)

10.9 Review

10.9.1 Key knowledge summary

Use this dot point summary to review the content covered in this topic.

10.9.2 Reflection

Reflect on your learning using the activities and resources provided.

 Resources

eWorkbook	Reflection (doc-31779)
	Crossword (doc-31780)
Interactivity	Factors affecting human wellbeing crossword (int-7677)

KEY TERMS

favela an area of informal housing usually located on the edge of many Brazilian cities. Residents occupy the land illegally and build their own housing. Dwellers often live without basic infrastructure such as running water, sewerage or garbage collection.

female infanticide the killing of female babies, either via abortion or after birth

gross domestic product (GDP) the value of all goods and services produced within a country in a given period, usually discussed in terms of GDP per capita (total GDP divided by the population of the country)

maternal mortality the death of a woman while pregnant or within 42 days of termination of pregnancy

micro-credit the provision of small loans to borrowers who usually would not be eligible to obtain loans due to having few assets and/or irregular employment

poverty rate the ratio of the number of people whose income is below the poverty line

regional and remote areas areas classified by their distance and accessibility from major population centres

rural relating to the country, rather than the city

sanitation facilities that safely dispose of human waste (urine, faeces and menstrual waste)

sex ratio the number of females per 1000 males

socioeconomic of, relating to or involving a combination of social and economic factors

spatial variation difference observed (in a particular measure) over an area of the Earth's surface

Sustainable Development Goals (SDGs) a set of 17 goals established by the United Nations Development Programme, which aim to end poverty, protect the Earth and promote peace, equality and prosperity

11 The impact of conflict on human wellbeing

11.1 Overview

When conflict occurs in a place, what happens to the people who live there? What would happen to us?

11.1.1 Introduction

People's wellbeing is put under stress when a country's development is threatened by pressures on society, politics, the economy and the environment. Sometimes the tension results in outbreaks of conflict.

Societies pressure governments for change. Improvements are sought in living conditions, and freedoms may be demanded. Tension can spill over into conflict. People can find themselves forced to fight or flee.

Over time, countries change, but always somewhere in the world there are people trapped by conflict.

on Resources

- ✓ **eWorkbook**   Customisable worksheets for this topic
- **Video eLesson** A better life (eles-1716)

LEARNING SEQUENCE

To access a pre-test and starter questions and receive immediate, **corrective feedback** and **sample responses** to every question, select your learnON format at www.jacplus.com.au.

11.2 Conflict across the world

11.2.1 Global conflicts

FIGURE 1 shows an uneven distribution of conflict affecting wellbeing across the world. By 2016 there were 31 active armed conflicts and numerous other conflicts of varying degrees of intensity.

FIGURE 1 World conflicts, 2017

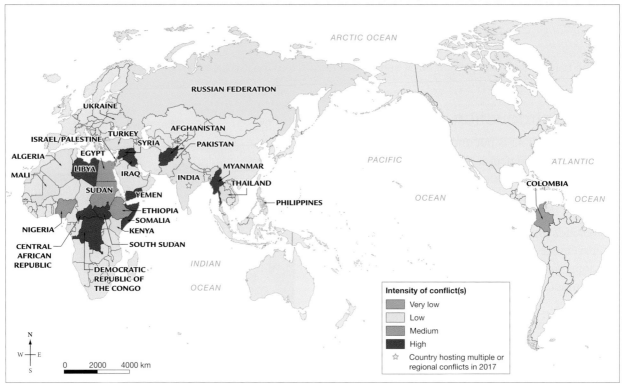

Source: Project Ploughshares.

Most conflicts are **civil wars**, where the victims are mostly the residents of the country. In World War I, less than 5 per cent of the casualties were civilians; however, in today's conflicts the figure is over 80 per cent.

Conflicts are very expensive for the countries in which they occur. The cost is not just financial, but also social and environmental. Natural and human environments can be devastated, and money is often drawn away from basic essential services, such as education and health, affecting wellbeing across the world. **FIGURE 2** shows the significant proportion of conflict-affected countries that are ranked low and medium on the **Human Development Index**.

Huge sums are spent not just in engaging in conflict but also in establishing and maintaining defence systems and armed forces. **FIGURE 3** shows global military spending for 2018.

FIGURE 2 Percentage of countries by Human Development Index (HDI) ranking hosting armed conflict, 2016

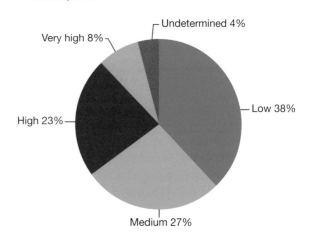

FIGURE 3 World military spending, 2018 (in US$ billions)

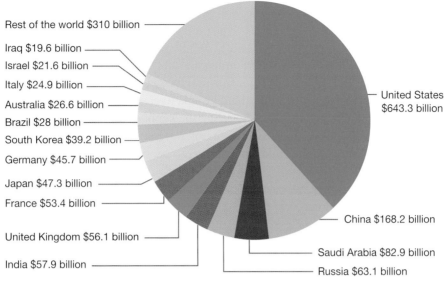

Rest of the world $310 billion

Iraq $19.6 billion

Israel $21.6 billion

Italy $24.9 billion

Australia $26.6 billion

Brazil $28 billion

South Korea $39.2 billion

Germany $45.7 billion

Japan $47.3 billion

France $53.4 billion

United Kingdom $56.1 billion

India $57.9 billion

United States $643.3 billion

China $168.2 billion

Saudi Arabia $82.9 billion

Russia $63.1 billion

Source: International Institute for Strategic Studies.

The role of diplomacy

Because of the devastating social, economic and environmental impacts of conflict, it is in everyone's interest that it be avoided where possible and resolved quickly when it occurs. **Diplomats** across the world strive to achieve this. They endeavour to negotiate successful outcomes between opposing groups to deter conflict from breaking out. In addition, soldiers in the field (as seen in **FIGURE 4**) often work with civilians to return to cooperation and peace, and improve the wellbeing of a country's population.

FIGURE 4 Working to avoid conflict

11.2.2 Ways to identify conflicts

Three approaches to **classification** that we can use to categorise conflicts are to look at the cause, length and scale of the conflict.

Cause of conflict

- Religious and cultural conflicts are based predominantly on characteristics of people or society. The break-up of Yugoslavia (1992–95) into Serbia, Bosnia-Herzegovina, Montenegro, Slovenia, Croatia and Macedonia saw the mass movement of ethnic groups to areas of safety.
- Economic conflicts involve monetary value. Securing the supply of oil from the Middle East, including shipping routes, has been important to the wellbeing of Americans. Large-scale deforestation and mining across Asia has destroyed the environment, forced people off the land and brought conflict between users of the land.
- Resource conflicts are those where resource distribution and use are the issue. A river crossing national borders is prone to manipulation of river flows, such as along the Nile River.
- Political conflicts can arise where people's political ideologies clash. The ongoing search for democracy across the Arab world signifies a break from dictatorships.
- Land conflicts, or territorial disputes, are often ongoing issues or revivals of past situations, such as the consequence of colonialism. Conflict in the Middle East between Palestine and Israel, for example, is an ongoing issue with British and French colonialism a key factor in its development.

Length of conflict

Short-term conflicts are those that last a limited time and have less ongoing impact on people. Long-term conflicts are those in which a resolution takes months or years to achieve, and even then there may be ongoing tensions.

Scale of conflict

International conflicts about the power to control land and civil conflict can destroy a nation. Conflict at this scale may become war. Small-scale or local conflicts are disagreements, generally over planning issues, that enter a dispute phase. Across Victoria the expansion of the wind farm industry is causing conflict within local communities.

11.2 INQUIRY ACTIVITY

On a blank world map, colour and label what you consider might be the areas of armed conflict in 2025. Justify your map. **Evaluating, predicting, proposing**

11.2 EXERCISES

Geographical skills key: GS1 Remembering and understanding **GS2** Describing and explaining **GS3** Comparing and contrasting **GS4** Classifying, organising, constructing **GS5** Examining, analysing, interpreting **GS6** Evaluating, predicting, proposing

11.2 Exercise 1: Check your understanding

1. **GS1** What is meant by the term *civilian*?
2. **GS1** Outline three classification systems (or ways) used to identify conflicts.
3. **GS5** Study the **FIGURE 1** map.
   (a) What classification system is used on the map?
   (b) Which continents show the highest intensity of conflict?
   (c) Which continents show a low intensity of conflict?
   (d) Which continents have no shown conflicts?
   (e) Describe the distribution pattern of conflict in 2017.
4. **GS2** Using **FIGURE 2**, suggest how the HDI ranking of countries affects the likelihood of conflict.
5. **GS1** Why is diplomacy important?

11.3 Land conflict and human wellbeing

11.3.1 A history of conflict in the Gaza Strip

Families have a strong bond with their homelands. Conflicts over land are about who the land belongs to. These conflicts stretch over long periods, from time to time erupting into hostilities. The question is: who has the right to the land?

The Gaza Strip is one of the most densely populated places in the world — more than 7500 people per square kilometre in an area 40 kilometres long and 8 kilometres wide. This strip of land on the south-eastern end of the Mediterranean Sea lies between the borders of Egypt to the south-west and Israel, defined by the ceasefire lines following the 1948 Arab–Israeli War. Palestinians came to this area as **refugees** when part of their traditional homeland was incorporated into the new state of Israel. More than 70 years later, hostilities between the Palestinians and Israelis continue.

FIGURE 1 The Gaza Strip

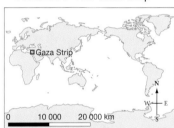

Source: Spatial Vision.

11.3.2 Lives fragmented by conflict

People's lives have been affected in many ways by the hostilities in the Gaza Strip.

- Unemployment levels are as high as 53 per cent (2018), with youth unemployment at 60 per cent.
- In 2018, 46 per cent of the population lived below the poverty line; over 70 per cent of the population relies on external aid.
- About 80 per cent of the population is dependent on food aid.
- Electricity is available for 8 hours per day; in 2016 availability was 4 hours per day.
- Medical services are hindered by drug shortages and by the lack of electricity required for complex medical equipment.
- Ninety-seven per cent of Gaza's drinking water is polluted by sewage or by salt infiltration to the aquifer.
- Buying fresh water costs six times more than the tap-water supply, which is available for only five hours per day.
- Israel restricts the entry of building materials for housing; 75 per cent of houses are in disrepair from the wars; 70 per cent of families cannot afford to rebuild their homes; 20 per cent live in overcrowded homes.
- Most of the overcrowded schools in Gaza run on double shifts; teachers are rarely paid.

FIGURE 2 shows other impacts and key features of the blockade.

FIGURE 2 Further impacts of the Israeli blockade of the Gaza Strip, which has been in place since 2007

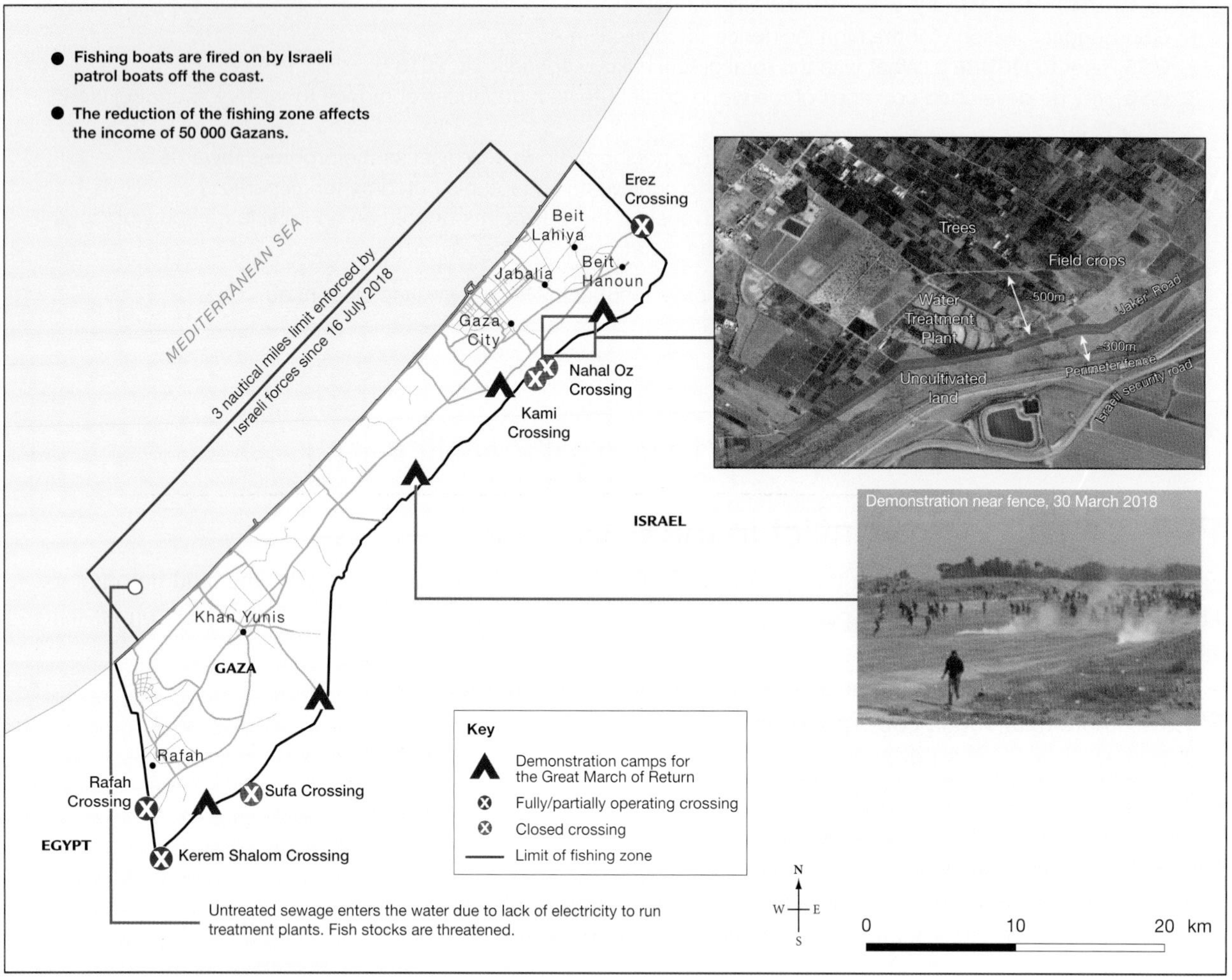

- Fishing boats are fired on by Israeli patrol boats off the coast.

- The reduction of the fishing zone affects the income of 50 000 Gazans.

MEDITERRANEAN SEA

3 nautical miles limit enforced by Israeli forces since 16 July 2018

Erez Crossing

Beit Lahiya

Jabalia

Beit Hanoun

Gaza City

Nahal Oz Crossing

Kami Crossing

ISRAEL

Trees

Field crops

Water Treatment Plant

500m

Uncultivated land

300m

Perimeter fence

Israeli security road

Baker Road

Demonstration near fence, 30 March 2018

Khan Yunis

GAZA

Rafah

Rafah Crossing

Sufa Crossing

EGYPT

Kerem Shalom Crossing

Untreated sewage enters the water due to lack of electricity to run treatment plants. Fish stocks are threatened.

Key

▲ Demonstration camps for the Great March of Return

⊗ Fully/partially operating crossing

⊗ Closed crossing

— Limit of fishing zone

N W E S

0   10   20 km

Source: United Nations Office for the Coordination of Humanitarian Affairs (OCHA).

11.3.3 The future

The opening and closing of border crossings impacts the daily wellbeing of the Gazan people. Relationships along the border fence see many skirmishes and, at times, open conflict. On 30 March 2018 Gazans began the Great March of Return, marking 11 years of being blockaded within the strip of land. The resulting cross-border clashes claimed over 200 lives and injured many thousands more. Still tensions continue, with no sign of resolution. In the meantime, Gazan families living near the sea try to maintain a sense of wellbeing with beach visits on weekends (**FIGURE 3**).

FIGURE 3 Families visit the beach on weekends in the Gaza Strip.

 Resources

Weblinks   Israel

UDHR

Google Earth The Gaza Strip

11.3 INQUIRY ACTIVITIES

1. Use the **Israel** weblink in the Resources tab to find out more about the key issues associated with the Gaza conflict. With a partner, make a list of some of the ways in which the daily lives of the people of this region would be affected. **Examining, analysing, interpreting**

2. Use the internet to research the impact that either electricity cuts of 12 to 16 hours per day or an irregular water supply have on Gazans' wellbeing. **Examining, analysing, interpreting**

3. Use the **UDHR** weblink in the Resources tab to access a version of the Universal Declaration of Human Rights. Identify those rights that you believe may be violated in this region (in both Israel and the Gaza Strip). **Examining, analysing, interpreting**

11.3 EXERCISES

Geographical skills key: GS1 Remembering and understanding **GS2** Describing and explaining **GS3** Comparing and contrasting **GS4** Classifying, organising, constructing **GS5** Examining, analysing, interpreting **GS6** Evaluating, predicting, proposing

11.3 Exercise 1: Check your understanding

1. **GS1** Outline the significance of border crossings for the Gazans.
2. **GS2** How are the lives of the Israeli people affected by the long-term conflict with the Gaza Strip?
3. **GS2** Use **FIGURE 2** to answer the following:
   (a) How does the denial of access to fishing grounds affect the Gazans?
   (b) Food security is affected by a lack of access to farmland. What activities prevent use of the *environment*?
4. **GS6** The **FIGURE 3** image seems unusual in a region of conflict. How *sustainable* is life in the Gaza Strip?
5. **GS6** Claiming territory (land) as a resource is at the heart of the conflict in the Gaza Strip. Why do you think this strip of land that is just 40 kilometres long and 8 kilometres wide is so important?

11.3 Exercise 2: Apply your understanding

1. **GS2** Explain the importance of foreign aid to the Gaza Strip.
2. **GS6** What potential forms of income can you suggest for a young person entering the workforce in the Gaza Strip?
3. **GS5** How does a lack of education impact on the future of the Gaza Strip?
4. **GS6** Why are border clashes in Gaza likely to continue in the next few years?
5. **GS6** What might be the 'stumbling blocks' to a peaceful resolution in the Gaza region?

Try these questions in learnON for instant, corrective feedback. Go to www.jacplus.com.au.

11.4 Minerals, wealth and wellbeing

11.4.1 Comparing Australia and the DRC

It is reasonable to say that any country rich in mineral resources could be expected to be highly developed, offering its people a high level of wellbeing. Clearly Australia can fit this category; regrettably this is not the case in many countries, including the Democratic Republic of Congo (DRC) in sub-Saharan Africa.

Human Development Index (HDI) comparisons

The Human Development Index takes a range of indicators of development and ranks countries according to its overall findings. **TABLE 1** shows the pronounced differences in the human development indicators for a selected range of indicators for Australia and the DRC.

TABLE 1 Comparison of selected HDI indicators for Australia and the DRC

Country	HDI ranking	GNI* per capita	Life expectancy (years)	Expected years of schooling	Child labour (% ages 5–17)	Percentage living below the poverty line ($1.90 per day)	Life satisfaction index (0 least satisfied; 10 most satisfied)
Australia	3	43 560	83.1	22.9	–	13.3	7.3
DRC	176	796	60	9.8	90.5	77.1	4.3

*GNI = **Gross national income** – the total income earned by a country's businesses and residents

Key mineral production comparisons

TABLE 2 looks at four selected minerals to show how the levels of output vary between the DRC and Australia. Diamonds, copper and gold are established mineral resources globally. Cobalt is a 'new' mineral resource. Most of these minerals are produced in greater quantities in the DRC than in Australia. However, the wellbeing indicators reflect that living conditions and human wellbeing in the DRC are considerably different to those in Australia.

TABLE 2 Mineral production, Australia and the DRC, 2016

Mineral	Unit of measurement	DRC — known quantity mined	Australia — quantity mined
Diamonds	Carat	15 559 447	13 958 000
Gold	Kilogram	30 664	288 000
Copper	Tonne	1 035 631	948 000
Cobalt	Tonne	69 038	5470 (2015–2016)

11.4.2 Riches from minerals in the DRC

The DRC has a wealth of untapped minerals, estimated in 2017 as being worth more than US$24 trillion. Many of these minerals are becoming increasingly important to us as technological change occurs. The key minerals in the DRC are gold, diamonds, tantalum (used in manufacturing mobile phones) and cobalt (used in battery technology in smartphones and electric cars). This wealth of minerals also perpetuates strife and ongoing poverty within the country.

11.4.3 The 'conflict minerals'

History shows that the colonial power of Belgium exploited the abundance of resources in the DRC. In 1960, Belgium abruptly gave independence to the country. As a result, the people had limited understanding of how to govern a country effectively. Corruption and civil war became rampant in the eastern part of the DRC; in particular, illegal trade in gold, diamonds and cobalt helped fund rebel groups.

Mining in the DRC is a significant 'artisanal' (traditional or non-mechanised) industry. People work long hours with their bare hands in poor working conditions to achieve very little financial reward. It is estimated that more than half of the gold miners operate with an armed group (**FIGURE 1**). Minerals are moved across the eastern borders of the DRC on the black market (hence the use of the term 'known quantity mined' in **TABLE 2**). Wealth from mining perpetuates the civil war or goes to international companies; this wealth could be used to improve the wellbeing of the country's people.

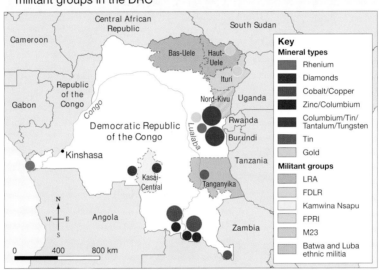

FIGURE 1 The interconnection between mining regions and militant groups in the DRC

Source: www.worldaware.com. Map by Spatial Vision.

Blood diamonds

'Blood diamonds' are uncut diamonds mined in areas of conflict and smuggled across borders; the funds generated are used for military-style activities that cause bloodshed, loss of life and a lack of wellbeing in the DRC. A cut-and-polished diamond prepared for the jewellery market cannot be identified as having its origins as a blood diamond. About 10 per cent of the DRC population relies on income from diamonds.

Gold

In some areas artisanal gold mines were forced to close in 2013 as international companies became involved in production. It is estimated that up to 10 000 miners were forcibly removed. As a result, a great deal of anger was vented on the international organisations by the displaced miners and their supporting communities. These localised conflicts have threatened production, often causing the mines to cease operation, at least for significant time periods.

Copper

Copper has become a sought-after mineral with the development of electrical vehicles and renewable energy projects that use four times more copper than traditional cars and energy production. Copper mining in the DRC is largely undertaken by international firms that struggle with the unreliable and limited power supply. The political instability of the country is also a challenge to these companies.

Cobalt

Cobalt mining remains an artisanal industry (see **FIGURE 2**). People seeking to improve their level of wellbeing work in the mines under the militia. In 2017, mine production of cobalt in the DRC totalled about 64 000 metric tons and placed the DRC as the dominant global producer (see **FIGURE 3**). Cobalt is an essential in the construction of modern jet engines, and the batteries that power our phones and electric cars.

FIGURE 2 Cobalt mining site near Kasulo, run by the DRC government

FIGURE 3 The world's key producers of cobalt, 2017

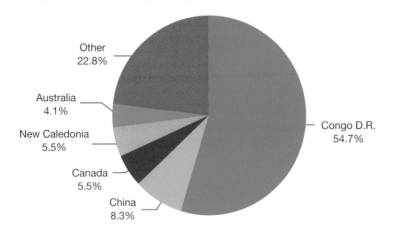

Other
22.8%

Australia
4.1%

New Caledonia
5.5%

Canada
5.5%

China
8.3%

Congo D.R.
54.7%

11.4.4 The future

The DRC is not a globally significant consumer of minerals or mineral fuels. Most of its mineral production leaves the country either illegally or through global mining organisations. In recent times the DRC has worked hard to improve the rights of the miners. In 2003 the government signed an agreement called the Kimberley Process, backed by the United Nations, trying to eradicate the sale of blood diamonds. In 2018 a new mining code was introduced, increasing the taxes and royalties for the international mining companies with the hope of returning more money to the DRC. Although the government is making strides towards achieving stability and the HDI ranking is slowly improving, it is the miners who continue to endeavour to improve their own wellbeing by working hard, and often illegally.

11.4 INQUIRY ACTIVITIES

1. Using the Human Development Index, research other key factors that show the discrepancy in wellbeing between Australia and the DRC. Write a reflection on what these indicators suggest to you about life in the DRC. **Examining, analysing, interpreting**
2. Use the internet to conduct research into the work life of child labourers. Record your findings as if you were one of these child labourers. **Examining, analysing, interpreting**

11.4 EXERCISES

Geographical skills key: GS1 Remembering and understanding **GS2** Describing and explaining **GS3** Comparing and contrasting **GS4** Classifying, organising, constructing **GS5** Examining, analysing, interpreting **GS6** Evaluating, predicting, proposing

11.4 Exercise 1: Check your understanding

1. **GS2** Explain how the mining of 'blood diamonds' limits benefits to the wellbeing of people in the DRC.
2. **GS2** Copper and cobalt are significant minerals of the twenty-first century. What global technological developments have increased their importance?
3. **GS2** Minerals are mined predominantly in the east of the DRC. Explain how this location might benefit the militia groups and those who export the minerals.
4. **GS1** What factors led to the development of corruption and unrest in the eastern DRC?
5. **GS2** Outline the status of the gold mining industry in the DRC.

11.4 Exercise 2: Apply your understanding

1. **GS5** Look at **TABLE 1**. Describe life in the DRC according to the data presented.
2. **GS2** Study **FIGURE 2**.
   (a) What surprises you most about this image?
   (b) Describe the conditions under which the miners work. Include the term 'artisan' or 'artisanal' in your response.
   (c) Do you think this work site meets international standards for workplace safety? Explain your answer.
3. **GS3** Compare the production of the four minerals listed in **TABLE 2** for the DRC and Australia.
4. **GS6** What is your assessment of the attitude of the miners to their work and to their wellbeing?
5. **GS6** Are you optimistic or pessimistic about DRC's future? Justify your answer.

Try these questions in learnON for instant, corrective feedback. Go to www.jacplus.com.au.

11.5 Fleeing conflict

11.5.1 The displacement of people

According to the Office of the United Nations High Commissioner for Refugees (UNHCR), global conflict saw more than 68.5 million people flee their homes in 2017 (see **FIGURE 1**); many have fled more than once. In 2017, one person was displaced every two seconds. These people feel they have no choice but to move. Each year the number of people on the move is different. Each year different places are in conflict. **FIGURE 2** shows where the majority of the world's refugees come from, and which countries are hosting them.

FIGURE 1 Persons of concern to UNHCR as refugees, end 2017

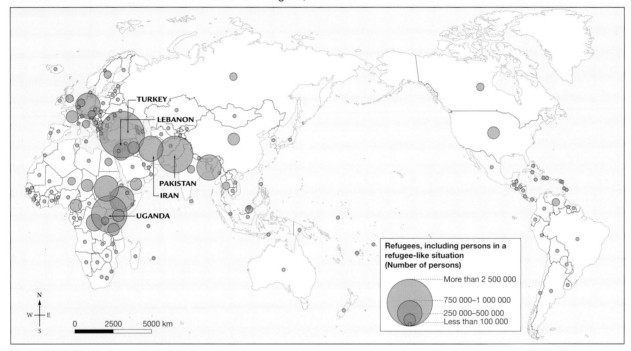

Source: UNHCR Statistics; The UN Refugee Agency.

FIGURE 2 Major refugee groups and their host countries, 2016–2017

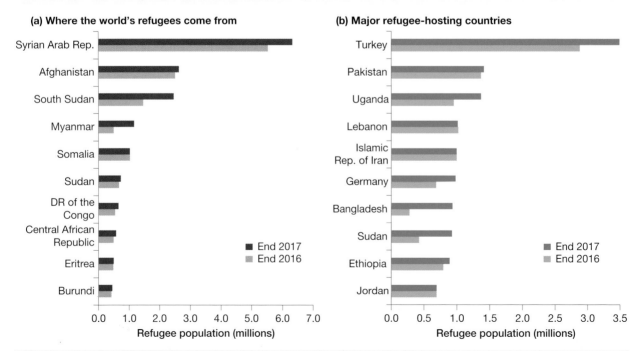

In 2017, the largest group of people on the move was the 40 million **internally displaced persons** (IDPs). In addition, some 25.4 million **refugees** left their country of origin, and 80 per cent of these arrived in a neighbouring country. Five countries — the Syrian Arab Republic, Afghanistan, South Sudan, Myanmar and Somalia — accounted for 68 per cent of the refugees in 2017. More than 10 million refugees have been **stateless people** for long periods of time.

Most of those who flee have experienced conflict, although some are 'environmental refugees', especially those escaping prolonged drought. Others flee as 'economic refugees', finding the living conditions of their country unacceptable and choosing to seek a better lifestyle.

11.5.2 Life as a refugee

People who flee are often forced to make the decision quickly. These people are distressed by the situation that they find themselves in and simply take with them possessions that can be carried — every family member carries something (see **FIGURE 3**). People walk to safety or cram into vehicles. Families and friends are torn apart.

FIGURE 3 Refugees arrive at Dadaab refugee camp, Kenya

The Office of the United Nations High Commissioner for Refugees (UNHCR) is the international organisation charged with leading and coordinating international action for the worldwide protection of refugees. The UNHCR monitors the movement of refugees. As the need arises, the UNHCR, governments and non-government organisations (NGOs) respond and establish camps across the borders from the conflict to accommodate people in the short term.

People of concern to the UNHCR are predominantly women (50 per cent of refugees) and children (52 per cent are under the age of 18). Camp life is basic. Women and children are at risk.

According to 2018 data, the three refugee camps with the largest populations in the world are:

- Kutupalong–Balukhali in Cox's Bazaar, Bangladesh (housing over 887 000 Rohingya from Myanmar)
- Bidi Bidi in Uganda (housing over 285 000 people, largely from South Sudan)
- Dadaab refugee complex in Kenya (housing over 235 000 people, mostly from Somalia).

Life in these refugee camps is no longer 'normal'. **FIGURE 4** shows some of the issues of living in these camps that affect human wellbeing.

Five million displaced persons returned home in 2017. Most of these people were IDPs. However, the rate of return was exceeded by new displacements.

FIGURE 4 Dadaab refugee camp, Kenya

- Space is restricted.
- Privacy is lacking.
- Only essential food items are provided.
- Cooking facilities are basic.
- Water is provided at a central location.
- Sanitation can be limited.
- Medical support is stretched to its limits.
- Education facilities are lacking.
- There is little to occupy people's time; there is little or no work.
- Family values need to be maintained.
- Violence against women and children can spread in the camp.

11.5 INQUIRY ACTIVITIES

1. Research Bidi Bidi or Dadaab refugee camps. Write an extended paragraph on what life is like in the camp for the thousands who live there. Include two images to illustrate your comments.

   Examining, analysing, interpreting

2. A mother in a refugee camp speaks to the media about the plight of her family. Working with a partner, create the interview questions and responses. You may like to create an audio or video recording of your interview to present to the class. **Classifying, organising, interpreting**

11.5 EXERCISES

Geographical skills key: GS1 Remembering and understanding **GS2** Describing and explaining **GS3** Comparing and contrasting **GS4** Classifying, organising, constructing **GS5** Examining, analysing, interpreting **GS6** Evaluating, predicting, proposing

11.5 Exercise 1: Check your understanding

1. **GS1** What is the difference between an internally displaced person and a stateless person?
2. **GS1** What is the role of the UNHCR?
3. **GS1** Which groups among refugees are of particular concern to the UNHCR?
4. **GS1** On which continents are the three largest refugee camps?
5. **GS1** In 2017, five million displaced persons returned home. What impact did this have on the total number of displaced persons? Explain.

11.6 CASE STUDY: Syria — the impact of conflict on wellbeing

11.6.1 The impact on the Syrian people

The civil war in the Syrian Arab Republic has become a long-term event. It began in 2011 as part of the uprising of its people against the government in the Arab Awakening (also known as the Arab Spring). Civil war does not mean that everyone living in the country is involved in the war, but everyone living in the country is affected by the war. Life and wellbeing is changed.

The Syrian people had four choices when government hostilities broke out against their protests in the Arab Awakening: join the Syrian Arab Republic's army, join the rebels, leave the fighting zones, or stay in their homes.

By 2018, 7.8 million Syrian people — especially women, children and young men — had fled areas of conflict to somewhere else within the Syrian Arab Republic, becoming internally displaced persons (IDPs) in their own country (see **FIGURE 1**). These people make up one in five of all IDPs globally — this is the largest displaced population worldwide.

FIGURE 1 The number of IDPs in Syria 2011–2017

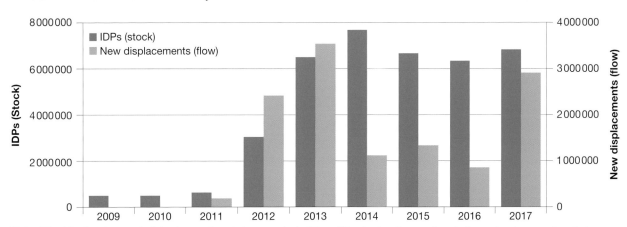

Note: 'Stock' refers to people living in a region during a period of time; 'Flow' refers to people entering or leaving a region during a period of time.

Source: © Internal Displacement Monitoring Centre.

In 2013, the Assad-led government declared 'surrender or starve' to its people and began sieges on key cities, particularly the capital city, Damascus, and large populations to the north of the country. Sieges 'lock' people within a city's boundaries, preventing easy movement out and denying entry to the city. In early 2016 it was estimated that between 390 000 and 1.9 million people were trapped in cities.

Multiple opposition groups formed in a wider context throughout the region and began to have a presence in the Syrian Arab Republic. Some of these groups have a religious base and others are terrorist cells. Since then the pressures of conflict in different areas have seen many IDPs flee again, often at night to avoid detection. Some of these people have crossed the border into surrounding countries to become refugees, massing in 'tent cities' on the border with Turkey, with many ultimately moving on either to other neighbouring countries or even further, into Europe (see **FIGURE 2**). The level of liveability for the Syrian people has declined.

FIGURE 2 The distribution of Syrian refugees

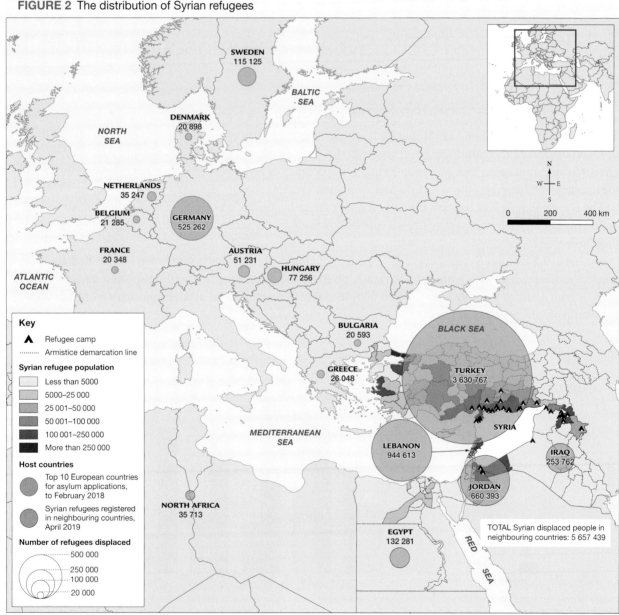

Source: Based on data from UNHCR, Government of Turkey.

11.6.2 The impact on housing

In **FIGURE 3**, Damascus shows significant change in liveability from 2008 to 2018. Living conditions have changed: safety in homes is at risk, there is food insecurity and children are traumatised. Global relief organisations estimate that more than 13 million people in Syria need humanitarian aid.

The street-to-street fighting that is a key element of civil war has destroyed buildings, including houses, in major cities such as Aleppo and Homs. Public services such as electricity, running water and gas supplies no longer operate. There is no transport system. Without oil, people rely on wood fires for heating and cooking, but this has brought about local deforestation. War continues to injure and kill local people who have remained in their homes. In late 2014 the United States, the United Kingdom and France began airborne bombing of cities to reduce the threat of rebel groups; Russia began air strikes in late 2015. Ongoing bombing strikes by the Syrian regime further destroy buildings.

FIGURE 3 Change in liveability score 2008–2018, showing the change in Syria

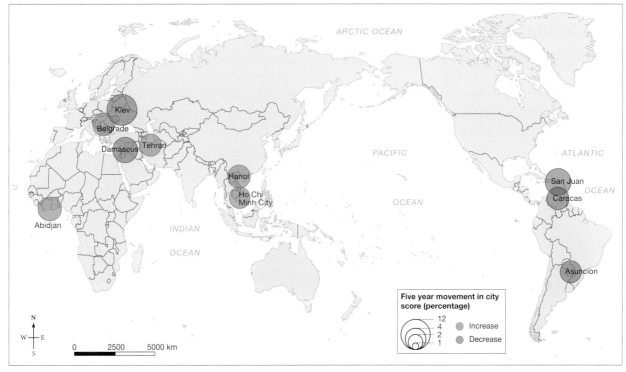

Source: Economist Intelligence Unit.

11.6.3 Feeding the Syrian people

Food insecurity is a daily issue for the war-torn areas of the Syrian Arab Republic. It is not safe to be outside for too long tending plants. Food transport cannot reach the besieged cities. Reports of malnourishment surfaced in 2016 at besieged Madaya (where 40 000 people were trapped) when social media reported that families were stripping the trees of leaves and boiling them to provide one meal a day. Aid organisations negotiated with the Assad government to be allowed to enter the city (free from attack) with a convoy of trucks bringing food, but this was only a short-term solution. As the conflict continues, food remains a major issue for the people trapped in the conflict zones of Syria.

FIGURE 4 A Syrian refugee makes a cooking fire.

11.6.4 The impact on children

Children in any war-torn area have their lives dramatically changed. The streets are no longer playgrounds. Education is disrupted or abandoned for months or years. Fear enters their lives — the sounds of aircraft, bombing and shooting punctuate their days and nights. Deafness in children becomes a problem. Families are torn apart, with some people fleeing and others staying. Children miss their friends. Young men are recruited for the fight by both sides of the conflict with blackmail, threats, fear and propaganda. Life is insecure, confusing and scary; children grow old before their time.

FIGURE 5 The destruction of their homes is just one of the many significant effects the war has had on the children of Syria.

11.6.5 Adapting to life in the besieged cities

The resilience of people is evident in the besieged cities as people become accustomed to a basic lifestyle. Innovation is required — static bicycles are pedalled to generate power for mobile phones; medicines are produced from home remedies; plastic is burned to extract oil derivatives; and rooftop gardens produce small amounts of vegetables.

FIGURE 6 Residents of Homs go about daily life among the destruction.

11.6.6 What are the costs to the Syrian Arab Republic?

International peace talks have brought ceasefires in the fighting, but will peace ever be achieved? In 2018–19, the Syrian armed forces continued to retake cities and push terrorist cells out of the country. The costs to the Syrian Arab Republic are immense. So many of its people have fled — more than 5.6 million are refugees and 7 million are IDPs, of whom nearly 3 million are in besieged cities or hard-to-reach locations. Some of those who fled will return to the Syrian Arab Republic, but they too have changed as a result of the experiences they have been through. And how will those who remained perceive the returnees and those who stay away? Families have been changed forever. The cities will take years to rebuild; more than a quarter of all housing has been destroyed. Services and food supplies will need to be re-established. Children will have years of schooling to catch up on. The country's soul has been irreversibly altered.

11.6 INQUIRY ACTIVITIES

1. Conduct an internet search to locate the latest Human Development Report.
   (a) Copy the table below and complete it using data from the report.
   (b) Using the indicators in the completed table, describe life in Syria.
   (c) From the table, is there an indication of why refugees opt to go to Turkey and Lebanon in preference to other cross-border countries? Support your answer using statistics.   **Examining, analysing, interpreting**

Country indicator	Syrian Arab Republic	Turkey	Jordan	Lebanon	Iraq	Egypt
HDI ranking						
GNI per capita						
Public expenditure on education (% of GDP)						
Local labour market (% answering 'good')						
Public health expenditure (% GDP)						
Internet users (% of population)						

2. As a small group activity, use the internet to create a photographic essay of at least six images of life in Syria during conflict. Choose the images carefully and include some qualitative and quantitative data by adding annotations to support your choice of images and to show your understanding of the situation and the impact on the wellbeing of Syrians.   **Classifying, organising, constructing**

11.6 EXERCISES

Geographical skills key: GS1 Remembering and understanding **GS2** Describing and explaining **GS3** Comparing and contrasting **GS4** Classifying, organising, constructing **GS5** Examining, analysing, interpreting **GS6** Evaluating, predicting, proposing

11.6 Exercise 1: Check your understanding

1. **GS1** Who are the sides in the civil war in the Syrian Arab Republic?
2. **GS1** What was the government policy that forced great hardship on the Syrians? Explain its implications.
3. **GS5** Study **FIGURE 1**.
   (a) Which time period saw the greatest movement of Syrians from their homes to another place in Syria?
   (b) When did the total number (stock) of IDPs peak?
   (c) Suggest reasons for the flow of IDPs from 2014 to 2017.
   (d) What does the variation in the IDP totals indicate about the location of Syrians impacted by the conflict?
4. **GS5 FIGURE 2** shows the cross-border movement of Syrian refugees.
   (a) Rank the neighbouring countries from highest to lowest in the number of Syrian refugees registered in each country in November 2017.
   (b) Is the distribution of Syrian refugees even across the neighbouring countries? In particular, refer to the situation in Turkey.
   (c) Suggest why the refugee camps are found along the borders.
5. **GS5** Using **FIGURE 3**, state how Syria's capital, Damascus, has faired in terms of liveability as a city.

▶

11.7 Seeking refuge

11.7.1 Refugees and asylum seekers

Refugees flee conflict and cross a border into another country to seek relief from the trauma of war and make a home elsewhere. All refugees (those who cannot return home due to a fear of persecution because of race, religion, nationality or membership of a social group) have been **asylum seekers**, but not all asylum seekers are found to be refugees. Asylum seekers who are not found to be refugees have either not satisfied the UNHCR criteria to be deemed a refugee or have gone outside of the formal process to seek a place to live.

11.7.2 The movement of people to Europe

Most of those arriving in Europe are fleeing the civil war in the Syrian Arab Republic, with other significant numbers arriving from the ongoing chaos in northern Africa since the Arab Awakening. Syrians have fled through Turkey to reach the shores of the Aegean Sea, from which on a clear day the Greek islands of Lesbos and Kos can be seen a mere 4 kilometres away (**FIGURE 1**). People from northern Africa come across the Mediterranean Sea, particularly from Libya. However, movement across these waters is treacherous in small boats and dinghies, and loss of life by drowning is high (see **FIGURE 3**).

On the eastern route, Greece's islands are the first point of arrival, where the refugees are fingerprinted, photographed and given a document allowing legal residency for 30 days in Greece. Greece does not accommodate the mass of people arriving on its shores. It is costly for the already poor country to rescue people from the seas and process their movement.

FIGURE 1 Refugees board a dinghy with all their possessions to make the short crossing from Turkey to Greece.

FIGURE 2 Tents of migrants and refugees in the port of Piraeus, Athens, Greece

People crossing from Africa reach Italy or Spain as the closest landfall places, or are picked up by the rescue ships in the middle of the Mediterranean Sea and taken either back to Africa or to a European country that will accept them.

11.7.3 Arranging the journey

It is estimated that 90 per cent of refugees have their journeys organised by criminal gangs, including individual people smugglers and migrant smuggling networks across Europe. Thousands of dollars are extorted for the risky sea crossings and for travel on trains within Europe. High prices are paid for accommodation and fake documents such as passports that allow refugees to apply for asylum elsewhere in Europe, especially in Germany and Sweden. People smugglers often instruct refugees that when a coast guard ship is in sight the boat or dinghy should be destroyed to ensure the refugees' rescue, a meal and health checks before arriving on European soil. **FIGURE 3** shows the number of asylum seekers moving to Europe over a four-year period.

FIGURE 3 Crossings of the Mediterranean Sea, 2015–2018

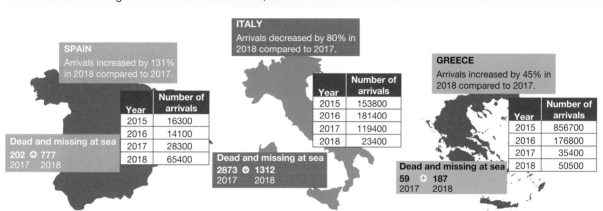

SPAIN
Arrivals increased by 131% in 2018 compared to 2017.

Year	Number of arrivals
2015	16300
2016	14100
2017	28300
2018	65400

Dead and missing at sea
202 ○ 777
2017   2018

ITALY
Arrivals decreased by 80% in 2018 compared to 2017.

Year	Number of arrivals
2015	153800
2016	181400
2017	119400
2018	23400

Dead and missing at sea
2873 ○ 1312
2017   2018

GREECE
Arrivals increased by 45% in 2018 compared to 2017.

Year	Number of arrivals
2015	856700
2016	176800
2017	35400
2018	50500

Dead and missing at sea
59 ↑ 187
2017   2018

11.7.4 Syrian refugees across Europe

FIGURE 2 in subtopic 11.6 shows the distribution of Syrian refugees throughout Syria's neighbouring countries and beyond, across Europe. After the sea crossing from Turkey, Syrians seeking asylum in Europe entered the region predominantly through Greece, although some entered via Bulgaria and then spread through neighbouring countries.

11.7.5 The European response that affects human wellbeing

In 2015, four times as many refugees arrived in Europe as in 2014. This was to be the peak in arrivals from the eastern Mediterranean route (see **FIGURE 3**). Germany, with its developed economy, high living standards and political compassion was targeted as a place to go. The German community initially showed open-minded goodwill and generosity (see **FIGURE 4a**), but in 2016 attitudes began to change; the numbers of migrants became overwhelming. By mid 2018, 1.4 million people had sought asylum in Germany. In 2018, the surge of asylum seekers from the western Mediterranean route through Spain and onward to Germany raised further concerns about the impact on the German way of life, with issues raised such as housing availability and infrastructure pressure, as well as how people with different languages and cultures would live together.

Sweden had a very open approach to asylum seekers, providing safety for people in need of protection (see **FIGURE 4b**). Permanent residency permits were offered to those with appropriate documents. Accommodation, a small daily allowance, health care and schooling were provided. Early in 2016, Sweden announced tougher rules as it felt that it had reached its limit regarding the numbers of asylum seekers that it could take. Some scenes of violence and criminal activity had changed Swedish attitudes and expulsion of asylum seekers began.

Hungary saw itself as a stepping stone for those moving north, but the sheer number of people moving through the country along disused railway lines, on roads and across paddocks struck fear within the government. In late 2015, a 4-metre-high wire fence was erected along the border with Serbia and patrolled by police with tear gas and water cannons (see **FIGURE 4c**), but refugees found gaps and cut holes to continue their movement north-west, or changed their path to go through Croatia.

FIGURE 4 (a) Welcome to Germany (b) Volunteers providing supplies in Sweden (c) The Hungarian fence (d) Sleeping at an Italian shelter (e) The Calais tent city before it was dismantled in 2016

Italy, with its influx of refugees from northern Africa, has given the task of caring for the refugees to charities, companies, cooperatives and individuals. Shelters are often substandard and overcrowded (see **FIGURE 4d**). Italy hopes that the people will move on from the southern regions, through Milan and on to other European countries. In 2018, Italy banned the rescue ships operating in the Mediterranean Sea from disembarking asylum seekers at its ports.

France has settled many of the northern African refugees within its cities. Most of these refugees speak French because of France's colonial dominance of northern Africa in the nineteenth century. Some refugees aimed to reach Britain by stowing away on ferries or on trucks travelling through the tunnel under the English Channel. Refugees established a tent camp city near Calais (see **FIGURE 4e**) while they waited to attempt a crossing. Authorities did not approve and in October 2016, amid protests and clashes, the camp was closed and dismantled. A program introduced in France in 2019 aims to assimilate the migrants via a volunteering scheme that will see people contribute to the public good, learn work skills, gain additional language lessons and receive a monthly payment.

11.7.6 A regional perspective on asylum seekers

The European countries tried to find a regional solution to the flood of migrants and found themselves bickering with each other over decisions made within one country that affected a neighbouring country — a domino effect. Greece, Italy and Spain, as major entry points, felt the pressure as other countries closed their borders and restricted the on-flow of migrants. Some countries felt they had taken their 'fair share' of the numbers of asylum seekers and began turning away those that couldn't prove their status.

By 2019 the European Union (EU) had taken some control of the situation. It claims there is no longer a crisis situation — arrivals in 2018 were at the lowest for five years. The EU works with countries of origin of asylum seekers, in particular Turkey, Syria and Libya, to provide financial assistance and assistance with processing arrivals at the EU borders. It also provides financial, operational and material support to its member states most impacted by the arrivals — Greece, Italy and Spain. Since 2015 two EU programs have assisted in the resettlement of 50 000 people across Europe.

DISCUSS

Using the criteria of fairness and two other criteria of your choice, evaluate the effectiveness of the responses of two European countries to the Syrian refugee crisis. **[Critical and Creative Thinking Capability]**

11.7 INQUIRY ACTIVITIES

1. Research the members of the European Union and the Schengen Area. In what ways may these organisations have contributed to the mass movement of people? **Examining, analysing, interpreting**

2. Conduct an internet search to find the current Human Development Report and answer the following questions.
   (a) Find the HDI ranking for each country in the table below.

Country	Greece	Italy	Hungary	Germany	France	Sweden	Spain
HDI ranking							

   (b) How do the HDI rankings for the 'first ports of call' of the refugees compare to the other countries listed?
   (c) How might the HDI rankings help to explain the movement of the asylum seekers through Europe?
   (d) According to the HDI rankings, which countries might be best placed to cater for the wellbeing of large numbers of people on the move? **Examining, analysing, interpreting**

3. Imagine you are a refugee moving from *place* to *place* on your journey across Europe seeking asylum. In small groups, write a series of tweets for the social media site Twitter that describe your wellbeing in a number of countries. **Evaluating, predicting, proposing**

11.8 Providing assistance for global human wellbeing

11.8.1 The Global Peace Index

The Global Peace Index (see **FIGURE 1**) uses 23 indicators and 30 other factors of wellbeing to assess a country's 'peacefulness'. Among the criteria used are elements of peace at home (government stability, democratic processes, community relations, security and trust between people) and peace in foreign relations (military spending levels, commitment to the United Nations and avoidance of war). Where countries experience low and very low states of peace, they are often greatly in need of assistance from other countries throughout the world in order to maintain their people's wellbeing.

11.8.2 Caring for human wellbeing

Developed countries across the world provide financial and personnel assistance to those who have their wellbeing pressured by environmental, social, political or economic factors, including in those countries where conflict disrupts lives and makes human wellbeing a struggle.

During the twentieth century, Australia accepted many people hoping to improve their wellbeing after conflict. Many thousands of migrants from Europe, in particular Italians and Greeks, arrived after World War II; the Vietnamese came as a result of the Vietnam War; and eastern Europeans came after the break-up of Yugoslavia. In the twenty-first century, conflict in the Middle East, western Asia and Africa has seen people of these cultures seek refuge in an ever-changing, multicultural Australia.

FIGURE 1 Global Peace Index, 2018

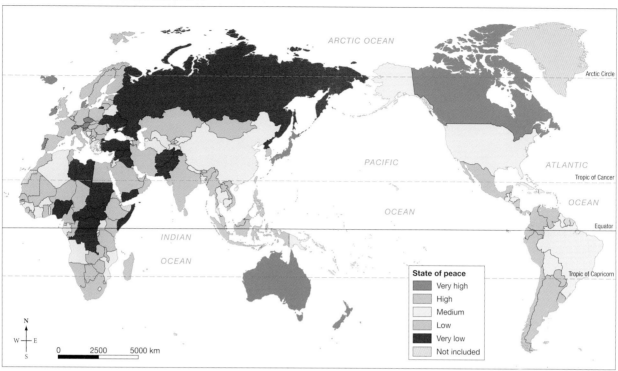

Source: Institute for Economics & Peace, Global Peace Index 2018: Measuring Peace in a Complex World, Sydney, June 2018. Available from: http://visionofhumanity.org/reports

11.8.3 The role of the Australian government today

The Australian government each year in its budget sets out a program of Official Development Assistance (ODA). The aim of this program is to promote sustainable economic growth and reduce poverty. The ODA budget is broken up into various investment priorities, which are shown in **FIGURE 2**. In the 2018–19 budget, some $4.2 billion was allocated for global assistance, the vast majority of which was earmarked for the Indo-Pacific region of which Australia is a part (see **FIGURE 3**).

Australian officials also sit on many international organisations providing a global perspective on issues, including events in conflict zones. For over 70 years Australia has been a member of the many United Nations peacekeeping and security groups providing support in conflict zones. In numerous other organisations, Australia is seen as a key driver of a change in attitude to ensuring civilians caught up in conflict are treated in a humane manner.

FIGURE 2 Distribution of Australia's ODA budget by investment priority, 2018–19

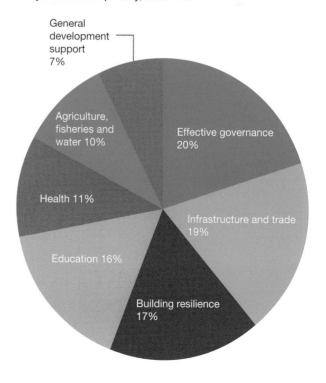

FIGURE 3 Australian Official Development Assistance (ODA) by region, 2018–19

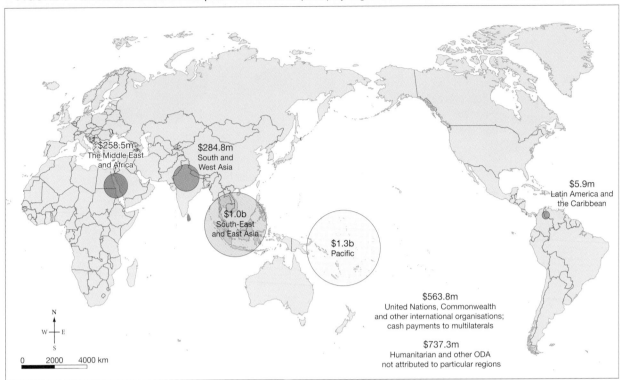

$258.5m
The Middle East
and Africa

$284.8m
South and
West Asia

$5.9m
Latin America and
the Caribbean

$1.0b
South-East
and East Asia

$1.3b
Pacific

$563.8m
United Nations, Commonwealth
and other international organisations;
cash payments to multilaterals

$737.3m
Humanitarian and other ODA
not attributed to particular regions

N
W E
S

0   2000   4000 km

Source: Data from © Commonwealth of Australia, DFAT, Australian Aid Budget Summary 2018–19. Map drawn by Spatial Vision.

FIGURE 4 It is in Australia's interests to promote peace and stability throughout the Indo–Pacific region. Australia has provided military personal for numerous peacekeeping missions; our role in East Timor was of particular importance.

In 2017, Australia was elected by the United Nations General Assembly to serve on the UN Human Rights Council from 2018 to 2020. This is the first time Australia had been chosen to serve on this council, and it is seen as a reflection of our commitment to the protection of human rights. In addition to this role, Australia is actively engaged with other entities that promote human rights, such as the Asia Pacific Forum of National Human Rights Institutions, of which the Australian Human Rights Commission was a founding member in 1996.

Among Australia's immigration statistics there is also a specified annual intake of humanitarian refugees. This was increased in 2015 to accept an additional 12 000 refugees from the Syrian crisis for the following five years. For the 2018–2019 period, it was expected that Australia's humanitarian refugee intake would be just under 19 000 people.

11.8.4 International NGOs working for human wellbeing

International **non-government organisations** (NGOs) assist the wellbeing of civilians caught up in conflicts. Three significant organisations are Médecins Sans Frontières, the International Red Cross and Red Crescent Movement, and the World Food Programme.

- Médecins Sans Frontières provides emergency medical care (see **FIGURE 5**). During conflict, local health systems often fail and hospitals close. In refugee camps, waterways may become contaminated, waste abounds and there is a lack of sanitation, all of which can lead to an outbreak of disease.
- The International Red Cross and Red Crescent Movement is the largest humanitarian network. It aims to alleviate human suffering, protect life and health, and uphold human dignity.
- World Food Programme (WFP) steps in when the distribution of food and other resources for the population is disrupted. WFP saves lives and protects livelihoods, reduces chronic hunger, and restores and rebuilds lives, especially for women and children.

FIGURE 5 Médecins Sans Frontières personnel help children in South Sudan.

DISCUSS

Hold a class debate on the contention 'Australia should do more to support global human wellbeing'.

[Ethical Capabililty]

Explore more with myWorldAtlas

Deepen your understanding of this topic with related case studies and questions.
- Investigate additional topics > Australia's links with the world > **Defence and peacekeeping**

Resources

🔗 **Weblink** Peacekeeping

11.8 INQUIRY ACTIVITIES

1. Using the internet, research an international NGO and show how it is working towards peace in areas of very low peacefulness. Focus on a country not studied in this topic.   **Examining, analysing, interpreting**
2. Write an essay to show how a country's HDI ranking and Peace Index levels indicate that people's wellbeing may **change**. Use examples from three countries that are in conflict.   **Describing and explaining**
3. Use the **Peacekeeping** weblink to learn more about global peacekeeping operations.

Examining, analysing, interpreting

11.8 EXERCISES

Geographical skills key: GS1 Remembering and understanding **GS2** Describing and explaining **GS3** Comparing and contrasting **GS4** Classifying, organising, constructing **GS5** Examining, analysing, interpreting **GS6** Evaluating, predicting, proposing

11.8 Exercise 1: Check your understanding

1. **GS1** What role do NGOs play in restoring wellbeing to a country?
2. **GS1** What assistance does the UN provide for a country to move towards peace?
3. **GS2** The Global Peace Index (2018) is mapped in **FIGURE 1**. Describe the distribution of **places** with a very low level of peacefulness and those with a high level of peacefulness.
4. **GS2** Explain why criteria for assessing peace levels of a country use indicators at home and in foreign relations.
5. **GS2** Explain how Australia has become a multicultural country.

11.8 Exercise 2: Apply your understanding

1. **GS5** The amount and distribution of Australian development assistance is shown in **FIGURES 2** and **3**.
   (a) To which regions of the world does Australia provide assistance?
   (b) Can you offer an explanation as to why Australia provides assistance to these regions?
   (c) More than half of the assistance is provided as governance, infrastructure, trade and education. Explain why each of these aspects is important to human wellbeing in the regions receiving assistance.
   (d) Suggest any aspects in which you would like to see Australia's priority in assistance expanded. Justify your response.
2. **GS6** From the conflicts discussed in this topic, which of the countries would you expect to be rated differently on the Global Peace Index in 2025?
3. **GS6** Australia took on a role in the UN Human Rights Council for 2018–2020. What do you think this should have meant in terms of our understanding of the wellbeing of refugees? Explain.
4. **GS6** Do you think Australia takes enough refugees? Explain your answer.
5. **GS6** How optimistic or pessimistic are you regarding the wellbeing of global citizens into the future? Explain your response.

Try these questions in learnON for instant, corrective feedback. Go to www.jacplus.com.au.

11.9 SkillBuilder: Debating like a geographer

What does debating like a geographer mean?

Debating like a geographer is being able to give the points for and against any issue that has a geographical basis, and supporting the ideas with arguments and evidence of a geographical nature.

Select your learnON format to access:

- an overview of the skill and its application in Geography (Tell me)
- a video and a step-by-step process to explain the skill (Show me)
- an activity and interactivity for you to practise the skill (Let me do it)
- questions to consolidate your understanding of the skill.

Affirmative speaker 1
(Introduces key ideas)
- Where is the Southern Ocean?
- Who is whaling?
- Which countries are involved in the issue?
- How far is it from Japan?
- Whale species
- Uses of whale meat
- The role of tradition
- Scientific research

Affirmative speaker 2
(Negates negative speaker 1 and expands on key ideas—provides the facts, statistics, emotional argument)
- Whale numbers
- Scientific research: what is research achieving?
- Importance of tradition

Affirmative speaker 3
(Negates negative speaker 2 and sums up key ideas)
- Emphasises that resource is well managed: whaling is not the only threat to species

Negative speaker 1
(Negates affirmative speaker 1 and introduces key ideas)
- Southern Ocean is a whale sanctuary
- Why don't the trawlers work closer to home?
- What is so important about the whale hunting that the benefits outweigh the costs?
- Global food chains affected
- Animal cruelty

Negative speaker 2
(Negates affirmative speaker 2 and expands on key ideas—provides the facts, statistics, emotional argument)
- Global food chains: facts
- How are whales caught? Is it humane?
- The work of Greenpeace, its actions, the conflict
- International Whaling Commission, its work, the global ban

Negative speaker 3
(Negates affirmative speaker 3 and expands on key ideas)
- Emphasises the resource is being degraded and conflict is rife

Resources

Video eLesson Debating like a geographer (eles-1762)

Interactivity Debating like a geographer (int-3380)

11.10 SkillBuilder: Writing a geographical essay

What is a geographical essay?

A geographical essay is an extended response structured like an essay, but it focuses on geographical facts and data, particularly data that can be mapped.

Introduction: A freeway should not go through ...
Theme 1: Noise levels from traffic. ...
Theme 2: House and land prices will decrease. ...
Theme 3: Animals will lose habitat and movement ...
Conclusion: If a road has to go through this area, it must be a tunnel under the parkland.

Select your learnON format to access:

- an overview of the skill and its application in Geography (Tell me)
- a video and a step-by-step process to explain the skill (Show me)
- an activity and interactivity for you to practise the skill (Let me do it)
- questions to consolidate your understanding of the skill.

Resources

Video eLesson Writing a geographical essay (eles-1763)

Interactivity Writing a geographical essay (int-3381)

11.11 Thinking Big research project: The displaced Rohingya children

SCENARIO

More than 380 000 Muslim Rohingya children have been displaced because of the conflict between their ethnic group and the Myanmar armed forces. You will create an annotated photographic essay detailing the daily wellbeing of these displaced children now living in the Kutupalong–Balukhali refugee camp in Bangladesh.

Select your learnON format to access:

- the full project scenario
- details of the project task
- resources to guide your project work
- an assessment rubric.

 Resources

 ProjectsPLUS Thinking Big research project: The displaced Rohingya children (pro-0220)

11.12 Review

11.12.1 Key knowledge summary
Use this dot point summary to review the content covered in this topic.

11.12.2 Reflection
Reflect on your learning using the activities and resources provided.

 Resources

 eWorkbook Reflection (doc-31781)

Crossword (doc-31782)

 Interactivity The impact of conflict on human wellbeing crossword (int-7678)

KEY TERMS

asylum seekers people who are awaiting confirmation of their refugee status

civil war a war between between two opposing groups within the one country

classification the categorisation of characteristics, changes, factors into distinctive groups

diplomats people who manage international relations

gross national income (GNI) the total income earned by a country's businesses and residents

Human Development Index (HDI) measures the standard of living and wellbeing by measuring life expectancy, education and gross national income

internally displaced persons (IDPs) people travelling within their country to 'safer' places, legally remaining under the control of their government

non-government organisation (NGO) an organisation that operates independently of government, usually to deliver resources or serve some social or political purpose

refugees people who flee through fear of persecution — for reasons of race, religion, nationality, membership of a social group, or because of a political opinion — and cross outside their home borders

stateless people people who frequently lack identification documents, live on the edges of society and are subjected to discrimination

GLOSSARY

absolute poverty experienced when income levels are inadequate to enjoy a minimum standard of living (also known as extreme poverty)

ageing population an increase in the number and percentage of people in the older age groups (usually 60 years and over)

algal bloom rapid growth of algae caused by high levels of nutrients (particularly phosphates and nitrates) in water

alluvial plain an area where rich sediments are deposited by flooding

alpha world city a city generally considered to be an important node in the global economic system

aquifers layers of rock which can hold large quantities of water in the pore spaces

asylum seekers people who are awaiting confirmation of their refugee status

atoll a coral island that encircles a lagoon

base flow water entering a stream from groundwater seepage, usually through the banks and bed of the stream

biocapacity the capacity of a biome or ecosystem to generate a renewable and ongoing supply of resources and to process or absorb its wastes

biodiversity the variety of plant and animal life within an area

biophysical environment all elements or features of the natural or physical and the human or urban environment, including the interaction of these elements

bioremediation the use of biological agents, such as bacteria, to remove or neutralise pollutants

booms floating devices to trap and contain oil

brackish (water) water that contains more salt than fresh water but not as much as sea water

carbon credits term for a tradable certificate representing the right of a company to emit one metric tonne of carbon dioxide into the atmosphere

carrying capacity the ability of the land to support livestock

child any person below 18 years of age

child soldier a child who is, or who has been, recruited or used by an armed force or armed group in any capacity, including but not limited to children, boys and girls, used as fighters, cooks, porters, messengers, spies or for sexual purposes. This term does not refer only to a child who is taking or has taken a direct part in hostilities.

chlamydia a sexually transmitted disease infecting koalas

civil war a war between between two opposing groups within the one country

classification the categorisation of characteristics, changes, factors into distinctive groups

climate change any change in climate over time, whether due to natural processes or human activities

coastal dune vegetation succession the process of change in the plant types of a vegetation community over time — moving from pioneering plants in the high-tide zone to fully developed inland area vegetation

conurbation an urban area formed when two or more towns or cities (e.g. Tokyo and Yokohama) spread into and merge with each other

Coriolis force force that results from the Earth's rotation. Moving bodies, such as wind and ocean currents, are deflected to the left in the southern hemisphere and to the right in the northern hemisphere.

cull selective reduction of a species by killing a number of animals

deltaic plain flat area where a river(s) empties into a basin

demographic transition model a graph attempting to explain how a country's population characteristics change as the level of wellbeing in a country improves over time

dependent population those in the under 15 years and over 60 years age groups. People in these age groups are generally dependent on those in the working age groups, either directly or indirectly for support.

desertification the transformation of land once suitable for agriculture into desert by processes such as climate change or human practices such as deforestation and overgrazing

developing nation a country whose economy is not well developed or diversified, although it may be showing growth in key areas such as agriculture, industries, tourism or telecommunications

development According to the United Nations, development is defined as 'to lead long and healthy lives, to be knowledgeable, to have access to the resources needed for a decent standard of living and to be able to participate in the life of the community'.

diplomats people who manage international relations

drainage area (or basin) an area drained by a river and its tributaries

dryland ecosystems characterised by a lack of water. They include cultivated lands, scrublands, shrublands, grasslands, savannas and semi-deserts. The lack of water constrains the production of crops, wood and other ecosystem services.

dyke an embankment constructed to prevent flooding by the sea or a river

dynamic equilibrium when the input of a coastal system such as winds and waves moving sediments onshore is equal to the output that moves sediments offshore, the system is said to be in a steady state. It is therefore not unstable and it has a dynamic equilibrium.

ecological footprint a measure of human demand on the Earth's natural systems in general and ecosystems in particular; the amount of productive land required by each person in the world for food, water, transport, housing, waste management and other purposes

ecological service the benefits to humanity from the resources and processes that are supplied by natural ecosystems

ecology the environment as it relates to living organisms

economic downturn a recession or downturn in economic activity that includes increased unemployment and decreased consumer spending

ecosystems systems formed by the interactions between the living organisms (plants, animals, humans) and the physical elements of an environment

enhanced greenhouse effect increasing concentrations of greenhouse gases in the Earth's atmosphere, contributing to global warming and climate change

environmental flows the quantity, quality and timing of water flows required to sustain freshwater ecosystems

environmental impact assessment a tool used to identify the environmental, social and economic impacts, both positive and negative, of a project prior to decision-making and construction

environmental world view varying viewpoints, such as environment-centred as opposed to human-centred, in managing ecological services

ephemeral describes a stream or river that flows only occasionally, usually after heavy rain

eutrophication a process where water bodies receive excess nutrients that stimulate excessive plant growth

exotic species species introduced from a foreign country

experienced wellbeing an individual's subjective perception of personal wellbeing

extreme poverty a state of living below the poverty line (US$1.90 per day), and lacking resources to meet basic life necessities (also known as absolute poverty)

favela an area of informal housing usually located on the edge of many Brazilian cities. Residents occupy the land illegally and build their own housing. Dwellers often live without basic infrastructure such as running water, sewerage or garbage collection.

female infanticide the killing of female babies, either via abortion or after birth

fertility rate the average number of children born per woman

floating settlements anchored buildings that float on water and are able to move up and down with the tides

flood mitigation managing the effects of floods rather than trying to prevent them altogether

fossil fuels carbon-based fuels formed over millions of years, which include coal, petroleum and natural gas. They are called non-renewable fuels as reserves are being depleted at a faster rate than the process of formation.

geothermal (power) describes power that is generated from molten magma at the Earth's core and stored in hot rocks under the surface. It is cost-effective, reliable, sustainable and environmentally friendly.

global warming the observable rising trend in the Earth's atmospheric temperatures, generally attributed to the enhanced greenhouse effect

grassroots movement action by ordinary citizens, as compared with the government, aid or a social organisation

green energy sustainable or alternative energy (e.g. wind, solar and tidal)

gross domestic product (GDP) the value of all goods and services produced within a country in a given period, usually discussed in terms of GDP per capita (total GDP divided by the population of the country)

gross national income (GNI) the total income earned by a country's businesses and residents

groundwater water held underground within water-bearing rocks or aquifers

groundwater salinity presence of salty water that has replaced fresh water in the subsurface layers of soil

hinterland the land behind a coast or shoreline extending a few kilometres inland

historical architecture urban environment that has significant value due to its unique form and history of development

Human Development Index (HDI) measures the standard of living and wellbeing by measuring life expectancy, education and gross national income

human–environment systems thinking using thinking skills such as analysis and evaluation to understand the interaction of the human and biophysical or natural parts of the Earth's environment

humanitarianism concern for the welfare of other human beings

humus decaying organic matter that is rich in nutrients needed for plant growth

icon sites six sites located in the Murray–Darling Basin that are earmarked for environmental flows. They were chosen for their environmental, cultural and international significance.

impervious a rock layer that does not allow water to move through it due to a lack of cracks and fissures

indicator a value that informs us of a condition or progress. It can be defined as something that helps us to understand where we are, where we are going and how far we are from the goal

Industrial Revolution the period from the mid 1700s into the 1800s that saw major technological changes in agriculture, manufacturing, mining and transportation, with far-reaching social and economic impacts

industrialised having developed a wide range of industries or having highly developed industries

infrastructure the basic physical and organisational structures and facilities (e.g. buildings, roads, power supplies) needed for the operation of a society

internally displaced persons (IDPs) people travelling within their country to 'safer' places, legally remaining under the control of their government

International Bill of Human Rights the informal name given to the Universal Declaration of Human Rights and the two International Covenants

International Covenants a multilateral treaty adopted by the United Nations General Assembly, in force from 1976. It commits those who have signed the Covenant to respect the civil and political rights of individuals and their economic, social and cultural rights.

invasive plant species commonly referred to as weeds; any plant species that dominates an area outside its normal region and requires action to control its spread

Kyoto Protocol an internationally agreed set of rules developed by the United Nations aimed at reducing climate change through the stabilisation of greenhouse gas emissions into the atmosphere

lagoon a shallow body of water separated from the sea by a sand barrier or coral reef

life expectancy the number of years a person can expect to live, based on the average living conditions within a country

mass wasting the movement of rock and other debris downslope in bulk, due to a destabilising force such as undermining compounded by the pull of gravity

maternal mortality the death of a woman while pregnant or within 42 days of termination of pregnancy

mediterranean (climate) characterised by hot, dry summers and cool, wet winters

medium-density housing a form of residential development such as detached, semi-attached and multi-unit housing that can range from about 25 to 80 dwellings per hectare

megacity a settlement with 10 million or more inhabitants

micro-credit the provision of small loans to borrowers who usually would not be eligible to obtain loans due to having few assets and/or irregular employment

micro hydro-dams produce hydro-electric power on a scale serving a small community (less than 10 MW). They usually require minimal construction and have very little environmental impact.

monoculture cultivating a single crop or plant species over a wide area over a prolonged period of time

mulch organic matter such as grass clippings

national park an area set aside for the purpose of conservation

natural increase the difference between the birth rate (births per thousand) and the death rate (deaths per thousand). This does not include changes due to migration.

non-government organisation (NGO) an organisation that operates independently of government, usually to deliver resources or serve some social or political purpose

Paris Agreement United Nations Framework Convention on Climate Change (UNFCCC) agreement outlining steps to reduce greenhouse gas emissions and tackle global warming

pastoral run an area or tract of land for grazing livestock

perennial describes a stream or river that flows permanently

population density the number of people within a given area, usually per square kilometre

population distribution the spread of people across the globe

population structure the number or percentage of males and females in a particular age group

poverty cycle circumstances whereby poor families become trapped in poverty from one generation to the next

poverty line an official measure used by governments to define those living below this income level as living in poverty

poverty rate the ratio of the number of people whose income is below the poverty line

qualitative indicators subjective measures that cannot easily be calculated or measured; e.g. indices that measure a particular aspect of quality of life or that describe living conditions, such as freedom or security

quantitative indicators objective indices that are easily measured and can be stated numerically, such as annual income or the number of doctors in a country

rainwater harvesting the accumulation and storage of rainwater for reuse before it soaks into underground aquifers

ratify to formally consent to and agree to be bound by a treaty, contract or agreement

recharge the process by which groundwater is replenished by the slow movement of water down through soil and rock layers

refugees people who flee through fear of persecution — for reasons of race, religion, nationality, membership of a social group, or because of a political opinion — and cross outside their home borders

regional and remote areas areas classified by their distance and accessibility from major population centres

relative poverty where income levels are relatively too low to enjoy a reasonable standard of living in that society

replacement rate the number of children each woman would need to have in order to ensure a stable population level — that is, to 'replace' the children's parents. This fertility rate is 2.1 children.

reservoir large natural or artificial lake used to store water, created behind a barrier or dam wall

ringbark remove the bark from a tree in a ring that goes all the way around the trunk. The tree usually dies because the nutrient-carrying layer is destroyed in the process.

river delta a landform composed of deposited sediments at the mouth of a river where it flows into the sea

river fragmentation the interruption of a river's natural flow by dams, withdrawals or transfers

river regime the pattern of seasonal variation in the volume of a river

Royal Commission a public judicial inquiry into an important issue, with powers to make recommendations to government

rural relating to the country, rather than the city

rural–urban fringe the transition zone where rural (country) and urban (city) areas meet

Sahel a semi-arid region in sub-Saharan Africa. It is a transition zone between the Sahara Desert to the north and the wetter tropical regions to the south. It stretches across the continent, west from Senegal to Ethiopia in the east, crossing 11 borders.

salinity an excess of salt in soil or water, making it less useful for agriculture

salt scald the visible presence of salt crystals on the surface of the land, giving it a crust-like appearance

sanitation facilities that safely dispose of human waste (urine, faeces and menstrual waste)

sex ratio the number of females per 1000 males

slum rundown area of a city with substandard housing

socioeconomic of, relating to or involving a combination of social and economic factors

spatial variation difference observed (in a particular measure) over an area of the Earth's surface

standard of living a level of material comfort in terms of goods and services available. This is often measured on a continuum; for example, a 'high' or 'excellent' standard of living compared to a 'low' or 'poor' standard of living.

stateless people people who frequently lack identification documents, live on the edges of society and are subjected to discrimination

stewardship the caring and ethical approach to sustainable management of habitats for the benefit of all life on Earth

STI sexually transmitted infection

storm surge a temporary increase in sea level from storm activity

subsidence the gradual sinking of landforms to a lower level as a result of earth movements, mining operations or over-withdrawal of water

Sustainable Development Goals (SDGs) a set of 17 goals established by the United Nations Development Programme, which aim to end poverty, protect the Earth and promote peace, equality and prosperity

terminal lake a lake where the water does not drain into a river or sea. Water can leave only through evaporation, which can increase salt levels in arid regions. Also known as an endorheic lake.

thermohaline relating to the combined influence of temperature and salinity

topsoil the top layers of soil that contain the nutrients necessary for healthy plant growth

training walls a pair of rock walls built at a river's mouth to force the water into a deeper and more stable channel. The walls improve navigation and reduce sand blockages.

turbid water that contains sediment and is cloudy rather than clear

Universal Declaration of Human Rights the first specific global expression of rights to which all human beings are inherently entitled

urban environment the human-made or built structures and spaces in which people live, work and recreate on a day-to-day basis

urban infilling the division of larger house sites into multiple sites for new homes

urban renewal redevelopment of old urban areas including the modernisation of household interiors

urban sprawl the spreading of urban developments into areas on the city boundary

water rights refers to the right to use water from a water source such as a river, stream, pond or groundwater source

water security the reliable availability of acceptable quality water to sustain a population

watertable upper level of groundwater; the level below which the earth is saturated with water

weathering the breaking down of rocks

weed any plant species that dominates an area outside its normal region and requires action to control its spread

weir wall or dam built across a river channel to raise the level of water behind. This can then be used for gravity-fed irrigation.

wellbeing a good or satisfactory condition of existence; a state characterised by health, happiness, prosperity and welfare

wetland an area covered by water permanently, seasonally or ephemerally. They include fresh, salt and brackish waters such as rivers, lakes, rice paddies and areas of marine water, the depth of which at low tide does not exceed 6 metres.

INDEX

Note: Figures and tables are indicated by italic *f* and *t*, respectively, following the page reference.

MOdulo Sperimentale Elettromeccanico (MOSE)   195
more economically developed countries (MECDs)   206, 260
mulch   48, 84
multilateral aid   213
multiple component index   210
Mumbai, growth of   187–90
Murray–Darling Basin
    declining river health   113
    features of   112*f*
    managing   112–15
    restore the balance of   113–14
    Royal Commission   58
    salinity in   56–8
    salt interception in   57*f*
Murray–Darling Basin Authority (MDBA)   113

N

Narre Warren   186–7
National Aboriginal and Torres Strait Islander Social Survey (NATSISS)   279
national parks   74, 84
National Rural Health Mission   261
native species and environmental change   74–8
    African elephant   75–7
    cull   74–5
    possums   77
    protecting biodiversity   74
natural environmental processes   136
natural increase   235, 257
natural systems, coasts   119
newly industrialised country (NIC)   206, 228
non-government organisations (NGOs)   213, 269, 311
North China Plain   102
nutrient-rich soil   41*f*
nutrients   42

O

ocean currents   148
    deep water currents   148
    downwellings currents   148–50
    role of   170
    surface currents   148
    upwellings currents   148–50
ocean processes   120–1
oceanographers   2*t*
official development assistance (ODA)   309
oil and water
    cleaning up   167–8
    impacts of   166–7
    marine pollution, sources of   165–6
    oil spill   166

P

Pacific, rising sea levels   129–30
Pacifying Police Units (UPP)   267
Papua New Guinea, AusAID work in   255–6
Paris Agreement   130, 146
pastoral run   51, 84
Paterson's Curse   70–1
people impacts, marine debris   157–8
perennial   86
place   8
political indicators   204
political rights   223*f*
population changing, India   244–6
population characteristics, Australia   247–50
population density   230, 257
population distribution   231–3, 257
population growth   238–9
population structure   239–41, 257
possible management solutions   137, 138*t*
possums   77
poverty
    defined   206–9, 216–19
    Sub-Saharan Africa   216–17
    Sustainable Development Goals   219–21
    world hunger   217–19
poverty cycle   254, 257
poverty line   216, 228
poverty rate   264
poverty, as factor in wellbeing
    effect of favela-living   266–7
    global progress   268–9
    haves and have-nots of Rio   265–8
    impacts of Rio de Janeiro's development   266
    improvements   267–8
    interconnection between water supply, sanitation   268–70
    success stories   269–70

Q

qualitative indicators   204
quantitative indicators   204, 228

R

rabbits   69–70, 72
ragpickers   188
ratify   131, 146
recharge of groundwater   100
refugees   289, 304, 314
    arranging the journey   305
    and asylum seekers   304
    European response   305–7
    movement of people to Europe   304–5
    regional perspective on asylum seekers   307–8
    Syrian across Europe   305

regional and remote areas   271, 285
relative poverty   216, 228
renewable energy   28–30
replacement rate   235, 257
reservoirs   89
restoring grasslands   66–8
    changing land use   66
    grazing bans   66
    moving people   66
rice   58
rill erosion   48–9
ringbark trees   70, 84
river deltas   194, 201
river flow   90–1
river fragmentation   89
river health   113
river regimes   112
rural–urban fringe   184–6
rural–urban fringe change   186–7
rural–urban wellbeing variation
    in Mumbai, India   276–8
    risk factor of   271–2
    in service accessibility   272–8
    within cities   275–6

S

Sahel   65, 84
salinity   54, 84
    dryland salinity   55
    irrigation salinity   55–6
    in Murray–Darling Basin   56–8
    Vietnam's Mekong Delta region   58–61
salt   54
salt scald   55, 84
sand bypass system   140*f*
sand storms   65
sanitation   261, 285
scale   12–13
sex ratio   262, 285
sexually transmitted infection (STI)   255, 257
sheep farming   43
sheet erosion   48
shifting sands on the Gold Coast   138–41
slum   188, 201
small-scale aid   214
small-scale solutions   95–6
social indicators   204
socioeconomic factors   278, 285
soil   48
    development and importance of   48
    formation   48
    gully erosion   49
    quality   71
    rill erosion   48–9
    sheet erosion   48
    tunnel erosion   49
    wind erosion   49–50